Environmental Remediation Technologies for Metal-Contaminated Soils

Hiroshi Hasegawa • Ismail Md. Mofizur Rahman •
Mohammad Azizur Rahman
Editors

Environmental Remediation Technologies for Metal-Contaminated Soils

Springer

Editors
Hiroshi Hasegawa
Institute of Science and Engineering
Kanazawa University
Kakuma, Kanazawa
Japan

Ismail Md. Mofizur Rahman
Department of Applied and Environmental Chemistry
Faculty of Science
University of Chittagong
Chittagong
Bangladesh

Mohammad Azizur Rahman
School of the Environment
University of Technology Sydney
Sydney, New South Wales
Australia

ISBN 978-4-431-55758-6 ISBN 978-4-431-55759-3 (eBook)
DOI 10.1007/978-4-431-55759-3

Library of Congress Control Number: 2015951462

Springer Tokyo Heidelberg New York Dordrecht London

Printed on acid-free paper

Springer Japan KK is part of Springer Science+Business Media (www.springer.com)

Preface

The contamination of soils by metals becomes a global concern due to the possibility of potential ecotoxic influence to plants and animals with the likely risk of accumulation in the human food chain. Soil with contaminants can either be left as is in the site or can be moved to a secure disposal site with continuous monitoring to limit any possible subsequent contamination. However, the distribution of contaminated sites around the world is increasing, e.g., (a) in China, about 50 thousands ha of the land area associated with mining activities are thought to be degraded each year with metal; (b) in the United States, 50 million m^3 of soil are estimated to be contaminated with metals; and (c) in Europe, several million ha of agricultural lands are reported to be polluted with metal. Hence, decontamination of the soil is considered as the economically beneficial option for the re-exploitation of the contaminated sites. The available remediation techniques include solidification, stabilization, flotation, soil washing, electro-remediation, bioleaching, and phytoremediation.

The book starts with an overview of the effects of metal intrusion on the natural properties of soils. The metal-loading extent in soils due to some notable anthropogenic activities is discussed in Chaps. 2 and 3. The test methods used to evaluate the metal content in the contaminated soils is discussed in Chap. 4. The following chapter provides a comparative discussion on the national and international legislative regulations so far proposed or being implemented to restrict the intrusion of toxic metal contaminants in soils. The remediation techniques in practice to manage the metal-contaminated soils are included as the core part of the book (Chaps. 6, 7, 8, 9, and 10). The final section of the book discusses the risk factors and cost modeling of the remediation options for the treatment of metal-contaminated soils.

The objective of this multi-authored book is to provide a compilation of the facts and issues that have been practiced and/or are required to be considered to meet the updated regulatory guidelines. The purpose of the book is to serve as reference material for both academic researchers and commercial-service professionals.

Kanazawa, Japan — Hiroshi Hasegawa
Chittagong, Bangladesh — Ismail Md. Mofizur Rahman
Sydney, Australia — Mohammad Azizur Rahman

Contents

Chapter 1
The Effects of Soil Properties to the Extent of Soil Contamination with Metals

Md. Alamgir

Abstract Heavy metal (HM) pollution of soils has been observed on local, regional, and global scales, and is likely to increase worldwide with growing industrial and agricultural activities. The HM pollution of soil is a significant environmental issue, because HM is responsible for causing adverse effect on human health through food chain contamination. The HM may originate and reach soils through pedogenic as well as anthropogenic processes. Once entered into the soil environment, the HM undergoes a number of chemical changes over time. The HM dynamics in soil is complex, and the bioavailability, mobility, and toxicity of metals in the soil fractions are influenced by different factors, including the properties of both the soil and the metal. This book chapter reviews the effect and significance of soil properties on the metal contamination of soils, which will help us to improve our understanding of the mechanisms involved in the transfer and mobilization of HM in soils.

Keywords Heavy metals • Soil properties • Adsorption • Soil pH • Soil texture • Clay minerals • Metal (hydro)oxides • Soil organic matter • Humus

1.1 Introduction

Metals are commonly defined as any element that has a silvery luster and is a good conductor of heat and electricity. Several terms are used to describe and categorize metals, including heavy metals, toxic metals, trace metals, transition metals, and micronutrients. Although the terms "heavy metals" or "trace metals" are poorly defined (Duffus 2002; Kabata-Pendias 2010; Steffan 2011), they are widely recognized and used to describe the widespread contaminants of terrestrial and freshwater ecosystems. Generally heavy metals refer to the group of metals and metalloids (semi metals) which have density greater than 5 g/cm^3 (Hackh et al. 1987; Morris 1992; Parker 1994) but the lower limit density may ranges from 3.5 to 7 g/cm^3 (Wild 1993; Duffus 2002). A total of 57 heavy metals/metalloids are known

M. Alamgir (✉)
Department of Soil Science, University of Chittagong, Chittagong 4331, Bangladesh
e-mail: mdalamgir@cu.ac.bd

H. Hasegawa et al. (eds.), *Environmental Remediation Technologies for Metal-Contaminated Soils*, DOI 10.1007/978-4-431-55759-3_1

(Vodyanitskii 2013), and among them lead (Pb), cadmium (Cd), mercury (Hg), arsenic (As), chromium (Cr), copper (Cu), selenium (Se), nickel (Ni), silver (Ag), zinc (Zn), aluminium (Al), cesium (Cs), cobalt (Co), manganese (Mn), molybdenum (Mo), strontium (Sr), uranium (U), Tin (Sn), Vanadium (V), and (Antimony (Sb) are considered as the most hazardous heavy metals/metalloids (McIntyre 2003; Vodyanitskii 2013). Any metal (or metalloid) species may be considered a "contaminant" if it occurs where it is unwanted, or in a form or concentration that causes a detrimental human or environmental effect (McIntyre 2003).

Heavy metals can enter the soil environment as a result of both pedogenic and anthropogenic processes. Heavy metals occur naturally in the soil environment from the pedogenetic processes of weathering of parent materials at levels that are regarded as trace (<1000 mg kg^{-1}) and have limited impact on soil except arsenic and selenium (Park et al. 2011; Wuana and Okieimen 2011). Unlike pedogenic sources, heavy metals in the soil from anthropogenic sources tend to be more mobile and can give rise to higher concentrations of metals into the environment. Soils act as a major sink for heavy metals released into the environment. However, when a soil's capacity to hold or retain the heavy metals is exceeded, the soil begins to act as a source for heavy metals (Selim 2013). The common sources of soil contamination with hazardous heavy metals/metalloids include: atmospheric deposition; organic manure, mineral fertilizers, and pesticides; industrial sewage discharge and industrial solid waste; municipal, agriculture, and food waste; coal ash, ore tailing dumps; logging and timber industry wastes; paints and other decorative materials; commodity impurities, etc. (Nriagu and Pacyna 1988; Yongsheng 2008; Zhang et al. 2011; Alloway 2012; Vodyanitskii 2013; Su et al. 2014). Heavy metals are highly hazardous to the environment and organisms. Unlike organic pollutants, heavy metals once introduced into the environment cannot be biodegraded. They persist indefinitely and cause pollution of soils, water, and air and can be enriched through the food chain. Intake of plants containing high levels of heavy metals might pose a serious risk to human health. Heavy metal dynamics in soils are complex, and the bioavailability, mobility, and toxicity of metals in the soil fractions are influenced by variety of factors including the properties of both the soil and the metal (Adriano et al. 2004; Buekers 2007; Naidu and Bolan 2008). Therefore, an understanding of the effects of soil properties on the behavior of heavy metals in the soil is essential for assessing the extent of the soil contamination with metals.

1.2 Pathways of Metal Movement in Soil

Once entered the soil, heavy metals can undergo a number of processes in soils: they may be retained in soil solution as free ions or complexed to inorganic or organic ligands; adsorbed on soil surfaces; precipitated such as oxides, hydroxides, and carbonates; or fixed chemically as solid compounds (Lasat 2000). The metals may also subject to plant uptake, transport through the vadose zone, and diffuse into

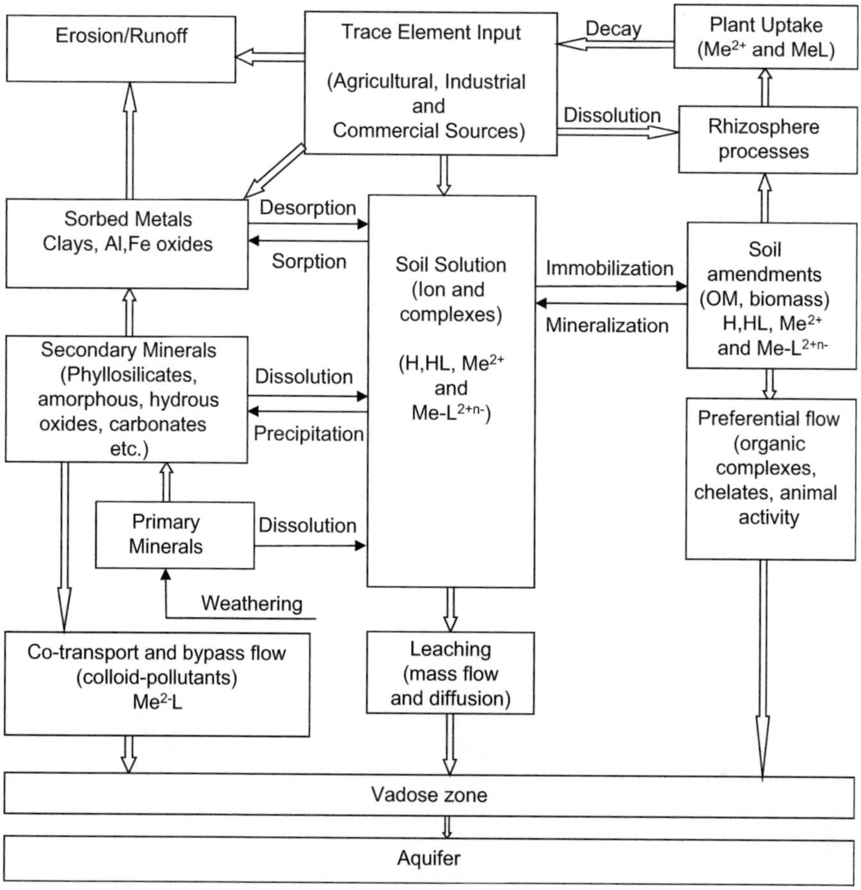

Fig. 1.1 Possible mechanisms and pathways of trace element movement in soils. Me represents trace element ion or atom, H hydrogen ion or atom, L ligand or other substance that combines with Me (Modified from Carrillo-Gonzalez et al. 2006 and Roberts et al. 2005)

porous material. The mechanisms and pathways of trace element mobility in soils have been reviewed by Carrillo-Gonzalez et al. (2006). Figure 1.1 illustrates the various chemical and physical pathways of a metal ion movement in soils.

The concentration of available forms of metals in soil is controlled by various physical and chemical processes such as ion exchange, adsorption and desorption, complexation, precipitation and dissolution, oxidation-reduction, sequestration and occlusion, diffusion and migration, metal competition, biological immobilization and mobilization, and plant uptake (Kabata-Pendias 2010; Wuana and Okieimen 2011). All the processes are not equally important for each element, but all of them are affected by soil pH and biological processes. Adsorption-desorption and precipitation-dissolution reactions directly affect the partitioning of metals between solid and aqueous phases, while complexation and oxidation-reduction reactions affect metal reactivity (e.g., solubility and bioavailability) (Allen et al. 1994). Ion exchange is a reversible process by which metal ions are exchanged between solid and liquid phases and between solid phases if in close contact with each other. The process of ion exchange on a clay surface can be illustrated as follows:

$$Me^{n+} + nNa^{+}\text{-clay} \leftrightarrow Me^{n+}\text{-clay} + nNa^{+}$$

where Me^{n+} is a metal cation with valence *n*. Sorption is considered a key process in soil environments that controls the mobility, fate, and bioavailability of metals in soils (Sparks 1995). Sorption is a general term that involves many different mechanisms (e.g. cation exchange or non-specific adsorption and specific adsorption) and refers to the general removal of a metal ion from solution and its subsequent association with the soil solid fraction. Specific adsorption is characterized by more selective and less reversible reactions including chemisorbed inner-sphere complexes, and nonspecific adsorption (or ion exchange), involves rather weak and less selective outer-sphere complexes (Sparks 1995; Bradl 2004). The reverse of sorption process, or the removal of a metal from a solid material and introduction into the soil solution, is termed desorption (Roberts et al. 2005). Desorption reactions are usually slower than sorption reactions. Through the process of precipitation, dissolved metals become insoluble, usually as metal hydroxides. A reaction in which a loss of electrons occurs is termed an oxidation reaction, while reduction is the gain of electrons (Sparks 1995). In reduction reaction, heavy metals ions (mono-, di-, or trivalent, M^{+}, M^{2+}, or M^{3+}) can be converted to metals ions with lower valency or zero-valent metal (M^{0}). On the other hand, in oxidation reaction heavy metals (with valency zero or higher) can be converted to higher valency (Chen 2012). A range of microorganisms can mobilize certain metals, metalloids, and organometallic compounds by reduction, oxidation, chelation by microbial metabolites and siderophores, and methylation which can result in volatilization. Similarly, many organisms can contribute to immobilization by sorption to cell components or exopolymers, transport and intracellular sequestration or precipitation as insoluble organic and inorganic compounds, e.g., oxalates, sulfides, or phosphates (White et al. 1997; Gadd 2007).

Heavy metals in soils exist in colloidal, ionic, particulate, and dissolved species depending on the particular environmental conditions. Colloidally bound metals are mainly associated with amorphous Fe and Mn oxyhydroxides or humic substances. Particulate metal forms can occur as discrete chemical forms (precipitates), adsorbed onto solid surfaces of hydrous oxides and organic substances, or as intermediates (coprecipitates). Dissolved metal species include the free unhydrated or hydrated metal ion (M^{n+}), as well as dissolved organic and inorganic complexes (Reuther 1999). The physico-chemical properties of metal ions that influence metal sorption rate include atomic weight, ionic radius, hydrated ion radius, electronegativities, reduction potential, and covalent binding.

Metal behavior in soils is a dynamic process and bioavailability of metals is regulated by physical, chemical and biological properties of soils. Numerous studies, investigations, and observations indicated that the principle soil parameters governing the binding of heavy metals and hence their bioavailability include pH, soil texture, cation exchange capacity (CEC), organic matter, oxides and hydroxides, mainly Fe, Mn, and Al, activity of microorganisms, occurrence and form of cations, content of macro and micronutrients, oxidation-reduction potential, sorption capacity, bioavailability for plants and animals, and resistance of the soil (Gadd 2004; Nouri et al. 2009; Kabata-Pendias 2010; Fijałkowski et al. 2012). Some recent studies indicate that there is a significant impact of carbonates on the sorption and retention of metals (Shirvani et al. 2006; Ahmed et al. 2008; Irha et al. 2009). However, the role of carbonates in affecting the fate of metals in soils is not yet fully clear. The mentioned parameters are often dependent to one another and fluctuation of parameters cannot change the total amount of heavy metals in soil but can significantly affect their bioavailability (Skłodowski et al. 2006).

1.3 The Role of Soil Properties on Metal Availability in Soils

As mentioned earlier, a range of physical, chemical, and biological properties of soil influence the retention and transport of heavy metals in the soils. Among these pH, soil texture, clay mineralogy, organic matter, redox potential, and cation exchange capacity (CEC) are important (Adriano 2001; Bolan et al. 2013; Selim 2013). The influence of different soil properties on metal sorption and desorption has been evaluated extensively but several studies indicate the possibility of the combined effects of soils properties on some metals sorption and desorption (Harter and Naidu 2001; Appel and Ma 2002; Dutta et al. 2011). In the following sections, an overview of current knowledge on the effects of soil pH, soil texture and clay mineralogy, SOM, CEC and redox potential on the extent of the soil contamination with metals has been provided.

1.3.1 Soil pH

Soil pH is defined as the negative logarithm of the hydrogen concentration. The pH scale goes from 0 to 14 with pH 7 as the neutral point. From pH 7 to 0 the soil is more acidic; from pH 7 to 14, the soil is increasingly more alkaline (basic). In general, soil pH value has the greatest effect of any single factor on the solubility or retention of metals in soils (Ghosh and Singh 2005; Alloway 2012). pH influence metal sorption through a range of mechanisms e.g. (I) changes in surface charge (Naidu et al. 1997), (II) competition for adsorption sites (Benjamin and Leckie 1981), (III) hydrolysis of metal species in solution (McBride 1989) and (IV) dissolution of metal complexing anions.

Basically, sorption of trace element in soils occurs at permanent and variable charged sorption sites of phyllosilicates, soil humic substances, carbonates, hydroxides, and oxyhydroxides of Fe, Al, Mn, and Ti, aluminosilicates such as allophanes and imogolite, but they differ greatly in their sorption capacities, in their ion-exchange capacities, and in the binding energies of their sorption sites (Violante et al. 2007; Proust 2015). Most minerals have reactive surface groups that are capable of binding or releasing protons which leads to the development of electrical charges at the surface. In all colloids, organic or inorganic, the surface charges are associated with OH groups, that are largely pH dependent. The OH groups or oxygen atoms are attached to iron and/or aluminum in the inorganic colloids (e.g., Al–OH) and to the carbon in humus (e.g., –C–OH). Broken edges of mineral colloids also generate pH-dependent charges. Most of the charges associated with humus, 1:1-type clays, the oxides of iron and aluminum and allophane are of this type (Brady and Weil 2008).

The mechanism of metal ion association with variable charge minerals involves an ion-exchange process in which the sorbed cations replace bound protons. At acidic pH, more protons (H^+) are available to saturate metal-binding sites; therefore, metals are less likely to form insoluble precipitates. Under basic conditions, metal ions can replace protons to form other species, such as hydroxo-metal complexes (Olaniran et al. 2013). Desorbing protons can leave negatively charged groups at the surface, which act as Lewis bases that coordinate metal ions. Variable-charge minerals often adsorb transition cations (Pb, Cu, Cr, Ni, Co, Zn, Al, Fe, Mn) as inner-sphere complexes more strongly than are alkaline earth cations (Violante et al. 2007). Adsorbed protons can form proton bonds between surface groups and metal complexes and can generate positive charges at the surface repelling or attracting respectively positively or negatively charged metal complexes (Selim and Kingery 2003).

Generally, metal sorption increases with increasing soil pH, and when pH falls to below 5, mobility is enhanced as a result of the increased proton concentration (McLaughlin et al. 2000; Paulose et al. 2007). The lower the pH value the more metal can be found in solution and thus more metal is mobilized. At low pH values, the dominant species in aqueous solution is the free aquo cation [e.g., $Zn(OH_2)_6^{3+}$], but with rising pH it is hydrolyzed to an hydroxylated cation and then to a simple uncharged complex [e.g., $ZnOH^+$ or $Zn(OH)_2$] and under alkaline conditions, to anionic hydroxy species [e.g., $Zn(OH)_3^-$ or $Zn(OH)_4^{2-}$] (Violante et al. 2007). Trace elements occur predominantly in cationic form [Pb, Cu, Zn, Ni, Cd, Hg, Cr(III), and Co], but some trace elements are present in anionic form [As, Se, Cr(VI), Mo, and B].

A number of authors have found positive correlations between pH and retention of Cu, Cd, Pb and Zn (Tyler and McBride 1982; Harter 1983; Christensen 1984; King 1988; Jopony and Young 1994; Chuan et al. 1996; Deurer and Bottcher 2007). The effect of pH on metal bioavailability varies between different metals. For example, a rapid decline in the concentrations of the free, ionic species of copper and zinc in minimal media was observed at pH values higher than 5, while the free,

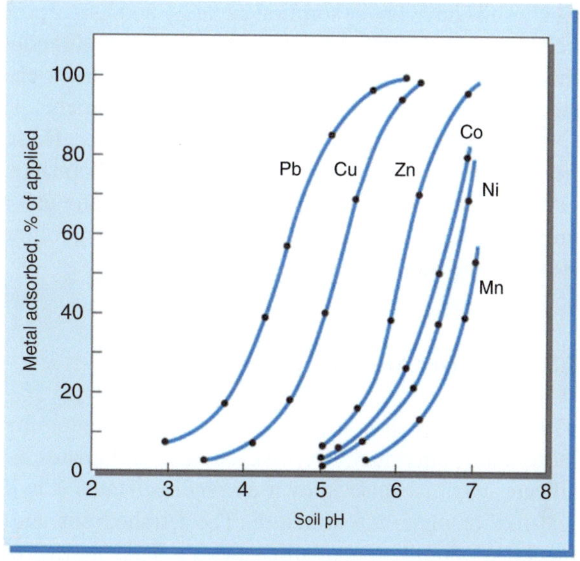

Fig. 1.2 The effect of pH on the adsorption of six heavy metal cations on goethite (After Basta et al. 2005)

ionic form of cobalt remains prevalent until the pH value is higher than 8 (Olaniran et al. 2013). It has been reported that cationic trace element adsorption by oxide surfaces may increases to almost 100 % with increasing pH (Fig. 1.2). Several studies have reported the interaction of pH with other properties such as organic matter, surface charge, and ionic strength responsible for metal sorption (Naidu et al. 1994; Sauvé et al. 1998; Bolan et al. 1999; Trancoso et al. 2007). Soil pH increases are often correlated with mineralogy and changes in solution chemistry, such as higher solubility in DOC and base cation concentration at high pH; high acidic cation concentration at lower pH and higher metal solubility.

1.3.2 Soil Texture and Clay Mineralogy

Both soil texture and mineral types play an important role in mobility of metals in soil. Soil texture reflects the relative amounts of sand, silt and clay particles in a soil. The clay fraction of the soil contains particles less than 0.002 mm in size. Particles less than 0.001 mm size possess colloidal properties and are known as soil colloids. The soil colloids are the most active portion of the soil and largely determine the physical and chemical properties of a soil. Clay fraction, which is mainly composed of clay minerals, has a high sorption capacity and a strong ability to bind metallic elements due to their large specific surface area, chemical and mechanical stability, layered structure and high cation exchange capacity (CEC). Generally soils having higher amounts of clay and humus also have high buffering capacity, the sorption capacity of soils which despite the increase in concentrations of contaminants do not cause adverse biological effects. Compared to clay soils, sandy soils have lower sorption capacity and larger pore size so they weakly absorb heavy metals, which lead to their movement to groundwater and surface water. Clay minerals may contain negligible amounts of trace elements as structural components, but their sorption capacities to trace elements play the most important role.

There are four major types of clay minerals. These include the layer silicates, iron and aluminum oxides, amorphous and allophanes, and humus. Layer silicate clays, iron and aluminum oxide clays, allophane and associated amorphous clays are inorganic colloids while humus is an organic colloid. Inorganic colloids usually make up the bulk of soil colloids.

1.3.2.1 Layer Silicate Clays

The mineralogical constituents of soil are predominantly layer silicate minerals (secondary minerals), oxy/hydroxides, carbonates and sulfates. The important silicate clays are also known as phyllosilicates (Phyllon – leaf) because of their leaf-like or plate-like structure. The tetrahedrons and octahedrons are the fundamental structural units of silicate clays. The tetrahedrons are composed of one Si^{4+} in the center surrounded by four O^{2-} in the corners giving a four-sided configuration. In octahedrons, a metal cation Me^{m+} (usually Al^{3+} or Mg^{2+}) is surrounded by six O^{2-} and/or OH^{-} giving an eight-sided building block termed octahedron. An interlocking array or a series of silica tetrahedra tied together horizontally by shared oxygen anions gives a tetrahedral sheet. Similarly, the octahedrons share oxygens to form a flat octahedral sheet (Selim and Kingery 2003; Brady and Weil 2008). The tetrahedral and octahedral sheets are bound together within the crystals by shared oxygen atoms into different phyllosilicate layers. The specific nature and combination of sheets in these layers vary from one type of clay to another and control the physical and chemical properties of each clay. Schematic representations of sheets, layers, and stacks of layers are given in Fig. 1.3.

Based on the number and arrangement of tetrahedral (Si) and octahedral (Al, Mg, Fe) sheets contained in the crystal units or layers, crystalline clays may be classed into two main groups: 1:1 silicate clays, in which each layer contains one tetrahedral and one octahedral sheet, and 2:1 silicate clays, in which each layer has one octahedral sheet sandwiched between two tetrahedral sheets. The 1:1 silicate clays include kaolinite, halloysite, nacrite, and dickite and 2:1 silicate clays include smectite (e.g., montmorillonite) and vermiculite group and fine-grained micas and the chlorites. The charge characteristics of clay minerals and oxides are categorized as permanent charge and variable charge. The variable-charge sites are located at the crystal edge surfaces and originate in the protonation or deprotonation of surface hydroxyl (SOH) groups. These sites promote pH-dependent adsorption attributed to surface complexation reactions with silanol and aluminol groups to form inner-sphere complexes. The permanent-charge sites are located at the interlayer basal surfaces of the clay minerals and result from the isomorphic substitutions of Al^{3+} for Si^{4+} in the silica tetrahedral sheet and Fe^{2+}/Mg^{2+} for Al^{3+} in the alumina octahedral sheet. These interlayer negative charges are neutralized by

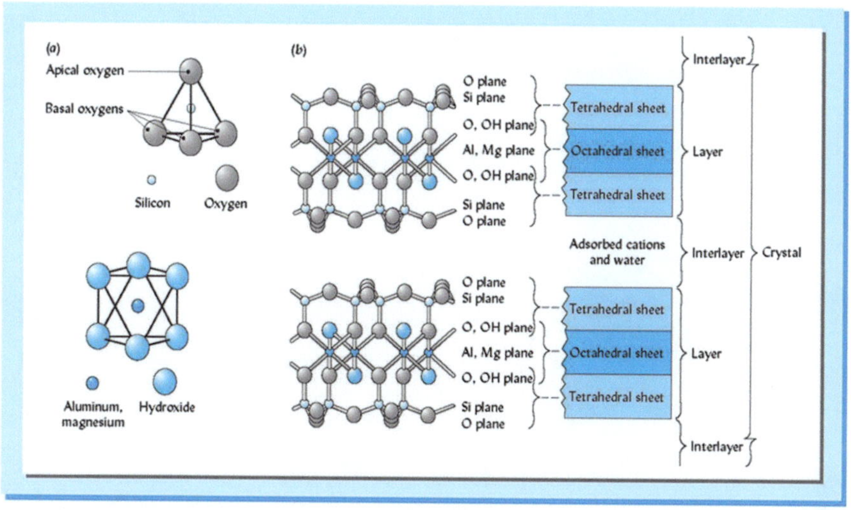

Fig. 1.3 The basic molecular and structural components of silicate clays: (**a**) a single tetrahedron and octahedron (**b**) different combinations of tetrahedral and octahedral sheets (Source: Brady and Weil 2008)

pH-independent adsorption of cations as outer-sphere complexes (cation exchange reactions) (Violante et al. 2007). When at least one water molecule of the hydration sphere is retained upon sorption, the surface complex is referred to as outer-sphere. When the ion is bound directly to the adsorbent without the presence of the hydration sphere, an inner-sphere complex is formed (Sposito 1984; Sparks 2003; Borda and Sparks 2008). A diagrammatic representation of the adsorption of ions on a colloid by the formation of outer-sphere and inner-sphere complexes has been shown in Fig. 1.4.

The main source of charge on clay minerals is isomorphous substitution which confers permanent charge on the surface of most layer silicates. Binding of metal cations increases with increase of their valence, atomic weight and ionic potential. Helios Rybicka et al. (1995) in their study found that 80 % and 70 % of the total concentration of Pb and Cu were adsorbed on illite, beidellite, and montmorillonite, whereas the amounts of Zn, Ni, and Cd ions adsorbed were smaller (40–50 %). As cited in Fijałkowski et al. (2012), the affinity of metal cations relative of clay minerals is arranged in a series of $Cu^{2+} > Cd^{2+} > Fe^{2+} > Pb^{2+} > Ni^{2+} > Co^{2+} > Mn^{2+} > Zn^{2+}$.

1.3.2.2 Iron, Aluminum, and Manganese Oxide Clays

These non-silicate clays consist of modified octahedral sheets with either iron, aluminum, or manganese in the cation positions. Examples of iron aluminum and manganese oxides common in soils are gibbsite ($Al_2O_3.3H_2O$) and goethite (Fe_2O_3.

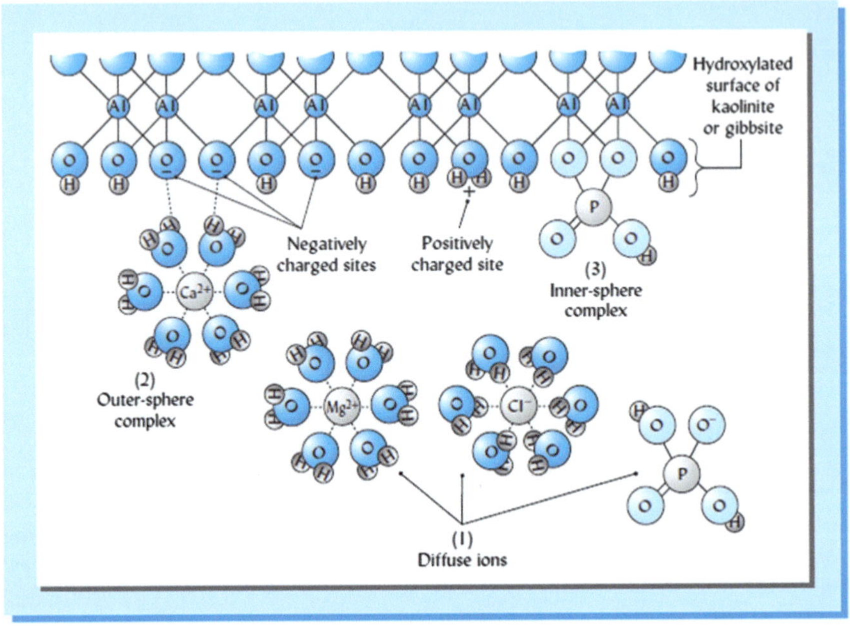

Fig. 1.4 A diagrammatic representation of the adsorption of ions on a colloid by the formation of outer-sphere and inner-sphere complexes (Source: Brady and Weil 2008)

H_2O). Lithiophorite (Al_2 Li $Mn_2^4Mn^3O_6(OH)_6$), lepidocrocite (γ-FeO(OH)), and maghemite (γ-Fe_2O_3), although less common, may also be present in soils (Schwertmann 1991; Dixon and Schulze 2002). Numerous studies indicate that oxides and hydroxides of Fe, Al, and Mn can markedly affect heavy metal retention, mobility, and bioavailability because of their ability to make covalent links with them (Korte et al. 1976; Hickey and Kittrick 1984; Tiller et al. 1984; King 1988; McBride 1994). Compared to Al oxides and other clay minerals, Fe and Mn oxides have a much greater adsorption capacity (Brown and Parks 2001). Isomorphic substitution ions are rare in these minerals and positive or negative charge on hydroxylated or hydrated surface is developed by sorption or desorption of H^+ or OH^- ions. The pH at which the net variable charge on the surfaces of these components is zero is called the point of zero charge (PZC). The reported PZC of Fe-oxides range from pH 7.0 to 9.5, whereas that of Al-oxides ranges from pH 8.0 to 9.2 as cited in Violante et al. (2010). In many soils, iron and aluminum oxide minerals are mixed with silicate clays. The oxides may form coatings on the external surfaces of the silicate clays, or they may occur as "islands" in the interlayer spaces of 2:1 clays substantially altering the colloidal behavior of the associated silicate clays (Brady and Weil 2008).

Heavy metal adsorption on iron oxides is generally accompanied by the release of protons at the surface oxy-hydroxyl groups, and the extent of the adsorption is strongly pH dependent (Silveira et al. 2003). Metals can be sorbed initially to the

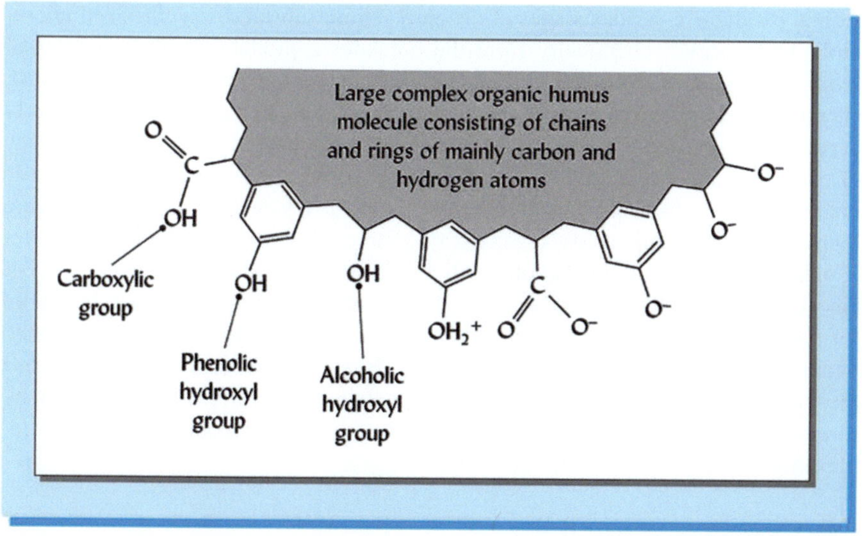

Fig. 1.5 A simplified diagram showing the carboxylic, phenolic, and alcoholic groups responsible for the high amount of negative charge on humus colloids (Source: Brady and Weil 2008)

iron oxide surface sites, but they may also diffuse to internal sorption sites, which are not readily accessible by the bulk solution (Ford et al. 1997). Metal retention by ferric hydroxide has been found to decrease in the order of: Pb > Cu > Zn > Cd (Landner and Reuther 2004) and $Pb^{2+} > Cu^{2+} > Zn^{2+} > Co^{2+} > Ni^{2+}$ (Dube et al. 2001).

1.3.2.3 Humus

The humus colloids are not crystalline. They are composed basically of carbon, hydrogen, and oxygen rather than of silicon, aluminum, iron, oxygen, and hydroxyl groups. The negative charges of humus are associated with partially dissociated enolic (–OH), carboxyl (–COOH), and phenolic groups; these groups in turn are associated with central units of varying size and complexity and act as main complexing functional groups (Takamatsu and Yoshida 1978). Bonding may be either covalent, where the metal and counter ion contribute one electron each, or coordinate, where the ligand provides both electrons. Figure 1.5 represents a simplified diagram showing major functional groups on humus colloids.

1.3.3 Soil Organic Matter

According to Schnitzer (1999), soil organic matter is the sum total of all organic carbon containing substances in the soil, which comprises of a mixture of plant and animal residues in various stages of decomposition, substances synthesized microbiologically and/or chemically from the breakdown products, and the bodies of living and dead microorganisms and their decomposing remains. The physical, chemical, and biological properties of soil and functions in the environment largely depend on soil organic matter (Balabane et al. 1999). Soil organic matter has been of particular interest in studies of heavy metal sorption by soils, because organic matter is known to form strong complexes with heavy metals and metals have a high affinity for humic acids, organo clays, and oxides coated with organic matter (Connell and Miller 1984; Elliott et al. 1986; Gaffney et al. 1996; Karaca 2004; Ghosh and Singh 2005). Organic matter in soils serves as a reactive adsorbent pool for trace metals, due to their high surface area and their high reactivity associated with various S-, O- and N-functional groups. Organic matter can reduce or increase the bioavailability of heavy metal in soil through immobilization or mobilization by forming various insoluble or soluble heavy metal-organic complexes. The complexation reaction follows the formula:

$$\text{R-L}^{l-} + \text{M}^{m+} \leftrightarrow \text{R-L-M}^{m-l}$$

where R is the C-chain, L the active group which actually binds, M the metal, and m and l are the valencies of metal and ligand, respectively. The effect of SOM on metals in soils depends on its amounts, composition, and dynamics.

Organic matter upon decomposition produces low molecular weight organic acids and stable humic substances. Humic substances may be classified according to their solubility. Humic acids (HA) are soluble only in alkali, while fulvic acids (FA) are soluble both in acid and alkali. Humic acids constitute a large fraction of humic substances in soil, which in turn may account for 70–80 % of all organic matter in soil. A wide range of organic acids acting as ligands for many metal ions has been found in soil. These include oxalic, citric, formic, acetic, malic, succinic, malonic, maleic, lactic, aconitic, and fumeric acids (Fijałkowski et al. 2012; Vranova et al. 2013). Several studies indicated that the reactions between organic acid and heavy metal are related to the amount and place of the carboxyl and hydroxyl groups (Shan et al. 2002; Gao et al. 2003; Schwab et al. 2008). Generally, citric acid is the most effective in terms of desorption of different metals (Cu, Hg, Pb, Cd, Zn, and ^{137}Cs), followed by malic > acetic > tartaric > oxalic acid as organic acid with more carboxyl group form more stable ligand (Vranova et al. 2013; Yan et al. 2014). The more stable of the ligand formed, the more difficult for it to be adsorbed by the soil and sediment, and thus the metal leaching is much easier (Gao et al. 2003). Metal sorption capacity of humic substances largely exceeds the mineral sorption capacity of the soil.

The binding of heavy metals by organic matter is a complex process, due to the diversity of its connections with the mineral phase (Harter and Naidu 2001; Lamb 2010). Soluble organic ligands may coat reactive mineral adsorption site, inactivating them and therefore inhibit metal adsorption, but in turn sorb on mineral surfaces and generate new sorbing sites and increase metal retention. Organic ligands on humic and fulvic acids behave as soft Lewis bases, while metals are soft acids and this is why they tend to form complexes.

1.3.4 Cation Exchange Capacity

Cation exchange capacity (CEC) is a dominant factor in heavy metals retention. CEC is defined simply as the sum total of exchangeable cations that a soil can adsorb or the number of cation adsorption sites per unit weight of soil expressed as centimoles per kg ($cmol_c/kg$). The CEC of soils depends on soil types, amounts, and types of different colloids present and on the CEC of the colloids. Fine-textured (clay) soils tend to have higher cation exchange capacity than sandy soils. CEC for clay soils usually exceeds 30 $cmol_c/kg$ while the value ranges from 0 to 5 for sandy soils. 2:1 type clays have higher CEC than 1:1 clay. The CEC values of clay vary in the following sequence: montmorillonite, imogolite > vermiculite > illite, chlorite > halloysite > kaolinite. Humus has very high CECs compared to the inorganic clays, especially kaolinite and Fe, Al oxides. CEC of some soils and soil materials has been shown in Fig. 1.6.

The capacity of the soils for adsorbing heavy metals is correlated with their CEC (Fontes et al. 2000; Harter and Naidu 2001). The greater the CEC values, the more exchange sites on soil minerals will be available for metal retention. Clay minerals, such as montmorillonite and vermiculite, have a high cation exchange capacity and have a high total capacity toward some heavy metals (Malandrino et al. 2006). Competing ions can have a marked effect on ion sorption by soils. In solution, metal cations such as Cu, Zn, Cd, and Pb compete with more abundant soil cations such as Ca^{2+} and Mg^{2+} for both nonspecific and specific exchange sites. Chen (2012) found that the presence of Pb did significantly reduce the adsorption maximum of Cd on soils. Mutual interactions between clay minerals, metal oxides, and organic matter can greatly alter the sorptive properties of these soil constituents for heavy metals because such interactions usually involve cation exchange sites (Cruz-Guzmán et al. 2003).

1.3.5 Oxidation-Reduction Potential

Oxidation-reduction potential (redox potential) is one of the critical factors regulating the speciation and bioavailability of metals in soils. Oxidation and reduction reactions are common in soils. Oxidation and reduction occur together as an electron cannot exist as an isolated entity; it is transferred from one species (the reductant) to another (the oxidant). The extent to which a soil is reduced or oxidized is generally assessed by the values Eh and pe. Eh is a redox potential and is expressed in terms of electrochemical energy (millivolts) and assumes a system

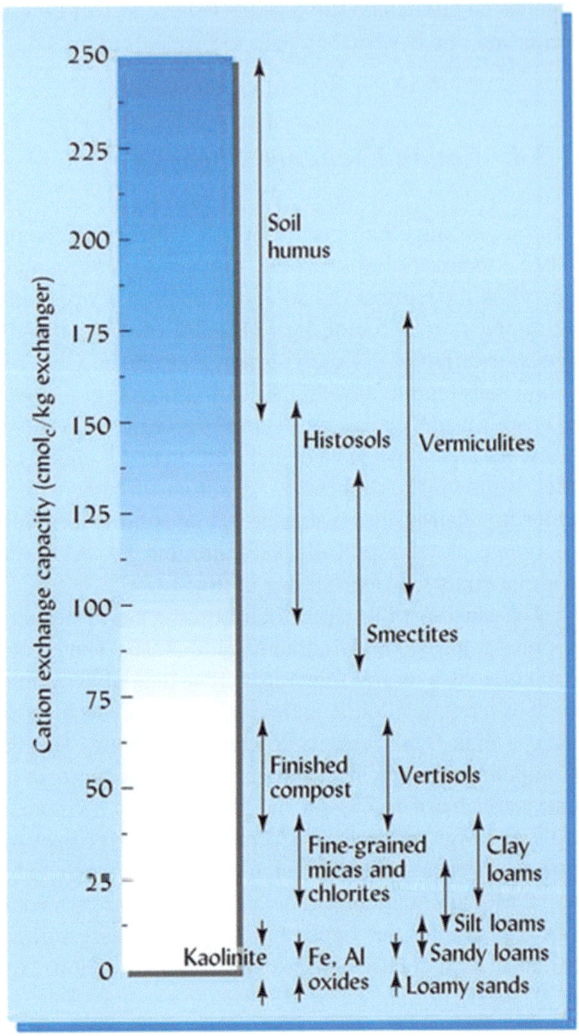

Fig. 1.6 Ranges in the cation exchange capacities (at pH 7) that are typical of a variety of soils and soil materials (Source: Brady and Weil 2008)

at thermodynamic equilibrium. Oxidized soils have values ranging from +400 to +700 mV while reduced soils may have values from −250 to −300 mV (Roberts et al. 2005). Redox reactions play a major role in the form and reactivity of some soil oxides (Fe and Mn) responsible for metal sorption. Redox also controls the chemical speciation of several metalloid contaminants (As, Cr, and Se) thus affecting sorption (McLaughlin et al. 2000). Generally reducing conditions cause a reduction in heavy metal mobility (Kabata-Pendias and Pendias 1991; Gonsior et al. 1997). Oxidation-reduction reactions may not only affect the partitioning of redox-active trace metals, like Cr or Mo, but also of redox stable metals like Zn, Cu, or Ni, in soil or aquatic environments (Landner and Reuther 2004).

1.4 Conclusions

Contamination of the soil with heavy metals is widespread and it poses a long-term risk for a quality of ecosystems. Environmental effects of metals can only be understood when their fate and interactions with the components of the environment are known. The soil properties that control the availability of metals in soils are interdependent and hence change in one value will also influence the other one. So the properties of soils and metals and their interactions should have to be taken into the account to achieve a comprehensive insight into complex processes and in the risk assessments and soil remediation strategies.

References

Adriano DC (2001) Trace elements in terrestrial environments: biogeochemistry, bioavailability, and risks of metals. Springer, New York

Adriano DC, Wenzel WW, Vangronsveld J, Bolan NS (2004) Role of assisted natural remediation in environmental cleanup. Geoderma 122:121–142

Ahmed IAM, Crout NMJ, Young SD (2008) Kinetics of Cd sorption, desorption and fixation by calcite: a long-term radiotracer study. Geochim Cosmochim Acta 72:1498–1512

Allen HE, Huang CP, Bailey GW, Bowers AR (1994) Metal speciation and contamination of soil. CRC Press, Boca Raton

Alloway BJ (2012) Heavy metals in soils. Springer, Dordrecht

Appel C, Ma L (2002) Concentration, pH, and surface charge effects on cadmium and lead sorption in three tropical soils. J Environ Qual 31:581–589

Balabane M, Faivre D, van Oort F, Dahmani-Muller H (1999) Mutual effects of soil organic matter dynamics and heavy metals fate in a metallophyte grassland. Environ Pollut 105:45–54

Basta NT, Ryan JA, Chaney RL (2005) Trace element chemistry in residual-treated soil: key concepts and metal bioavailability. J Environ Qual 34:49–63

Benjamin MM, Leckie JO (1981) Multiple-site adsorption of Cd, Cu, Zn, and Pb on amorphous iron oxyhydroxide. J Colloid Interface Sci 79:209–221

Bolan NS, Naidu R, Khan MAR, Tillman RW, Syers JK (1999) The effects of anion sorption on sorption and leaching of cadmium. Aust J Soil Res 37:445–460

Bolan NS, Makino T, Kunhikrishnan A, Kim PJ, Ishikawa S, Murakami M, Naidu R, Kirkham MB (2013) Cadmium contamination and its risk management in rice ecosystems. Adv Agron 119:183–273

Borda MJ, Sparks DL (2008) Mobility of trace elements in soil environments. In: Violante A, Huang PM, Gadd GM (eds) Biophysico-chemical processes of metals and metalloids in soil environments. Wiley, Hoboken, pp 97–168

Bradl HB (2004) Adsorption of heavy metal ions on soils and soils constituents. J Colloid Interface Sci 277:1–18

Brady N, Weil R (2008) The nature and properties of soils, 14th edn. Prentice Hall, Harlow

Brown GE, Parks GA (2001) Sorption of trace elements on mineral surfaces: modern perspectives from spectroscopic studies, and comments on sorption in the marine environment. Int Geol Rev 43:963–1073

Buekers J (2007) Fixation of cadmium, copper, nickel and zinc in soil: kinetics, mechanisms and its effect on metal bioavailability. PhD, Department of Microbial and Molecular Systems, Katholieke Universiteit Leuven, Flanders, Belgium

Carrillo-Gonzalez R, Simunek J, Sauve S, Adriano D (2006) Mechanisms and pathways of trace element mobility in soils. Adv Agron 91:111–178
Chen JP (2012) Decontamination of heavy metals: processes, mechanisms, and applications. CRC Press, Boca Raton
Christensen TH (1984) Cadmium soil sorption at low concentrations: II. Reversibility, effect of changes in solute composition, and effect of soil aging. Water Air Soil Pollut 21:115–125
Chuan MC, Shu GY, Liu JC (1996) Solubility of heavy metals in a contaminated soil: effects of redox potential and pH. Water Air Soil Pollut 90:543–556
Connell DW, Miller GJ (1984) Chemistry and ecotoxicology of pollution. Wiley, New York
Cruz-Guzmán M, Celis R, Hermosin M, Leone P, Negre M, Cornejo J (2003) Sorption-desorption of lead (II) and mercury (II) by model associations of soil colloids. Soil Sci Soc Am J 67:1378–1387
Deurer M, Bottcher J (2007) Evaluation of models to upscale the small scale variability of Cd sorption in a case study. Geoderma 137:269–278
Dixon JB, Schulze DG (2002) Soil mineralogy with environmental applications. Soil Science Society of America, Madison
Dube A, Zbytniewski R, Kowalkowski T, Cukrowska E, Buszewski B (2001) Adsorption and migration of heavy metals in soil. Pol J Environ Stud 10:1–10
Duffus JH (2002) "Heavy metals" a meaningless term? Pure Appl Chem 74:793–807
Dutta SK, Singh D, Sood A (2011) Effect of soil chemical and physical properties on sorption and desorption behavior of lead in different soils of India. Soil Sediment Contam 20:249–260
Elliott H, Liberati M, Huang C (1986) Competitive adsorption of heavy metals by soils. J Environ Qual 15:214–219
Fijałkowski K, Kacprzak M, Grobelak A, Placek A (2012) The influence of selected soil parameters on the mobility of heavy metals in soil. Inżynieria i Ochrona środowiska 15:81–92
Fontes F, Matos MP, Teixeira A, da Costa ML, Neves L, César J (2000) Competitive adsorption of zinc, cadmium, copper, and lead in three highly-weathered Brazilian soils. Commun Soil Sci Plant Anal 31:2939–2958
Ford RG, Bertsch PM, Farley KJ (1997) Changes in transition and heavy metal partitioning during hydrous iron oxide aging. Environ Sci Technol 31:2028–2033
Gadd GM (2004) Microbial influence on metal mobility and application for bioremediation. Geoderma 122:109–119
Gadd G (2007) Transformation and mobilization of metals, metalloids, and radionuclides by microorganisms. In: Violante A, Huang PM, Gadd GM (eds) Biophysico-chemical processes of heavy metals and metalloids in soil environments. Wiley, Hoboken, pp 53–96
Gaffney JS, Marley NA, Clark SB (1996) Humic and fulvic acids: isolation, structure, and environmental role. American Chemical Society, Washington, DC
Gao Y, He J, Ling W, Hu H, Liu F (2003) Effects of organic acids on copper and cadmium desorption from contaminated soils. Environ Int 29:613–618
Ghosh M, Singh S (2005) A review on phytoremediation of heavy metals and utilization of it's by products. Asian J Energy Environ 6:18
Gonsior SJ, Sorci JJ, Zoellner MJ, Landenberger BD (1997) The effects of EDTA on metal solubilization in river sediment/water systems. J Environ Qual 26:957–966
Hackh IWD, Grant RL, Grant C (1987) Grant & Hackh's chemical dictionary. McGraw-Hill, New York
Harter RD (1983) Effect of soil pH on adsorption of lead, copper, zinc, and nickel. Soil Sci Soc Am J 47:47–51
Harter RD, Naidu R (2001) An assessment of environmental and solution parameter impact on trace-metal sorption by soils. Soil Sci Soc Am J 65:597–612
Helios Rybicka E, Calmano W, Breeger A (1995) Heavy metals sorption/desorption on competing clay minerals; an experimental study. Appl Clay Sci 9:369–381
Hickey M, Kittrick J (1984) Chemical partitioning of cadmium, copper, nickel and zinc in soils and sediments containing high levels of heavy metals. J Environ Qual 13:372–376

Irha N, Steinnes E, Kirso U, Petersell V (2009) Mobility of Cd, Pb, Cu, and Cr in some Estonian soil types. Est J Earth Sci 58:209–214
Jopony M, Young S (1994) The solid–solution equilibria of lead and cadmium in polluted soils. Eur J Soil Sci 45:59–70
Kabata-Pendias A (2010) Trace elements in soils and plants, 4th edn. CRC Press, Boca Raton
Kabata-Pendias A, Pendias H (1991) Trace elements in soils and plants, vol V, 2nd edn. CRC Press, Boca Raton
Karaca A (2004) Effect of organic wastes on the extractability of cadmium, copper, nickel, and zinc in soil. Geoderma 122:297–303
King LD (1988) Retention of metals by several soils of the southeastern United States. J Environ Qual 17:239–246
Korte N, Skopp J, Fuller W, Niebla E, Alesii B (1976) Trace element movement in soils: influence of soil physical and chemical properties. Soil Sci 122:350–359
Lamb DT (2010) Heavy metal phytotoxicity and bioavailability in contaminated soils. University of South Australia, Adelaide
Landner L, Reuther R (2004) Metals in society and in the environment, vol 8, Environmental pollution. Springer, Dordrecht
Lasat M (2000) Phytoextraction of metals from contaminated soil: a review of plant/soil/metal interaction and assessment of pertinent agronomic issues. J Hazard Subst Res 2:1–25
Malandrino M, Abollino O, Giacomino A, Aceto M, Mentasti E (2006) Adsorption of heavy metals on vermiculite: influence of pH and organic ligands. J Colloid Interface Sci 299:537–546
McBride MB (1989) Reactions controlling heavy metal solubility in soils. In: Advances in soil science, vol 10. Springer, New York, pp 1–56
McBride MB (1994) Environmental chemistry of soils. Oxford University Press, New York
McIntyre T (2003) Phytoremediation of heavy metals from soils. Adv Biochem Eng Biotechnol 78:97–123
McLaughlin MJ, Hamon RE, McLaren RG, Speir TW, Rogers SL (2000) Review: a bioavailability-based rationale for controlling metal and metalloid contamination of agricultural land in Australia and New Zealand. Aust J Soil Res 38:1037–1086
Morris CG (1992) Academic Press dictionary of science and technology. Academic, San Diego
Naidu R, Bolan N (2008) Contaminant chemistry in soils: key concepts and bioavailability. Chem Bioavailab Terr Environ 9:38
Naidu R, Bolan NS, Kookana RS, Tiller K (1994) Ionic-strength and pH effects on the sorption of cadmium and the surface charge of soils. Eur J Soil Sci 45:419–429
Naidu R, Kookana RS, Sumner ME, Harter RD, Tiller KG (1997) Cadmium sorption and transport in variable charge soils: a review. J Environ Qual 26:602–617
Nouri J, Khorasani N, Lorestani B, Karami M, Hassani AH, Yousefi N (2009) Accumulation of heavy metals in soil and uptake by plant species with phytoremediation potential. Environ Earth Sci 59:315–323
Nriagu JO, Pacyna JM (1988) Quantitative assessment of worldwide contamination of air, water and soils by trace metals. Nature 333:134–139
Olaniran AO, Balgobind A, Pillay B (2013) Bioavailability of heavy metals in soil: impact on microbial biodegradation of organic compounds and possible improvement strategies. Int J Mol Sci 14:10197–10228
Park JH, Lamb D, Paneerselvam P, Choppala G, Bolan N, Chung JW (2011) Role of organic amendments on enhanced bioremediation of heavy metal(loid) contaminated soils. J Hazard Mater 185:549–574
Parker SP (1994) McGraw-Hill dictionary of scientific and technical terms. McGraw-Hill, New York
Paulose B, Datta SP, Rattan RK, Chhonkar PK (2007) Effect of amendments on the extractability, retention and plant uptake of metals on a sewage-irrigated soil. Environ Pollut 146:19–24

Proust D (2015) Sorption and distribution of Zn in a sludge-amended soil: influence of the soil clay mineralogy. J Soils Sediments 15:607–622
Reuther R (1999) Trace metal speciation in aquatic sediments: methods, benefits, and limitations. In: Mudroch A, Azcue JM, Mudroch P (eds) Manual of bioassessment of aquatic sediment quality. CRC Press, Boca Raton, pp 1–54
Roberts D, Nachtegaal M, Donald Sparks D (2005) Speciation of metals in soils. In: Tabatabai MA, Sparks DL (eds) Chemical processes in soils, vol Soil science Society of America book series, no. 8. Soil Science Society of America, Madison
Sauvé S, McBride M, Hendershot W (1998) Lead phosphate solubility in water and soil suspensions. Environ Sci Technol 32:388–393
Schnitzer M (1999) A lifetime perspective on the chemistry of soil organic matter. Adv Agron 68:1–58
Schwab AP, Zhu DS, Banks MK (2008) Influence of organic acids on the transport of heavy metals in soil. Chemosphere 72:986–994
Schwertmann U (1991) Solubility and dissolution of iron oxides. Plant Soil 130:1–25
Selim HM (2013) Transport and retention of heavy metal in soils: competitive sorption. Adv Agron 119:275–308
Selim HM, Kingery WL (2003) Geochemical and hydrological reactivity of heavy metals in soils. CRC Press, Boca Raton
Shan XQ, Lian J, Wen B (2002) Effect of organic acids on adsorption and desorption of rare earth elements. Chemosphere 47:701–710
Shirvani M, Kalbasi M, Shariatmadari H, Nourbakhsh F, Najafi B (2006) Sorption-desorption of cadmium in aqueous palygorskite, sepiolite, and calcite suspensions: isotherm hysteresis. Chemosphere 65:2178–2184
Silveira MLA, Alleoni LRF, Guilherme LRG (2003) Biosolids and heavy metals in soils. Sci Agric 60:793–806
Skłodowski P, Maciejewska A, Kwiatkowska J (2006) The effect of organic matter from brown coal on bioavailability of heavy metals in contaminated soils. In: Soil and water pollution monitoring, protection and remediation. Springer, Dordrecht, pp 299–307
Sparks DL (2003) Environmental soil chemistry, 2nd edn. Academic Press, San Diego
Sparks DL (1995) Environmental soil chemistry. Academic Press, San Diego
Sposito G (1984) The surface chemistry of soils. Oxford University Press, New York
Steffan L (2011) Transfer of heavy metals from contaminated soil into organs of soybean crops [Glycine max (L.) Merr.] in Córdoba, Argentina. University of Hohenheim, Stuttgart
Su C, Jiang L, Zhang W, Ferrarini A (2014) A review on heavy metal contamination in the soil worldwide: situation, impact and remediation techniques. Environ Skept Crit 3:24–38
Takamatsu T, Yoshida Y (1978) Determination of stability constants of metal-humic acid complexes by potentiometric titration and ion-selective electrodes. Soil Sci 125:377–386
Tiller KG, Gerth J, Brümmer G (1984) The relative affinities of Cd, Ni and Zn for different soil clay fractions and goethite. Geoderma 34:17–35
Trancoso MA, Dos Santos MMC, Goncalves MLS (2007) Lead sorption to selected Portuguese soils. Eur J Soil Sci 58:854–863
Tyler LD, McBride MB (1982) Mobility and extractability of cadmium, copper, nickel, and zinc in organic and mineral soil columns. Soil Sci 134:198–205
Violante A, Krishnamurti G, Pigna M (2007) Factors affecting the sorption-desorption of trace elements in soil environments. In: Violante A, Huang PM, Gadd GM (eds) Biophysico chemical processes of heavy metals and metalloids in soil environments. Wiley, Hoboken, pp 169–214
Violante A, Cozzolino V, Perelomov L, Caporale AG, Pigna M (2010) Mobility and bioavailability of heavy metals and metalloids in soil environments. J Soil Sci Plant Nutr 10:268–292
Vodyanitskii YN (2013) Contamination of soils with heavy metals and metalloids and its ecological hazard (analytic review). Eurasian Soil Sci 46:793–801

Vranova V, Rejsek K, Formanek P (2013) Aliphatic, cyclic, and aromatic organic acids, vitamins, and carbohydrates in soil: a review. Scientific World Journal 2013:1–15

White C, Sayer JA, Gadd GM (1997) Microbial solubilization and immobilization of toxic metals: key biogeochemical processes for treatment of contamination. FEMS Microbiol Rev 20:503–516

Wild A (1993) Soils and the environment: an introduction. Cambridge University Press, New York

Wuana RA, Okieimen FE (2011) Heavy metals in contaminated soils: a review of sources, chemistry, risks and best available strategies for remediation. ISRN Ecol 2011:402647

Yan Y, Gao J, Wu J, Li B (2014) Effects of inorganic and organic acids on heavy metals leaching in contaminated sediment. In: Sui W, Sun Y, Wang C (eds) An interdisciplinary response to mine water challenges. China University of Mining and Technology Press, Xuzhou

Yongsheng Q (2008) Study on the influences of combined pollution of heavy metals Cu and Pb on soil respiration. J Anhui Agric Sci 36:1117–1128

Zhang W, Jiang F, Ou J (2011) Global pesticide consumption and pollution: with China as a focus. Proc Int Acad Ecol Environ Sci 1:125–144

Chapter 2
Heavy Metals Accumulation in Coastal Sediments

S.M. Sharifuzzaman, Hafizur Rahman, S.M. Ashekuzzaman, Mohammad Mahmudul Islam, Sayedur Rahman Chowdhury, and M. Shahadat Hossain

Abstract Heavy metals, such as arsenic (As), cadmium (Cd), chromium (Cr), copper (Cu), lead (Pb), mercury (Hg) and zinc (Zn), are major environmental problem due to their toxic nature, nonbiodegradability and accumulative behaviors. Once in the estuarine/coastal and marine environments, sources predominantly form industrial, agricultural and hydrocarbon-related activities, scrap metal recycling, commercial ports and sewage, these contaminants accumulate in sediments and soils. Thus, heavy metals concentrations in coastal areas around shipyards, ports and industrial sites with refineries, smelters and milling facilities are often far exceed their background values or standard limits that can be toxic. The toxicity of heavy metals may negatively affects marine biodiversity as higher concentration is detected in fish and other organisms. Due to their persistence, through bioaccumulation and biomagnification along the aquatic food chain, heavy metals contamination ultimately affects human health. Here, the sources and impacts of heavy metals pollution in living systems are discussed.

Keywords Sediment • Heavy metals • Toxic effect • Ship breaking • Food chain • Pollution

---

S.M. Sharifuzzaman (✉) • S.R. Chowdhury • M.S. Hossain
Institute of Marine Sciences and Fisheries, University of Chittagong, Chittagong 4331, Bangladesh
e-mail: sharifuzzaman@cu.ac.bd; sharifuzaman@yahoo.com

H. Rahman
Department of Environmental Science, Independent University, Bangladesh (IUB), Dhaka, Bangladesh

S.M. Ashekuzzaman
School of Engineering and Built Environment, Glasgow Caledonian University, Glasgow G4 0BA, Scotland, UK

M.M. Islam
Department of Coastal and Marine Fisheries, Sylhet Agricultural University, Sylhet 3100, Bangladesh

H. Hasegawa et al. (eds.), *Environmental Remediation Technologies for Metal-Contaminated Soils*, DOI 10.1007/978-4-431-55759-3_2

2.1 Introduction

Achieving millennium developmental goals (MDGs) and sustainable development have been a major concern of the world in the recent decades. In particular, developing countries are facing threats due to environmental pollution arising to meet the objectives of MDGs and economic growth. Populated and agro-based developing countries of the Asian and Pacific region are heavily relying on chemical fertilizers and inorganic pesticides to meet increasing demand of food supply that eventually contaminating the natural environment including human food chain (Akoto et al. 2008). Moreover, major industries of many countries are situated either in the big cities or on the banks of major rivers and coastal areas, directly disposing effluents into environment without treatment (Kawser et al. 2011).

Over the last few decades, land-based metal polluting industries, such as textiles, coal and energy bases, cement plant, tannery and ship breaking/recycling, are expanding progressively as this type of enterprises have become an important factor for both macro and micro perspectives of economic growth. In particular, poor compliance of the environmental and safety legislations along with low labor costs and complimentary geographical locations have made the ship-breaking industry highly lucrative in a few developing countries (Sarraf et al. 2010). Although dismantling ships release considerable amounts of heavy metals into the surrounding environment, the scraps are used as raw materials for re-rolling mills and some other purposes, suggesting both opportunities and potential threats to a progressively more globalized economy. However, by now, over 40 % of the world's oceans have been affected by human activities to a large extent but, undoubtedly the coastal areas are suffering from the worst impacts (Lotze et al. 2006; Halpern et al. 2008).

Heavy metals are toxic and cause serious problems to environment as they are long persistent and not easily oxidized, degraded, removed, or converted to less harmful components through biological or chemical processes. Although no fixed limit definition, heavy metals are commonly characterized by their density, atomic weight, atomic number, or periodic table position (Duffus 2002). Density of heavy metals ranged from 3.5 to 7.0 g/cm^3, atomic weight from 23 to 40, and atomic numbers from 20 to 92 (Hawkes 1997). Heavy metals being nondegradable, unlike most pollutants (both organic and inorganic), can be accumulated in aquatic habitats and undergo global ecological cycle via natural waters as main pathway (Siddiquee et al. 2009). Sediments served as excellent indicator of metal pollution in costal environment as large inputs (>90 %) of heavy metals ultimately find their way to the estuarine zone and on the continental shelf (Yeasts and Bewers 1983; Satpathy et al. 2012); therefore, concentrations in sediments are often 10–100 times higher than those in ambient solution (Temara et al. 1998; Clark 1999).

At contaminated sites, commonly occurring metals are lead (Pb), chromium (Cr), zinc (Zn), cadmium (Cd), copper (Cu), mercury (Hg), arsenic (As) and nickel (Ni), and their total concentration in soils persists for a long time after their introduction and thus, biogeochemical activity of soil is affected including organisms living therein, to a varying degree (Bryan and Langston 1992). Indeed, anthropogenic metal contamination in coastal and marine environments has been a huge concern, because it can result in possible toxic effects on aquatic life forms, degradation of natural ecosystems and serious threats to human health via consumption of contaminated seafood (Ruilian et al. 2008; Pan and Wang 2012; Hu et al. 2013; Naser 2013).

2.2 Anthropogenic Sources of Heavy Metals in Coastal Sediments

As naturally occurring constituents of the earth's crust, the concentrations of metals in marine ecosystems are usually expected to be in the low range between nanogram to microgram per liter (liquid phase) or per gram (solid phase). Standard guidelines of safety limits for metal concentrations in marine sediment as set by the United States Environmental Protection Agency (US EPA) are summarized in Table 2.1. Incidentally, in the last few decades, the concentrations of heavy metals (e.g., Zn, Pb, Cd, Cu and Hg) have been increased in marine sediments by five or ten times higher compared to concentrations recorded in 50 or 100 years ago (Cardoso et al. 2001; Mashiatullah et al. 2013). Mainly rapid industrialization and economic development in recent decades have raised the metal contamination into marine environment at global scale. Major routes for metals to be released into aquatic systems are industrial effluents, municipal and domestic sewage discharges, ship breaking and agriculture activities, exploration and production of oil/gas, petroleum refinery and metalliferous mining and smelting, Fig. 2.1 (Pan and Wang 2012; Ra et al. 2014). Once metals incorporated into sediments they are not readily available to aquatic habitats, but changes in physicochemical conditions like pH, temperature, salinity, redox potential and organic ligand concentrations can help dissolution of metals from a solid phase. Thus, environmental conditions of an area largely determine the bioavailability, mobility and toxicity of metals (Nobi et al. 2010; Cukrov et al. 2011; Mashiatullah et al. 2013; Sany et al. 2013).

Table 2.1 Standard guideline applicable for heavy metals in marine sediment

	Sediment quality (μg/g)			
Metal	Average crustal	Non-polluted	Moderately polluted	Heavily polluted
As	1.8	<3	3–8	>8
Cu	55	<25	25–50	>50
Cr	100	<25	25–75	>75
Ni	75	<20	20–50	>50
Pb	12.5	<40	40–60	>60
Zn	70	<90	90–200	>200

After Pazi (2011)

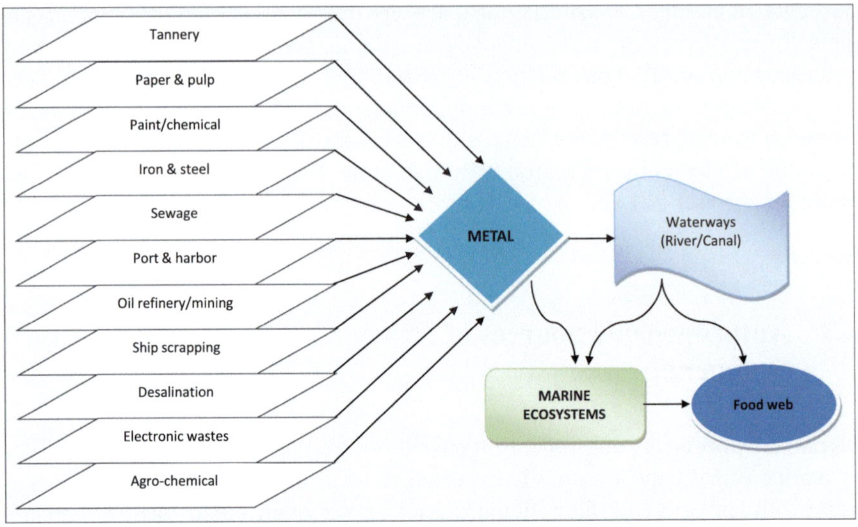

Fig. 2.1 Major anthropogenic sources of heavy metals into the marine environment

2.2.1 *Ship Scrapping and Repairing*

Ship scrapping activity is a good source of raw material for the steel mills and also generates substantial employment in several countries such as Bangladesh, India, China, Pakistan and Turkey. As a result, the coasts along the ship breaking area of these countries are getting polluted from ship scrapping (Reddy et al. 2004, 2005; Siddiquee et al. 2009; Neser et al. 2012). For example, concentrations of Hg, Cd, Pb, Cr, Cu, Zn, Mn and Ni around Allaga (Turkey) ship breaking facilities were higher in comparison to respective background levels. The metal levels (μg/g) were ranged from 0.32–7.02 for Hg, 0.06–3.94 for Cd, 26–751 for Pb, 65–264 for Cr, 20–703 for Cu, 86–970 for Zn, 283–1,192 for Mn, 28–240 for Ni, 32,390–54,666 for Fe and 4,006–41,962 for Al (Neser et al. 2012). Moreover, severe metal accumulation was observed at an abandoned shipyard site in Hong Kong, where the concentrations of Cu (653 ppm), Pb (1485 ppm) and Zn (1622 ppm) exceeded environmental standards (Chiu et al. 2006). Incidence of much higher concentration of metal pollutants in sediments of Sitakund (Bangladesh) was also due to presence of ship breaking yards at the area (Aktaruzzaman et al. 2014).

2.2.2 *Industrial Effluent*

Heavy metals enrichment in coastal and estuarine ecosystems due to direct discharge of wastewaters or solid wastes from various industrial activities including mining, metal smelting, printing, marine transport and port, desalination plants, electrical and electronic wastes (e-waste) recycling have been widely reported (Youssef and El-Said 2011; Pan and Wang 2012; Wang et al. 2013; Ra et al. 2014). Every year, an estimated 160,000 factories dump between 41,000 and 57,000 tons of toxic organic chemicals and 68,000 tons of toxic metals into coastal waters (UNEP and UN-HABITAT 2005). Recently, Sany et al. (2013) investigated temporal and spatial distribution of heavy metals in sediments of Port Klang (Malaysia) and observed significantly higher concentrations of As, Cd, Hg and Pb than that of background values at which these metals are considered hazardous to marine organisms and health. Main sources of metal contamination in Port Klang were industrial wastewater from cement manufacturing, oil/electrical-based power generation, palm oil, port activities (e.g., boat docking and corrosion of ships) and agricultural activities such as application of organic insecticides (lead-arsenate), pesticides, fertilizers. Metal contamination in coastal sediments of the Arabian Gulf countries due to effluents from desalination plants has also been reported by number of studies (de Mora et al. 2004; Naser 2012, 2013). For example, Sadiq (2002) confirmed an accumulation of high levels of Cd, Co, Cu, Hg, V, Fe, P and Zn in sediments at the proximity of a seawater desalination plant in Saudi Arabia. This fact is well supported by Abdul-Wahab and Jupp (2009) who documented localized pollution of Cu, Cr, Fe and Ni in coastal sediments of Oman due to effluent discharges from power/desalination plants. Due to discharges from printed circuit board (e-waste), electroplating, metal and textile industries, high levels of Cu, Pb, Zn, Cd and Hg were reported in sediments from coastal areas of Hong Kong (reviewed by Wang et al. 2013). Similarly, industrial pollutant is a major source of heavy metals contamination in the coast of Bangladesh. There are thousands of industrial establishments along the coastline and riverbank dealing with jute, paper and pulp, textiles, fertilizers, rubber and plastic, tannery, food and beverages, sugar, pharmaceuticals, tobacco, paint, distilleries, cement clinker, refinery, ship breaking, etc., and their untreated effluents and solid wastes contribute to metal pollution in coastal waters (Islam and Hossain 1986; Kashem and Singh 1999; Rasul et al. 2006). The degree of heavy metals accumulation in coastal sediments because of industrial activities can be understood from Table 2.2.

2.2.3 *Sewage*

Discharge of sewage is another important source of coastal pollution worldwide. About 67 % of sewage from urban areas is discharged untreated into lakes, rivers and coastal waters, and yearly 5.9 trillion gallons of sewage is discharged into coastal waters by the sewage treatment facilities (UNEP and UN-HABITAT 2005). The composition of sewage effluents includes biological and chemical pollutants with heavy metals and trace elements (Table 2.3) that may deteriorate the quality of receiving coastal and marine environments (Al-Muzaini et al. 1999; Singh et al. 2004; Naser 2013). In China, the annual discharge of municipal sewage was reported to increase from 45 billion tons in 2003 to 57 billion tons in 2008 which is

Table 2.2 Concentration range of heavy metals in sediments of coastal areas around the world

Major activity	Location	Heavy metals concentration (μg/g)								References
		As	Cu	Cr	Cd	Ni	Pb	Hg	Zn	
Industry	Jinzhou Bay, North China	20.4–820 (396.5)	9–1227 (417)	44.0–72 (60.6)	n.d.	n.d.	21.8–1828 (753.2)	n.d.	89–13,933 (6419)	Zhang et al. (2008)
	Fujian coast, Southeast China	2.4–15.1 (9.1)	2–64 (22)	4.5–279 (57.4)	0.01–1.2 (0.08)	2.1–53.3 (27.4)	7.7–144 (37.1)	0.00–0.22 (0.05)	4–319 (96)	Zhang (2008)
	Hong Kong, Southeast China	n.d.	1–4000 (119)	5.0–560 (48.9)	0.10–5.3 (0.3)	5.0–220 (24.7)	9.0–260 (53.6)	0.10–8.00 (0.19)	17–790 (148)	Zhou et al. (2007)
	Ulsan Bay, Korea	7.0–34 (15.8)	12.5–384.7 (95.6)	17.8–90.6 (64.7)	0.09–1.64 (0.4)	7.1–47 (32.2)	18.1–1088 (90.7)	0.01–1.01 (0.16)	91.6–4356 (361.9)	Ra et al. (2014)
	Karachi coast, Pakistan	n.d.	n.d.	12.0–319.8 (96.8)	n.d.	1.5–58.9 (31.4)	9.0–49.5 (23.2)	n.d.	15.6–666.3 (204.8)	Mashiatuallah et al. (2013)
Port	Port Klang, Malaysia	34.1–112.8 (60.4)	11.4–40.6 (17.4)	30.2–74.8 (46.4)	0.28–1.55 (0.83)	6.2–17.8 (11.4)	47.5–85.9 (59.4)	0.17–0.35 (0.23)	32.8–126.9 (51.1)	Sany et al. (2013)
	Raysut Port, Oman	n.d.	0.83	31.2	0.20	2.49	0.73	0.002	1.57	de Mora et al. (2004)
Desalination plant	Muscat, Oman	n.d.	3.1–15.9 (10.0)	27.1–223.5 (95.1)	0.0–1.2 (0.6)	75–790 (299.2)	0.1–3.3 (1.7)	n.d.	<0.02–41.9 (13.9)	Abdul-Wahab and Jupp (2009)
	Ras Tanajib, Saudi Arabia	n.d.	<0.01–32.2 (3.5)	2.7–23.2 (8.3)	0.7–31.6 (3.3)	3.3–37.2 (12.1)	4.7–46.3 (14.8)	0.01–0.15 (0.06)	2.1–33.6 (7.4)	Sadiq (2002)

Number within a bracket indicates average value
n.d. not determined

Table 2.3 Concentrations of heavy metals and trace elements in sewage

Metals in sewage	Concentration (mg/l)
As	<0.1
Cd	<0.02
Cr	0.1–0.5
Cu	0.2–0.5
Pb	0.08–0.4
Hg	–
Ni	<0.02
Ag	<0.02
Zn	0.4–0.7

Adapted from Islam and Tanaka (2004)

expected to follow a continuous temporal increase (NBSC 2004–2009; Pan and Wang 2012). Incidentally, total heavy metal loading of raw wastewater in a conventional wastewater treatment process usually ends up in the sludge or remains in the final effluent, because such treatment systems are mainly designed for solids/liquids separation, reduction of organic matter and nutrients. For example, in an investigation on the occurrence and fate of heavy metals during wastewater treatment process operated under activated sludge mode, Karvelas et al. (2003) reported that 47–63 % of daily input of Cr, Pb, Ni, Cd, Zn and Fe end up in the treated effluent that ultimately sink to sediments through recipient aquatic systems. Moreover, sewage from most coastal cities in Bangladesh, particularly in Chittagong find their way into the coastal environment, thus playing role in metal enrichment. Studies showed that untreated domestic sewage from residential source contaminate aquatic environments and soils with certain heavy metals including Cr, Ni, Zn and Pb (O'Connor et al. 2003).

2.2.4 Hydrocarbons and Related Activity

Measurable quantity of various metals, such as Ni, Cu, Cr, Cd, Pb, Fe, Zn, vanadium (V), Mn and Co are normal constituents of crude oil or hydrocarbons (Onojake et al. 2011; Onojake and Frank 2013). Therefore, discharge of petroleum refinery wastewaters into marine environment can contribute to an increased concentration of heavy metals (Table 2.4). Certainly, an increase in the concentrations of Cu, Hg, Pb and Zn was recorded in sediments nearby the BAPCO (Bahrain Petroleum Company) oil refinery in Bahrain, suggesting localized anthropogenic inputs of heavy metals in the environment (reviewed by Freije 2015). In addition, an oil spillage was partially responsible for the concentrations of Ni, Cu and Pb in oil-polluted soils in Nigeria (Onojake and Frank 2013).

Moreover, offshore oil/gas exploration and production activities could lead to possible sediment contamination with metals (Table 2.4) by generating considerable quantities of drill cuttings and drilling wastes/fluids onto the seafloor of

Table 2.4 Accumulation of heavy metals in coastal sediments due to hydrocarbons and related activity

Activity	Location	Heavy metals concentration (μg/g)								References
		Cd	Cr	Cu	Hg	Ni	Pb	V	Zn	
Oil production platform	[a]North Sea, Europe	<0.1–10.3 (n/a)	38.4–426 (50)	75.5–374 (39)	<0.1–32.6 (n/a)	26.3–137 (32)	62.3–4785 (63)	180–523 (64)	553.5–2511 (155)	Reviewed by Breuer et al. (2004)
	California, USA	1.20	85	30	–	41	19	71	290	
Oil spills	Nsukka, Nigeria	4.8	11.7	18.7	–	24.3	5.7	–	223	Adesodun and Mbagwu (2008)
	Taiwan	25.3	2294	1460	0.34	1278	2795	–	2983	Lai 210 et al. (2010)
Oil refinery	BAPCO, Bahrain	0.18	41.8	<8.3	0.22	23.2	99	–	52.2	de Mora et al. (2004)

n/*a* not applicable

(–): not measured

[a]Metal data from several North Sea oil drilling platforms located at Scotland and Norway, and number within a bracket refers to reported mean background value for mud (0–1 cm)

surrounding oil platforms (Newbury 1979; Agwa et al. 2013). The composition of drill cuttings include: (1) drilling mud; (2) speciality chemicals, such as viscosifiers, emulsifiers, lubricants, wetting agents, corrosion inhibitors, surfactants, detergents, caustic soda, salts and organic polymers; and (3) fragments of reservoir rock that carry metals, i.e., barium (Ba), Cr, Cu, Ni, Pb and Zn, and hydrocarbons. Estimates suggest that 7–12 million m^3 of drill cuttings have been accumulated on the seabed of North Sea due to oil/gas drilling activity and thus elevated concentrations of Cr, Cu, Ni, Pb and Zn were reported nearby drill sites (reviewed by Breuer et al. 2004). Indeed, due to presence of crude-oil production terminal, significant metal pollution, i.e., high enrichment factors for Pb and Fe, was observed in soils and sediments at Forcados, in the Niger Delta of Nigeria (Obiajunwa et al. 2002). Apart from oil refineries and terminals (i.e., offshore oil rigs, underwater pipelines), accidental oil spillage from (tanker) ships and oil residues from ballast waters and land-based units (i.e., industrial and shipbuilding units) are causing metal pollution by introducing petroleum hydrocarbons into the marine environment (Naser 2013).

2.2.5 *Dredging and Reclamation*

Due to major housing, recreational and economic developments, including artificial islands, ports, power/desalination plants, hotels, fisheries harbors, aquaculture projects, the coasts surrounding Arabian Gulf have been brought under massive construction activities that often involved with conversion of coastal areas, intensive dredging and land reclamation actions (Naser 2013; Freije 2015). This sort of scenario is also true for many other countries of the world. Dredging and reclamation processes, including mining for construction materials cause temporary and/or permanent physical (e.g., anoxia), chemical (e.g., toxicity) or biological (e.g., loss of species) alterations in marine environment that may cause reduced biodiversity, richness, abundance and biomass of marine organisms (Smith and Rule 2001). Indeed, an elevated level of heavy metals was reported to mobilize in aquatic environment due to such activities, Table 2.5 (Guerra et al. 2009; Hedge et al. 2009) that may enter into important food web components, posing threats to human health. In Bangladesh, periodic dredging is done to maintain navigable shipping channels, nautical depth in port and river waterways for ferry crossing,

Table 2.5 Metal concentrations in coastal sediments occurring from dredging and reclamation activity

		Heavy metals concentration (μg/g)								
Activity	Location	Cd	Cr	Cu	Hg	Mn	Ni	Pb	Zn	References
Dredging and reclamation	Hong Kong estuarine	1.5	80	65	1.0	–	40	75	200	Evans (1994)
	Victoria Harbour, Hong Kong	32	80	11	–	196	14	–	67	Tanner et al. (2000)

but no scientific data is available on the extent of heavy metals contamination resulting from such activity.

2.3 Assessment of Metals Pollution

Monitoring sediment quality for spatial distribution, accumulation and seasonal variation of metals is important in order to identify and understand anthropogenic input of metals into the marine environment. Many studies have suggested an increase of metals content in coastal sediment than that of their natural background concentration. The assessment of heavy metals contamination/pollution is usually done by calculating enrichment factor, pollution load index, geo-accumulation index and contamination factor (Youssef and El-Said 2011; Satpathy et al. 2012; Hasan et al. 2013; Hu et al. 2013; Mashiatullah et al. 2013; Ra et al. 2014).

Enrichment factor (EF) is commonly used to distinguish metals originating from anthropogenic and natural sources. For data normalization purpose, the sample metal concentrations are initially normalized to reference elements (e.g., iron or aluminum) to determine whether a sediment sample is enriched with metals in comparison to the sample's background conditions. To determine EF values (Eq. 2.1), iron (Fe) is commonly selected as normalizing element because it is a major sorbent phase for trace metals and is a quasi-conservative tracer of the natural metal-bearing phases in fluvial and coastal sediments (Schiff and Weisberg 1999; Turner and Millward 2000). According to this, EF can be calculated as the sample metal (X) to Fe concentration ratio divided by the background metal/Fe concentration ratio and expressed as:

$$\mathrm{EF} = \frac{\left(\frac{\mathrm{X}}{\mathrm{Fe}}\right)_{\mathrm{sample}}}{\left(\frac{\mathrm{X}}{\mathrm{Fe}}\right)_{\mathrm{background}}} \qquad (2.1)$$

The average crustal abundance data can be used to select background metal values, although regional background values have been suggested to be more appropriate (Christophoridis et al. 2009; Rubio et al. 2000; Hu et al. 2013). EF value of nearly unity denotes the elements that are naturally derived, while EF values of several orders indicate elements of anthropogenic origin. Table 2.6 shows the classification of EF value according to Taylor (1964) to determine the degree of metal contamination.

Geo-accumulation index (I_{geo}), which was introduced by Müller (1969), can be used to determine the degree of heavy metals pollution in sediment. The formula to calculate I_{geo} value is expressed as:

$$I_{\mathrm{geo}} = \log_2\left(\frac{C_{\mathrm{n}}}{1.5\ B_{\mathrm{n}}}\right) \qquad (2.2)$$

Table 2.6 The degree of metal pollution according to seven enrichment factor (EF) classes

EF value	Designation of sediment quality
<1	No enrichment
1–3	Minor enrichment
3–5	Moderate enrichment
5–10	Moderately severe enrichment
10–25	Severe enrichment
25–50	Very severe enrichment
>50	Extremely severe enrichment

After Taylor (1964)

Table 2.7 Classification of the geo-accumulation index (I_{geo})

I_{geo} value	Class	Quality of sediment
<0	0	Unpolluted
0–1	1	Unpolluted to moderately polluted
1–2	2	Moderately polluted
2–3	3	Moderately to strongly polluted
3–4	4	Strongly polluted
4–5	5	Strongly to extremely polluted
>5	6	Extremely polluted

After Müller (1981)

where C_n is the concentration of metal (n) and B_n is the geochemical background concentration of metal (n). The factor 1.5 is a background matrix correction that was introduced to minimize the effect of possible variations in the background values due to lithogenic effects. The sediment quality was graded into seven classes according to I_{geo} value as defined by Müller (1981), Table 2.7.

Pollution load index (PLI) is another parameter to evaluate metal pollution in marine environment, and it can be calculated from the following equation given by Tomlinson et al. (1980).

$$\text{PLI} = (\text{CF}_1 \times \text{CF}_2 \times \cdots \times \text{CF}_n)^{1/n} \qquad (2.3)$$

where *CF* is contamination factor and *n* is the number of metals investigated.

$$\text{CF} = \frac{\text{Metal concentration in sediment}}{\text{Background value of metal}}$$

A PLI value above one (>1) indicates that an area is polluted, whereas values <1 indicates no pollution or only background levels of pollutants are present (Chakravarty and Patgiri 2009; Mashiatullah et al. 2013). While an estimation of PLI can be used to identify whether a site is collectively polluted or non-polluted by metals, the value of CF can be used to identify the contamination of an individual metal in a basin. According to Hakanson (1980), CF was classified into four groups: CF <1 = low contamination factor, 1 ≤ CF <3 = moderate contamination factor, 3 ≤ CF <6 = considerable contamination factor and CF >6 = very high

Table 2.8 Sediment quality guidelines (SQGs) for heavy metals in marine sediments

	Metals concentration (μg/g)							
SQGs	As	Cu	Cr	Cd	Ni	Pb	Hg	Zn
TEL	7.2	18.7	52	0.68	15.9	30.2	0.13	124
PEL	41.6	108	160	4.2	42.8	112	0.7	271
ERL	8.2	34	81	1.2	21	47	0.15	150
ERM	70	270	370	9.6	52	218	0.71	410

Adapted from Grecco et al. (2011), Hu et al. (2013), and Dimitrakakis et al. (2014)
TEL threshold effects level, below which adverse biological effects are not expected to occur, *PEL* probable effects level, above which adverse biological effects are expected to occur, *ERL* effects range low, *ERM* effects range medium

contamination factor (Pazi 2011). In addition, sediment quality guidelines (SQGs), which was established by the National Oceanic and Atmospheric Administration (NOAA), are widely used to determine whether metals available in sediments pose any threat to aquatic ecosystems or lead to an adverse biological impacts, Table 2.8 (Grecco et al. 2011; Hu et al. 2013; Sany et al. 2013). The effects range-low (ERL) refer to the concentration at which small percentage of biota is affected, whereas the effects range-medium (ERM) point to greater percentage of adverse effects resulting from metal exposure equal to or greater than this concentration level (Long et al. 1995; Breuer et al. 2004).

2.4 Heavy Metals in Coastal Sediments: Example from Ship Breaking Area

Ocean-going vessels, after a life span of 20–30 years, are principally dismantled in India, Bangladesh, Pakistan and China (Sujauddin et al. 2015). Specially, due to long and flat uniform coastal intertidal zone and supply of cheap labour with lax environmental regulations, ship breaking activity has flourished tremendously in Bangladesh. As a result over 25-years period, the longitudinal extent of the shore used for ship breaking yards has expanded from about 3 km in 1989 to nearly 15 km in 2014 (Fig. 2.2). However, scrapping activities generate various hazards for the coastal and marine environments by releasing loads of pollutants, including toxic waste, harmful chemicals, i.e., poly-chlorinated biphenyls (PCBs), polyvinyl chloride (PVC), polycyclic aromatic hydrocarbons (PAHs), tin-organic compounds (TBT), oils, gas, asbestos, heavy metals (i.e., Hg, Pb, Cd, etc.) and fumes (dust, fume/gas components: dioxins, isocyanates, sulphurs) into the water, seabed/ground and the air (Sinha 1998). Heavy metals are found in many parts of ships such as in paints, coatings, anodes and electrical equipment. Of the metals, mercury, lead and cadmium are of greatest concern because of their ability to travel long distances in the atmosphere. Several studies have demonstrated higher

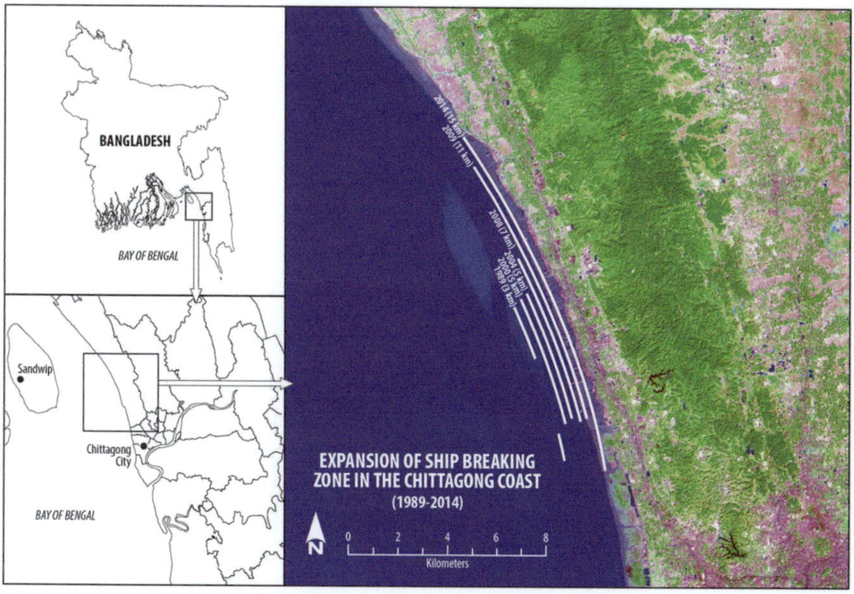

Fig. 2.2 Growth of ship breaking area (1989–2014) in the Chittagong coast, Bangladesh

concentration of heavy metals in the sediments along ship breaking areas (Reddy et al. 2004; Neşer et al. 2008).

For example, Tewari et al. (2001) observed 25–15, 500 % higher concentrations of metals, i.e., Fe, Mn, Co, Cu, Zn, Pb, Cd, Ni and Hg at Alang (India) ship breaking point when compared to a control site. In another study, Reddy et al. (2004) determined 19 times higher concentrations of heavy metals in the intertidal zone of Alang-Sosiya (India) ship scrapping yard than a reference point, located 60 km away. In particular, the contamination level was most intensive with Zn, Mn and Pb. Moreover, sediments surrounding Aliağa (Turkey) ship recycling zone were contaminated with Hg, Cd, Pb, Cr, Cu, Zn, Mn, Ni, and in particular, the levels of Pb, Cr, Cu, Zn, Ni in sediments were 'heavily polluted' as per the sediment quality guidelines (Neşer et al. 2012). Similarly, higher concentrations of Cd and Pb were recorded in the sediments of intertidal zone around the ship breaking area of Chittagong (Bangladesh) due to discharge of untreated effluents from ship breaking yards (Khan and Khan 2003). In addition, Mavrakis et al. (2004) reported much higher average concentrations of Cu (319–898 μg/g), Fe (34–74 mg/g), Mn (733–6560 μg/g) and Ni (98–126 μg/g) in the sediments of Skaramaga (Greece) ship yards than those measured at a remote sampling point (i.e., Cu = 44–351 μg/g; Fe = 24–29 mg/g; Mn = 399–1308 μg/g, and Ni = 84–112 μg/g). They concluded that coastal sediment linked to shipyards and scrap metal yards tend to hold nearly double quantity of heavy metals contaminant compared to the center of gulf/bay. Because of such heavy metal pollution, the coastal area nearby ship breaking zone is vulnerable to serious environmental catastrophe with regard to destruction of marine biodiversity and long-term effect on human health. Table 2.9 shows levels

Table 2.9 Levels of metal contaminations in sediments in the vicinity of ship breaking area

	Sitakund of Chittagong, Bangladesh		Alang-Sosiya, India	Aliağa, Turkey	Bakopoulos, Greece	
Metals (ppm)	Siddiquee et al. (2009)	Hasan et al. (2013)	Tewari et al. (2001)	Neşer et al. (2012)	Mavrakis et al. (2004)	Standard value
Cd	0.6–0.9	0.01–1.16	24	0.06–3.94	–	0.11[a, b]
Cr	23–87	311–1,232	–	65.5–264	–	77.2[a]
Cu	21–40	6–1,635	8	19.6–703	107–206	33[b]
Fe	11,933–41,362	22,890–131,110	75,088	32,390–54,666	26,000–60,000	41,000[b, c]
Hg	0.01–0.12	–	1.6	0.32–7.02	–	0.02[a]
Mn	2–8	334–2,524	1488	283–1,192	1,048–2,016	770[c]
Ni	23–49	8–45	117	27.6–240	75–116	56.1[a]
Pb	37–148	16–22	220	91.3–751	–	19[c]
Zn	84–143	58–978	51	86.4–970	–	95[b, c]

(–) not measured
[a]IAEA (1989)
[b]GESAMP (1982)
[c]Salomons and Froster (1984)

of heavy metals in the sediments of ship breaking area compared to recommended values of unpolluted sediments.

2.5 Impacts of Heavy Metals on Living Systems

Above the threshold concentrations, depending on type of metal, animal species and environment, all heavy metals can be toxic to aquatic life. Although most marine organisms have the ability to store, remove (through feces, eggs or molting) or detoxify (with metallothioneins) heavy metals, many still accumulate such pollutant when contamination level is acute (Clark 1999). Indeed, several studies have reported certain heavy metals concentrations in coastal sediments to exceed the ERM values by an order of magnitude that may be associated with biological effects (de Mora et al. 2004; Reddy et al. 2004; Neşer et al. 2012). The incidence of adverse effects increased with increasing concentrations of metals (Long et al. 1995); however, in the order of decreasing toxicity, Hg, Cd, Ni, Pb, Cu, Cr, As and Zn are considered most poisonous to marine life (Freije 2015).

Certainly, an accumulation of heavy metals in the tissue of aquatic organisms, such as plankton, algae, molluscs, crustaceans and fish, has been reported from coastal areas contaminated with metals (Hossain and Khan 2001; de Mora et al. 2004; Naser 2013; Freije 2015). Exposure of marine organisms to elevated levels of metal contaminants can damage tissue and DNA, interfere with damage tissue regeneration process and inhibit growth (Kennish 1996). Adverse effects on fish, both in natural and under laboratory conditions, were reported due to heavy metals toxicity that may include reduced fertility, problematic reproduction, hatching delay, damaged kidney, slower growth and development, organ deformities, abnormal behavior and even death. The most common deformities can be seen in vertebral column, swim bladder, cephalic region, fins and lateral line. Metabolic processes (i.e., developmental retardation, morphological and functional anomalies) in developing fish, and in particular, embryonic and larval stages are very vulnerable and sensitive to metal toxicity. Moreover, fish is expected to have damaged gills, gut and sensory systems under elevated levels of toxic metals (reviewed by Sfakianakis et al. 2015). Hg accumulation in fish may harm their predators in food chain, principally birds and aquatic mammals (e.g., whales, dolphins) that feed on fish are more exposed to Hg toxicity than others inhabiting in a particular ecosystem (UNEP 2002).

Along the food chain, accumulated metals in the tissue of marine organisms (Habashi 1992; Peplow 1999) can be bioaccumulated and then biomagnified (e.g., grazer → primary consumer → secondary consumer → top predator) to various extents (Fig. 2.3). However, a few studies have confirmed biomagnification of heavy metals and only mercury (as methylmercury) and arsenicum showed high affinity for organic tissues (Claridge 2004). Humans, in turn, are exposed to heavy metals by consuming contaminated seafood and this has been known to cause a wide range of toxic effects, such as carcinogenicity, mutagenicity and teratogenicity.

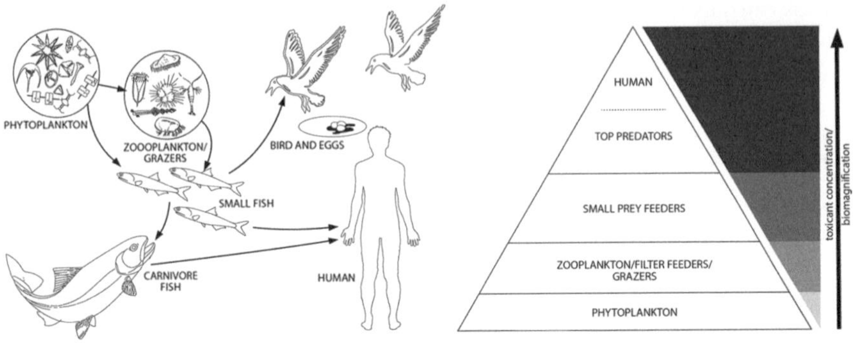

Fig. 2.3 Illustration of a typical marine food chain (*left*) showing feeding connections between organisms, and a trophic pyramid (*right*) demonstrating how toxicant becomes more concentrated as consumers/predators prey on smaller ones or producers

However, high-risk populations can be classified depending on multiple scenarios of exposure to specific metal or metal mixtures. Generally, children and pregnant women are vulnerable because of their physiological makeup and behavior. Metal susceptibility in pregnant women is even higher due to trans-placental shift of metals in maternal blood (Morse et al. 1979). Heavy metals are known to be persistent in the human body, with elimination half-lives last for decades. Risk assessment of human health to heavy metals can be efficiently done using biomarkers such as whole blood, hair, bone, hair, etc. (Barbosa et al. 2005).

The main threats to human health from heavy metals are associated with exposure to As, Cd, Cr, Hg and Pb (Järup 2003). As and Cr are classified as priority pollutants by the US EPA with a carcinogenicity classification A (human carcinogen), while Cd and Pb are classified in the same list with a carcinogenicity classification B, i.e., probable human carcinogen (Pekey 2006). Although a particular metal exhibit specific signs of its toxicity, there are some general signs including gastrointestinal disorders, diarrhea, stomatitis, tremor, hemoglobinuria causing a rust-red color to stool, ataxia, paralysis, vomiting and convulsion, depression and pneumonia (McCluggage 1991). Cadmium is toxic even at extremely low levels and ingesting very high levels severely irritates the stomach, leading to vomiting and diarrhea. Long-term exposure to lower levels leads to kidney damage and fragile bones. Long-term high cadmium exposure may cause skeletal damage, for instance, in Japan, outbreaks of itai-itai (ouch-ouch) disease in the late 1950s was due to cadmium poisoning (Järup 2003; Young 2005).

People who have been exposed to lead for a long time may suffer from memory deterioration, prolonged reaction time and reduced ability to understand. Studies have demonstrated that long-term low-level lead exposure in children may also lead to diminished intellectual capacity (Järup 2003; Udedi 2003). Long-term exposure of lead in adults can cause decreased performance in some tests that measure functions of the nervous system, weakness (in fingers, wrists or ankles), minor increase in blood pressure and anemia. Exposure to high lead levels can severely damage the brain and kidneys and ultimately cause death. In pregnant women, high levels of exposure to lead may cause miscarriage. High level exposure in men can damage the organs responsible for sperm production. In case of adults, several convincing population studies have shown a positive association between blood lead level and risk of death (Ogwuebgu and Muhanga 2005; Rossi 2008).

Exposure of the general population to mercury primarily occurs via food, fish being a major source of methyl mercury exposure (WHO 1990). Therefore, high mercury level, which is hypothesized to increase the risk of coronary heart disease (Salonen et al. 1995; Hossen et al. 2001), correlates positively with high dietary intake of contaminated fish (Järup 2003). High doses of mercury may lead to death, usually 2–4 weeks after onset of symptoms. The minamata disease in Japan in 1956 was caused by methyl mercury poisoning from fish contaminated by mercury being discharged into the surrounding sea. The nervous system is very sensitive to all forms of mercury and exposure to high levels can permanently damage the brain, kidneys and developing fetuses. Effects on brain functioning may result in irritability, shyness, tremors, changes in vision or hearing and memory problems (Järup 2003; Lenntech 2004).

Inorganic arsenic is acutely toxic, carcinogenic and intake of large quantities leads to gastrointestinal complexities, severe disturbances of the cardiovascular and central nervous systems, and eventually death. In less severe cases, bone marrow depression, hemolysis, hepatomegaly, melanosis, polyneuropathy and encephalopathy may be observed. Ingestion of inorganic arsenic may also induce peripheral vascular disease, which in its extreme form leads to gangrenous changes. Long-term low level exposure can cause a darkening of the skin and the appearance of small "corns" or "warts" on the palms, soles and torso (Järup 2003; Ogwuegbu and Ijioma 2003). An excess amount of zinc can cause system dysfunctions that result in impairment of growth and reproduction (INECAR 2000). The clinical signs of zinc toxicosis have been reported as vomiting, diarrhea, bloody urine, icterus (yellow mucus membrane), liver failure, kidney failure and anemia (Fosmire 1990).

2.6 Conclusion

Chronic inputs of heavy metals from ship breaking, hydrocarbons and related activity, seaport and industrial activities are developing highly contaminated sediments in the coastal environment. Thus, metal pollution raised concerns about toxic effects in marine organisms/ecosystems and the potential for bioaccumulation down the food chain with possible human health risks. However, enterprises center on marine environments are playing vital role in the economy of many countries, therefore, such activities cannot be stopped but sustainable approaches may be developed to reduce/combat polluting contaminants. Adequate legislation and proper management may prevent indiscriminate discharge of large quantity of toxic heavy metals into the environment. Moreover, proper investigations and research should be carried out to understand the sources, chemistry and potential toxicity of heavy metals in the contaminated ecosystem in order to undertake necessary measures such as to select appropriate management and remediation method(s). Indeed, it demands precautionary measures against the environmental and health threats due to heavy metals pollution from different human activities.

References

Abdul-Wahab SA, Jupp BP (2009) Levels of heavy metals in subtidal sediments in the vicinity of thermal power/desalination plants: a case study. Desalination 244:261–282

Adesodun JK, Mbagwu JSC (2008) Distribution of heavy metals and hydrocarbon contents in an alfisol contaminated with waste-lubricating oil amended with organic wastes. Bioresour Technol 99:3195–3204

Agwa A, Leheta H, Salem A, Sadiq R (2013) Fate of drilling waste discharges and ecological risk assessment in the Egyptian Red Sea: an aquivalence-based fuzzy analysis. Stoch Environ Res Risk A 27:169–181

Akoto O, Ephraim JH, Darko G (2008) Heavy metals pollution in surface soils in the vicinity of abundant railway servicing workshop in Kumasi, Ghana. Int J Environ Res 2:359–364

Aktaruzzaman M, Chowdhury MA, Fardous Z, Alam K, Hossain MS, Fakhruddin ANM (2014) Ecological risk posed by heavy metals contamination of ship breaking yards in Bangladesh. Int J Environ Res 8:469–478

Al-Muzaini S, Beg M, Muslamani K, Al-Mutairi M (1999) The quality of marine water around a sewage outfall. Water Sci Technol 40:11–15

Barbosa F Jr, Tanus-Santos JE, Gerlach RF, Parsons PJ (2005) A critical review of biomarkers used for monitoring human exposure to lead: advantages, limitations, and future needs. Environ Health Perspect 113:1669–1674

Breuer E, Stevenson AG, Howe JA, Carroll J, Shimmield GB (2004) Drill cutting accumulations in the Northern and Central North Sea: a review of environmental interactions and chemical fate. Mar Pollut Bull 48:12–25

Bryan GW, Langston WJ (1992) Bioavailability, accumulation and effects of heavy metals in sediments with special reference to United Kingdom estuaries: a review. Environ Pollut 76:89–131

Cardoso A, Boaventura G, Silva E, Brod J (2001) Metal distribution in sediments from the Ribiera bay, Rio de Janeiro Brazil. J Braz Chem Soc 12:767–774

Chakravarty M, Patgiri AD (2009) Metal pollution assessment in sediments of the Dikrong River, NE India. J Hum Ecol 27:63–67

Chiu SW, Ho KM, Chan SS, So OM, Lai KH (2006) Characterization of contamination in and toxicities of a shipyard area in Hong Kong. Environ Pollut 142:512–520

Christophoridis C, Dedepsidis D, Fytianos K (2009) Occurrence and distribution of selected heavy metals in the surface sediments of Thermaikos Gulf, N. Greece. Assessment using pollution indicators. J Hazard Mat 168:1082–1091

Claridge PN (2004) Fishes in estuaries. J Fish Biol 64:1462–1463

Clark RB (1999) Marine pollution, 4th edn. Clarendon, Oxford

Cukrov N, Frančišković-Bilinski S, Hlača B, Barišič D (2011) A recent history of metal accumulation in the sediments of Rijeka harbor, Adriatic Sea, Croatia. Mar Pollut Bull 62:154–167

de Mora S, Fowler S, Wyse E, Azemard S (2004) Distribution of heavy metals in marine bivalves, fish and coastal sediments in the Gulf and Gulf of Oman. Mar Pollut Bull 49:410–424

Dimitrakakis E, Hahladakis J, Gidarakos E (2014) The "Sea Diamond" shipwreck: environmental impact assessment in the water column and sediments of the wreck area. Int J Environ Sci Technol 11:1421–1432

Duffus JH (2002) Heavy metals – a meaningless term? Pure Appl Chem 74:793–807

Evans NC (1994) Effects of dredging and dumping on the marine environment of Hong Kong. Terra et Aqua 57:15–25

Fosmire GJ (1990) Zinc toxicity. Am J Clin Nutr 51:225–227

Freije AM (2015) Heavy metal, trace element and petroleum hydrocarbon pollution in the Arabian Gulf: review. J Assoc Arab Univ Basic Appl Sci 17:90–100

GESAMP(IMO/FAO/IJNESCO/WMO/IAEA/UN/IJNEP) (1982) The health of the oceans. Rep Stad GESAMP 15:108 and UNEP Res Seas Rep Stud 16:108

Grecco LE, Gómez EA, Botté SE, Marcos AO, Marcovecchio JE, Cuadrado DG (2011) Natural and anthropogenic heavy metals in estuarine cohesive sediments: geochemistry and bioavailability. Ocean Dyn 61:285–293

Guerra R, Pasteris A, Ponti M (2009) Impacts of maintenance channel dredging in a northern Adriatic coastal lagoon. I: effects on sediment properties, contamination and toxicity. Estuar Coast Shelf Sci 85:134–142

Habashi F (1992) Environmental issues in the metallurgical industry – progress and problems, environmental issues and waste management in energy and mineral production. Balkama, Rotherdam, pp 1143–1153

Hakanson L (1980) Ecological risk index for aquatic pollution control, a sedimentological approach. Water Res 14:975–1001

Halpern BS, Walbridge S, Selkoe KA, Kappel CV, Micheli F, D'Agrosa C, Bruno JF, Casey KS, Ebert C, Fox HE (2008) A global map of human impact on marine ecosystems. Science 319:948–952

Hasan AB, Kabir S, Reza AHMS, Zaman MN, Ahsan A, Rashid M (2013) Enrichment factor and geo-accumulation index of trace metals in sediments of the ship breaking area of Sitakund Upazilla (Bhatiary-Kumira), Chittagong, Bangladesh. J Geochem Explor 125:130–137

Hawkes SJ (1997) What is a heavy metal? J Chem Edu 74(11):1374

Hedge L, Knott A, Johnston E (2009) Dredging related metal bioaccumulation in oysters. Mar Pollut Bull 58:832–840

Hossain MS, Khan YSA (2001) Trace metals in Penaeid shrimp and Spiny lobster from the Bay of Bengal. Sci Asia 27:165–168

Hossen E, Mokhtar G, El-Awady M, Ali I, Morsy M, Dawood A (2001) Environmental exposure of the pediatric age groups in Cairo City and its suburbs to cadmium pollution. Sci Total Environ 273:135–146

Hu B, Li J, Zhao J, Yang J, Bai F, Dou Y (2013) Heavy metal in surface sediments of the Liaodong Bay, Bohai Sea: distribution, contamination, and sources. Environ Monit Assess 185:5071–5083

IAEA (1989) Trace elements in marine sediments, reference sheet SDM-2/TM. IAEA, Vienna

INECAR, Institute of Environmental Conservation and Research (2000) Position paper against mining in Rapu-Rapu. Published by INECAR, Ateneo de Naga University, Philippines. http://www.adnu.edu.ph/Institutes/Inecar/pospaper1.asp. Accessed 26 Jan 2015

Islam KL, Hossain MM (1986) Effect of ship scrapping on the soil and sea environment in the coastal area of Chittagong, Bangladesh. Mar Pollut Bull 17:462–463

Islam MS, Tanaka M (2004) Impacts of pollution on coastal and marine ecosystems including coastal and marine fisheries and approach for management: a review. Mar Pollut Bull 48:624–649

Järup L (2003) Hazards of heavy metal contamination. Br Med Bull 68:167–182

Karvelas M, Katsoyiannis A, Samara C (2003) Occurrence and fate of heavy metals in the wastewater treatment process. Chemosphere 53:1201–1210

Kashem MA, Singh BR (1999) Heavy metal contamination of soil and vegetation in the vicinity of industries in Bangladesh. Wat Air Soil Pollut 115:347–361

Kawser A, Monika D, Monirul I, Salma A (2011) Water pollution of Bangladesh. World Appl Sci J 12:152–159

Kennish MJ (1996) Practical handbook of estuarine and marine pollution. CRC Press, Boca Raton

Khan MAA, Khan YSA (2003) Trace metals in littoral sediments from the north east coast of the Bay of Bengal along the ship-breaking area, Chittagong, Bangladesh. J Biol Sci 3:1050–1057
Lai HY, Hseu ZY, Chen TC, Chen BC, Guo HY, Chen ZS (2010) Health risk-based assessment and management of heavy metals-contaminated soil sites in Taiwan. Int J Environ Res Public Health 7:3595–3614
Lenntech (2004) Water treatment and air purification. Lenntech, Rotterdamseweg
Long ER, MacDonald DD, Smith SL, Calder FD (1995) Incidence of adverse biological effects within ranges of chemical concentrations in marine and estuarine sediments. Environ Manag 19:81–97
Lotze HK, Lenihan HS, Bourque BJ, Bradbury RH, Cooke RG, Kay MC, Kidwell SM, Kirby MX, Peterson CH, Jackson JBC (2006) Depletion, degradation, and recovery potential of estuaries and coastal seas. Science 312:1806–1809
Mashiatullah A, Chaudhary MZ, Ahmad N, Javed T, Ghaffar A (2013) Metal pollution and ecological risk assessment in marine sediments of Karachi Coast, Pakistan. Environ Monit Assess 185:1555–1565
Mavrakis A, Theoharatos G, Asimakopoulos DN, Christides A (2004) Distribution of trace metals in the sediments of Elefsis Gulf. Mediterr Mar Sci 5:151–158
McCluggage D (1991) Heavy metal poisoning. NCS Magazine, Published by The Bird Hospital, Co., USA. http://www.cockatiels.org/articles/Diseases/metals.html. Accessed 16 Jan 2015
Morse DL, Harrington JM, Housworth J, Landrigan PJ (1979) Arsenic exposure in multiple environmental media in children near a smelter. Clini Toxicol 14:389–399
Müller G (1969) Index of geo-accumulation in sediments of the Rhine River. Geojournal 2:108–118
Müller G (1981) The heavy metal pollution of the sediments of Neckars and its tributary: a stocktaking. Chemiker-Zeitung 105:157–164
Naser HA (2012) Metal concentrations in marine sediments influenced by anthropogenic activities in Bahrain, Arabian Gulf. In: Shao HB (ed) Metal contaminations: sources, detection and environmental impacts. NOVA Science Publishers, Inc., New York, pp 157–175
Naser HA (2013) Assessment and management of heavy metal pollution in the marine environment of the Arabian Gulf: a review. Mar Pollut Bull 72:6–13
NBSC (2004–2009) National Bureau of Statistics of China. China statistical yearbook. Beijing
Neşer G, Ünsalan D, Tekoğul N, Stuer-Lauridsen F (2008) The shipbreaking industry in Turkey: environmental, safety and health issues. J Clean Prod 16:350–358
Neşer G, Kontas A, Ünsalan D, Uluturhan E, Altay O, Darılmaz E, Küçüksezgin F, Tekoğul N, Yercan F (2012) Heavy metals contamination levels at the Coast of Aliağa (Turkey) ship recycling zone. Mar Pollut Bull 64:882–887
Newbury TK (1979) Possible accumulation of heavy metals around offshore oil production facilities in the Beaufort Sea. Arctic 32:42–45
Nobi EP, Dilipan E, Thangaradjou T, Sivakumar K, Kannan L (2010) Geochemical and geo-statistical assessment of heavy metal concentration in the sediments of different coastal ecosystems of Andaman Islands, India. Estuar Coast Shelf Sci 87:253–264
O'Connor CS, Leppi NW, Edwards R, Sunderland G (2003) The combined use of electro-kinetic remediation and phytoremediation to decontaminate metal-polluted soils: laboratory-scale feasibility study. Environ Monit Assess 84:141–158
Obiajunwa EI, Pelemo DA, Owolabi SA, Fasasi MK, Johnson-Fatokun FO (2002) Characterisation of heavy metal pollutants of soils and sediments around a crude-oil production terminal using EDXRF. Nucl Instrum Method Phys Res Sect B Beam Interact Mater Atoms 194:61–64
Ogwuegbu MO, Ijioma MA (2003) Effects of certain heavy metals on the population due to mineral exploitation. In: International conference on scientific and environmental issues, University of Ado Ekiti, Ekiti State, pp 8–10
Ogwuegbu MOC, Muhanga W (2005) Investigation of lead concentration in the blood of people in the Copperbelt Province of Zambia. J Environ 1:66–75

Onojake MC, Frank O (2013) Assessment of heavy metals in a soil contaminated by oil spill: a case study in Nigeria. Chem Ecol 29:246–254

Onojake MC, Oforka NC, Osuji LC (2011) Trace metals geochemistry of crude oils from Umutu/Bomu fields in south west Niger Delta Nigeria. Energy Environ Res 1:139–146

Pan K, Wang W-X (2012) Trace metal contamination in estuarine and coastal environments in China. Sci Total Environ 421(422):3–16

Pazi I (2011) Assessment of heavy metal contamination in Candarli Gulf sediment, Eastern Aegean Sea. Environ Monit Assess 174:199–208

Pekey H (2006) The distribution and sources of heavy metals in Izmit Bay surface sediments affected by a polluted stream. Mar Pollut Bull 52:1197–1208

Peplow D (1999) Environmental impacts of mining in Eastern Washington. Centre for Water and Watershed Studies Fact Sheet, University of Washington, Seattle

Ra K, Kim J-K, Hong SH, Yim UH, Shim WJ, Lee S-Y, Kim Y-O, Lim J, Kim E-S, Kim K-T (2014) Assessment of pollution and ecological risk of heavy metals in the surface sediments of Ulsan Bay, Korea. Ocean Sci J 49:279–289

Rasul MG, Faisal I, Khan MMK (2006) Environmental pollution generated from process industries in Bangladesh. Int J Environ Pollut 28:144–161

Reddy MS, Basha S, Kumar VGS, Joshi HV, Ramachandraiah G (2004) Distribution, enrichment and accumulation of heavy metals in coastal sediments of Alang-Sosiya ship scrapping yard, India. Mar Pollut Bull 48:1055–1059

Reddy MS, Basha S, Joshi HV, Ramachandraiah G (2005) Seasonal distribution and contamination levels of total PHCs, PAHs and heavy metals in coastal waters of the Alang-Sosiya ship scrapping yard, Gulf of Cambay, India. Chemosphere 61:1587–1593

Rossi E (2008) Low level environmental lead exposure – a continuing challenge. The clinical biochemist reviews. Aust Assoc Clin Biochem 29:63–70

Rubio B, Nombela MA, Vilas F (2000) Geochemistry of major and trace elements in sediments of the Ria De Vigo (NW Spain): an assessment of metal pollution. Mar Pollut Bull 40:968–980

Ruilian YU, Xing Y, Yuanhui Z, Gongren HU, Xianglin TU (2008) Heavy metal pollution in intertidal sediments from Quanzhou Bay, China. J Environ Sci 20:664–669

Sadiq M (2002) Metal contamination in sediments from a desalination plant effluent outfall area. Sci Total Environ 287:37–44

Salomons W, Forstner U (1984) Metals in the hydro cycle. Springer, Berlin

Salonen JT, Seppanen K, Nyyssonen K, Korpela H, Kauhanen J, Kantola M, Tuomilehto J, Esterbauer H, Tatzber F, Salonen R (1995) Intake of mercury from fish, lipid peroxidation, and the risk of myocardial infarction and coronary, cardiovascular, and any death in eastern Finnish men. Circulation 91:645–655

Sany SBT, Salleh A, Sulaiman AH, Sasekumar A, Rezayi M, Tehrani GM (2013) Heavy metal contamination in water and sediment of the Port Klang coastal area, Selangor, Malaysia. Environ Earth Sci 69:2013–2025

Sarraf M, Steur-Lauridsen F, Dyoulgerov M, Bloch R, Wingfield S, Watkinson R (2010) Ship breaking and recycling industry in Bangladesh and Pakistan. Report No 58275-SAS, World Bank. http://siteresources.worldbank.org/SOUTHASIAEXT/Resources/223546-1296680097256/Shipbreaking.pdf. Accessed 10 Jan 2015

Satpathy KK, Mohanty AK, Prasad MVR, Natesan U, Sarkar SK (2012) Studies on the variations of heavy metals in the marine sediments off Kalpakkam, East Coast of India. Environ Earth Sci 65:89–101

Schiff KC, Weisberg SB (1999) Iron as a reference element for determining trace metal enrichment in southern California coastal shelf sediments. Mar Environ Res 48:161–176

Sfakianakis DG, Renieri E, Kentouri M, Tsatsakis AM (2015) Effect of heavy metals on fish larvae deformities: a review. Environ Res 137:246–255

Siddiquee NA, Parween S, Quddus MMA, Barua P (2009) Heavy metal pollution in sediments at shipbreaking area of Bangladesh. Asian J Wat Environ Pollut 6:7–12

Singh K, Mohan D, Sinha S, Dalwani R (2004) Impact assessment of treated/untreated wastewater toxicants discharged by sewage treatment plants on health, agricultural, and environmental quality in wastewater disposal area. Chemosphere 55:227–255

Sinha S (1998) Ship scrapping and the environment – the buck should stop! Marit Pol Manag 25:397–403

Smith S, Rule M (2001) The effects of dredge-spoil dumping on a shallow water soft sediment community in the Solitary Island Marine Park, NSW, Australia. Mar Pollut Bull 42:1040–1048

Sujauddin M, Koide R, Komatsu T, Hossain MM, Tokoro C, Murakami S (2015) Characterization of ship breaking industry in Bangladesh. J Mater Cycles Waste Manag 17:72–83

Tanner PA, Leong LS, Pan SM (2000) Contamination of heavy metals in marine sediment cores from Victoria Harbour, Hong Kong. Mar Pollut Bull 40:769–779

Taylor SR (1964) Abundance of chemical elements in the continental crust: a new table. Geochim Cosmochim Acta 28:1273–1285

Temara A, Skei JM, Gillan D, Warnau M, Jangoux M, Dubois P (1998) Validation of the asteroid *Asterias rubens* (Echinodermata) as a bioindicator of spatial and temporal trends of Pb, Cd, and Zn contamination in the field. Mar Environ Res 45:341–356

Tewari A, Joshi HV, Trivedi RH, Sravankumar VG, Raghunathan C, Khambhaty Y, Kotiwar OS, Mandal SK (2001) Studies the effect of ship scraping industry and its associated waste on the biomass production and biodiversity of biota in situ condition at Alang. Mar Pollut Bull 42:462–469

Tomlinson DL, Wilson JG, Harris CR, Jeffrey DW (1980) Problems in the assessment of heavymetal levels in estuaries and the formation of a pollution index. Helgol Mar Res 33:566–575

Turner A, Millward GE (2000) Particle dynamics and trace metal reactivity in estuarine plumes. Estuar Coast Shelf Sci 50:761–774

Udedi SS (2003) From Guinea Worm Scourge to metal toxicity in Ebonyi State. Chem Nigeria New Millenn Unfold 2:13–14

UNEP (2002) Global mercury assessment. Geneva. http://www.chem.unep.ch/mercury/Report/Chapter6.htm. Accessed 12 Jan 2015

UNEP, UN-HABITAT (2005) Coastal area pollution – the role of cities. http://www.unep.org/urban_environment/PDFs/Coastal_Pollution_Role_of_Cities.pdf. Accessed 01 Jan 2015

Wang S-L, Xu X-R, Sun Y-X, Liu J-L, Li H-B (2013) Heavy metal pollution in coastal areas of South China: a review. Mar Pollut Bull 76:7–15

WHO (1990) Methyl mercury. Environmental health criteria, vol 101. World Health Organization, Geneva

Yeast PA, Brewers JM (1983) Potential anthropogenic influences on trace metal distribution in the North Atlantic. Aquat Sci 40:245–249

Young RA (2005) Toxicity profiles: toxicity summary for cadmium. Risk Assessment Information System, RAIS, University of Tennessee. http://www.rais.ornl.gov/tox/profiles/cadmium.html. Accessed 26 Dec 2014

Youssef DH, El-Said GF (2011) Assessment of some heavy metals in surface sediments of the Aqaba Gulf, Egypt. Environ Monit Assess 180:229–242

Zhang KB (2008) Characteristics of the elemental abundance and content of near-shore neritic sediment in Fujian Province. Geol Fujian 4:402–408 (in Chinese)

Zhang YF, Wang LJ, Huo CL, Guan DM (2008) Assessment on heavy metals pollution in surface sediments in Jinzhou Bay. Mar Environ Sci 2:178–181 (in Chinese)

Zhou F, Guo HC, Hao ZJ (2007) Spatial distribution of heavy metals in Hong Kong's marine sediments and their human impacts: a GIS-based chemometric approach. Mar Pollut Bull 54:1372–1384

Chapter 3
Radionuclides Released from Nuclear Accidents: Distribution and Dynamics in Soil

Seiya Nagao

Abstract The Chernobyl nuclear power plant accident in 1986, the Mayak reprocessing plant accident in 1957, and the Fukushima nuclear power plant accident in 2011 have released various radionuclides. Spatial distribution of radionuclides in surface soil constitutes fundamental information related to radiation exposure and to post-accident sources for secondary dispersion of radionuclides. The spatial and vertical distributions of ^{137}Cs, ^{90}Sr, and Pu from those three nuclear accidents were reviewed to elucidate the fate of the radionuclides in the soil environment. This chapter specifically presents information about the dynamics of radiocesium (^{134}Cs and ^{137}Cs) in the soil of contaminated areas because of its long half-life and its major contribution to the overall external radiation dose. Along with an examination of global fallout, for soils of various types, the existence forms and geochemical behavior of radiocesium in surface soil after the Fukushima accident are reviewed, as are similar results from the Chernobyl and Mayak nuclear accidents.

Keywords ^{134}Cs • ^{137}Cs • Clay minerals • Organic matter • Fukushima nuclear accident

3.1 Distribution of Radionuclides Released from Nuclear Weapon Tests and Severe Nuclear Accidents in Chernobyl and Mayak

Nuclear weapon tests conducted during 1950–1960s have emitted huge amounts of U, Pu, and their fission products (e.g., ^{90}Sr, ^{131}I, ^{137}Cs) into the environment. These radionuclides were distributed as aerosols and gases across large distances, producing an anthropogenic radioactive background on ground surfaces, especially in the Northern Hemisphere. Aoyama et al. (2006) estimated the global fallout of ^{137}Cs as 765 ± 79 PBq based on global measurements in rain, seawater, and soil, as

S. Nagao (✉)
Low Level Radioactivity Laboratory, Institute of Nature and Environmental Technology, Kanazawa University, Kanazawa, Ishikawa, Japan
e-mail: nagao37@staff.kanazawa-u.ac.jp

H. Hasegawa et al. (eds.), *Environmental Remediation Technologies for Metal-Contaminated Soils*, DOI 10.1007/978-4-431-55759-3_3

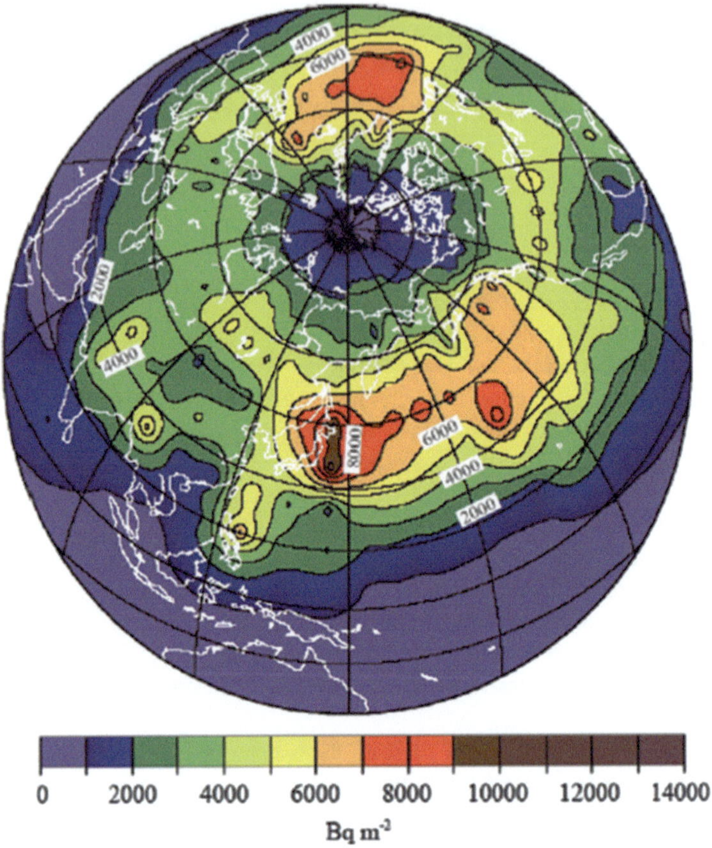

Fig. 3.1 Reconstructed deposition (Bq m^{-2}) of global fallout ^{137}Cs as of 1 January, 1979 from Aoyama et al. (2006)

data from 10° × 10° grids. Figure 3.1 portrays the precise spatial distribution of global ^{137}Cs fallout as reported by Aoyama et al. (2006). A typical feature of geographical distribution is that two high global ^{137}Cs fallout areas exist in the northern hemisphere, where the highest ^{137}Cs fallout was observed in the earth surface. These areas correspond to crossovers of areas where larger precipitation amounts were observed and where higher stratosphere–troposphere exchange is expected.

The International Nuclear Event Scale (INES) is useful for promptly communicating to the public in consistent terms the safety significance of events reported at nuclear installations. It was designed by an international group of experts convened jointly in 1989 by the International Atomic Energy Agency (IAEA) and the Nuclear Energy Agency (NEA) of the Organization for Economic Co-operation and Development (IAEA 2014). The nuclear accident at the Chernobyl Nuclear Power Plant (NPP) in 1986 was classified as level 7, a major accident, because of widespread

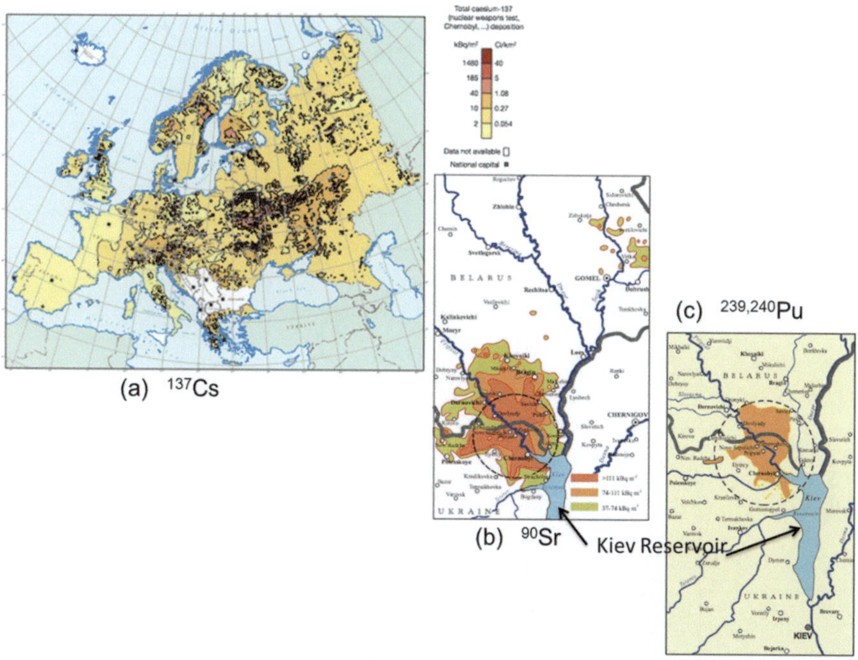

Fig. 3.2 Surface ground deposition of ^{137}Cs (**a**), ^{90}Sr (**b**), and ^{239,240}Pu (**c**) as a result of the Chernobyl accident. The maps of ^{137}Cs and ^{90}Sr-^{239,240}Pu were referred respectively from International Advisory Committee (1991) and de Cort (1998). The scale of radioactivity in **c** is described in **b**

environmental and human health effects. The 1957 accident (Kyshtym accident) at the Kyshtym reprocessing plant led to the release of large amounts of radionuclides at Mayak in the Ural region. Based on the off-site impact, it is classified as level 6, a severe accident. The Fukushima nuclear accident, which occurred in 2011, is classified as level 7 based on an estimate of the total amount of radioactivity released to the environment from the NPP. Details of the Fukushima accident will be presented in the next section of this chapter.

The Chernobyl NPP is located in the northeast region of Ukraine, about 130 km of north of Kiev. The nuclear accident occurred on 26 April, 1986, during a test of emergency preparedness for a loss of offsite power. Chernobyl-derived radionuclides affected a vast area of Europe (de Cort 1998; United Nations Scientific Committee on the Effects of Atomic Radiation (UNSCEAR) 2000; Terada et al. 2005). Figure 3.2a depicts the surface ground deposition of ^{137}Cs in Europe in 1986. Total ^{137}Cs activity of about 64 TBq was deposited on European territory (de Cort 1998). The three countries most heavily affected by the accident were 23 % of total deposition of Europe at Belarus (23 %), 30 % at the Russian Federation, and 18 % at Ukraine. The ^{90}Sr deposition was much more spatially confined to areas near the Chernobyl NPP than that of ^{137}Cs (Fig. 3.2b) because it is less volatile than Cs. The amounts of ^{239,240}Pu deposited on soil are also presented in Fig. 3.2c.

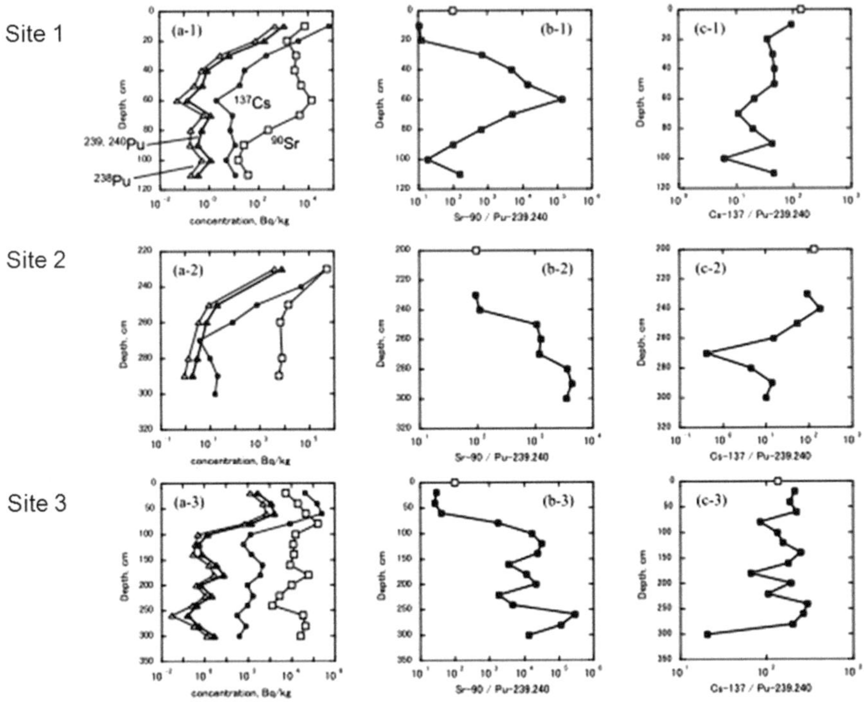

Fig. 3.3 Depth profiles of radionuclide concentrations (Bq kg^{-1} dry-soil) and their activity ratios in selected locations (Sites 1, 2, and 3) in the Exclusion Zone of Chernobyl area (Matsunaga and Nagao 2009). The radioactivity was corrected to the time of accident on 26 April, 1986. In sub-plots b and c, open squares denote abundance in the reactor core inventory at the time of the accident

Nearly all areas with $^{239,240}Pu$ deposits above 3.7 kBq m^{-2} are within the Chernobyl Exclusion Zone, a highly contaminated area to which access has been controlled by Ukrainian authorities. The IAEA and World Health Organization (WHO) released reports including the latest analyses of the influence of the accident including its environmental, radiological, and social aspects (IAEA 2006a, b; WHO 2006). These reports have stated that the radiological effects on people are mostly attributable to radionuclides of I, Cs, and Sr.

Figure 3.3 shows vertical profiles of Chernobyl-derived radionuclides in selected sites in the accident area from data of 1999–2003 of the Japanese Ministry of Education, Culture, Sports, Science, and Technology (MEXT) (Matsunaga and Nagao 2009). The $^{239,240}Pu$ contamination density was 40–100 kBq m^{-2}. Sites 2 and 3 were mutually close, 2 km west of the NPP, in an area affected strongly by the West Plume, and which has been called the "Red Forest." Highly contaminated trees and other wastes produced by the accident were buried at site 2. Leaching from those buried materials has led to high radioactivity in deep soil layers. The top layer (0–40 cm) of site 3, as well as other plots in this area, was covered with uncontaminated river sand for mitigation of radioactivity after the accident. The Chernobyl-derived $^{239,240}Pu$ is distributed predominantly in the upper soil horizons in the case of undisturbed soil (Pavlotskaya et al. 1991; Knatko et al. 1996). Most of the $^{239,240}Pu$ was confined to within 60 cm of the surface soil in the studied sites in the Exclusion Zone, as presented in Fig. 3.2a, c.

Three important contamination events have occurred at Mayak Production Association (PA) facilities (Joint Norwegian–Russian Expert Group (JNREG) 1997). During 1949–1956, the direct release of radionuclides to the Techa River, which is located in south Ural and a tributary of the Iset River, at the early stage of the Mayak operations, led to the contamination of its floodplain and radiation exposure of the local population residing in the river shore area. An explosion in a tank containing high-level liquid waste occurred in Kyshtym in September 1957. About 74 PBq was released in a plume that presumably reached an altitude of 1 km and became dispersed by the wind in a north-northeastward direction to form the East Ural Radioactive Trace (EURT) (JNREG 1997). Initially, the primary exposure pathway was external exposure. The main source of exposure gradually changed, becoming internal exposure to ^{90}Sr as a result of radioactive decay of most short-lived radionuclides and ^{90}Sr transfer through the food chain (Peremylova et al. 2004). The schematic contamination map for ^{90}Sr in soil is portrayed in Fig. 3.4. Measurements by governmental hydro-meteorological services were conducted during 1992–1997 using radiochemical and gamma-spectrum analysis methods (Inter-Departmental Commission on Radiation Monitoring (IDCRM) 2000). Results indicate that activity levels varied among the sites surrounding Mayak PA. The surface Pu contamination levels tended to decrease with distance from Reservoir 11 and the Asanov Swamp area.

Figure 3.5 exhibits soil profiles of ^{90}Sr and ^{137}Cs in surface soil in 1999 after the Mayak nuclear accident. Spring overflows of the Techa River and particularly the flood of 1951 contributed to intensive radioactive concentration in riverside areas. The ^{90}Sr profiles characteristically show depletion of the upper 0–10 cm layer. The presence of the maximum was found at the depletion of 5–40 cm with a dramatic drop in the contaminants in deeper layers (Kostyuchenko et al. 2012). This result derives from washing of the upper soil layer with surface water. However, vertical profiles of ^{137}Cs differ from those of ^{90}Sr because of different physicochemical and morphological properties of the floodplain soil.

Figure 3.6 presents vertical profiles of Pu isotopes in bottom sediments from Reservoir 10 (Skipperud et al. 2005), which is portrayed in Fig. 3.4, and which has received radioactive waste of various origins over several decades. $^{239,240}Pu$ was concentrated in the upper 0–8 cm and the middle (15–20 cm) layers of the station 4 sediment. Variable activity densities reflected the inhomogeneous spatial distribution of radionuclides in reservoir sediments after damming and subsequent flooding of the area. The vertical distribution of $^{239,240}Pu$ in the sediments was at its highest in the upper 20 cm layer of the profiles. The $^{238}Pu/^{239,240}Pu$ activity ratios were 0.46–1.83 at 0–10 cm depth. The $^{240}Pu/^{239}Pu$ atom ratios also varied spatially and vertically in the reservoir sediments. The differences reflect the

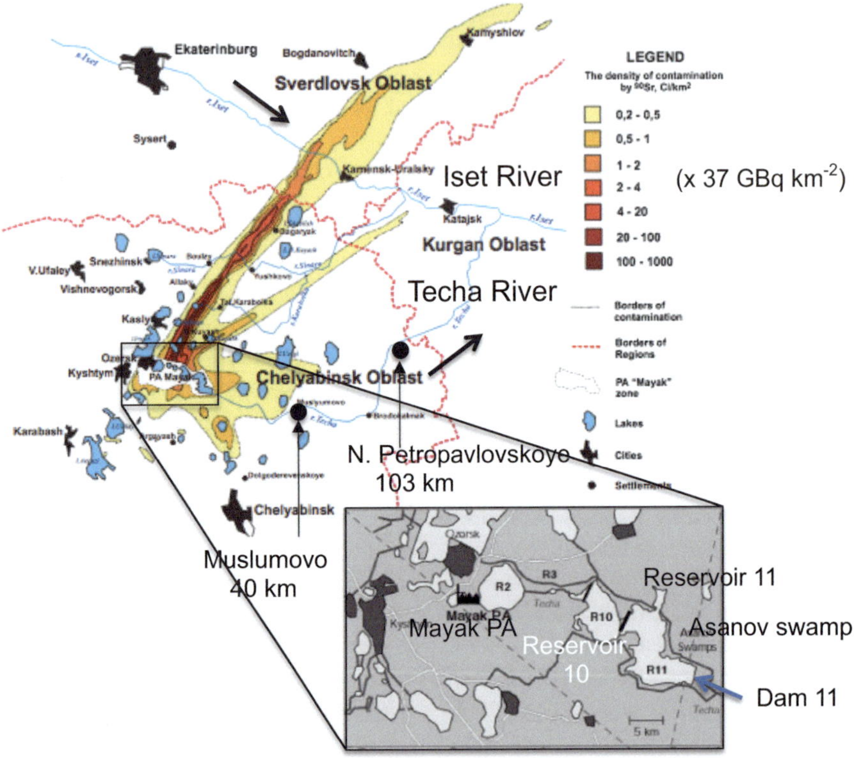

Fig. 3.4 Schematic map of density contamination by ^{90}Sr in 1997 of the territories of the southern Ural region (IDCRM 2000). The enlarged view around reservoirs near Mayak PA is referred from a report by Peremyslova et al. (2004). The sampling transect (Station 1 to station 7 for Pu isotope analysis) of Reservoir 10 is marked with a line. *Arrows indicate* the river water flow direction of at the Techa River and the Iset River

differences in sources of Pu isotopes before and after the 1957 Kysthym accident such as low to high burn-up of spent fuel.

3.2 Contamination of Radionuclides from the Fukushima Nuclear Power Plant Accident

The nuclear accident at the Fukushima Dai-ichi NPP of the Tokyo Electric Power Company (TEPCO) occurred after the 11 March, 2011, Tohoku Earthquake and Tsunami. Operating reactors Units 1–3 shut down automatically, but all off-site power supplies were lost. Units 1–3 overheated without controlling or cooling the reactors so that severe damage to the fuel occurred: melt-down or melt-through (TEPCO 2011a; Hirano et al. 2012). Major radionuclides released from the

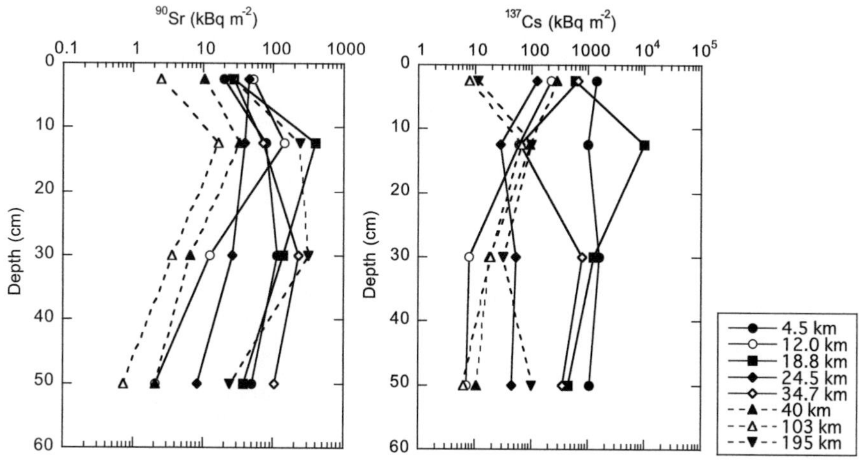

Fig. 3.5 Vertical distribution of ^{90}Sr (**a**) and ^{137}Cs (**b**) in floodplain soil over the length of the Techa River. Data were referred from Kostyuchenko et al. (2012). *Numbers* denote distances from dam 11 in Reservoir 11. Musumovo is located at 40 km; N. Petropavlovskoye is 103 km as shown in Fig. 3.4

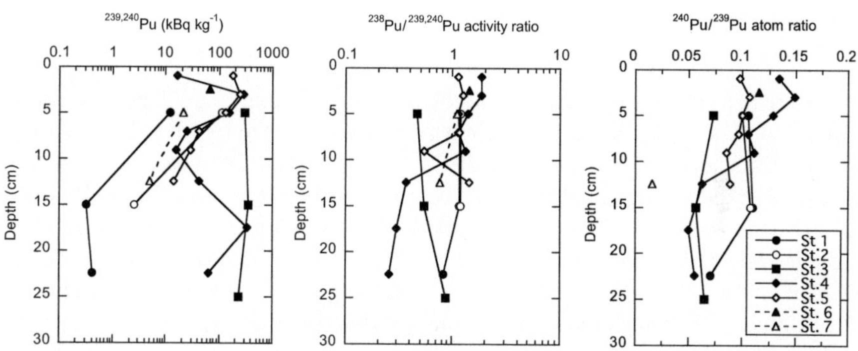

Fig. 3.6 Vertical distribution of ^{239,240}Pu radioactivity (**a**), ^{238}Pu/^{239,240}Pu activity ratio (**b**), and ^{240}Pu/^{239}Pu atom ratio (**c**) in sediments collected from Reservoir 10 (Data were referred from Skipperud et al. 2005)

Fukushima Dai-ichi NPP were noble gases (^{133}Xe), iodine (^{131}I, ^{133}I), tellurium (^{129m}Te, ^{132}Te), and cesium (^{134}Cs, ^{136}Cs, ^{137}Cs) by hydrogen explosion and vent operations (Nuclear Emergency Response Headquarters (NERH) 2011; Endo et al. 2012; Thakur et al. 2013). All of these radionuclides have a short half-life (20.8 h to 33.6 day) except for ^{134}Cs and ^{137}Cs. The physical half-life is 2.06 year for ^{134}Cs and 30.07 year for ^{137}Cs. About 6–20 PBq of ^{134}Cs and ^{137}Cs was released from the Fukushima Dai-ichi NPP (UNSCEAR 2014).

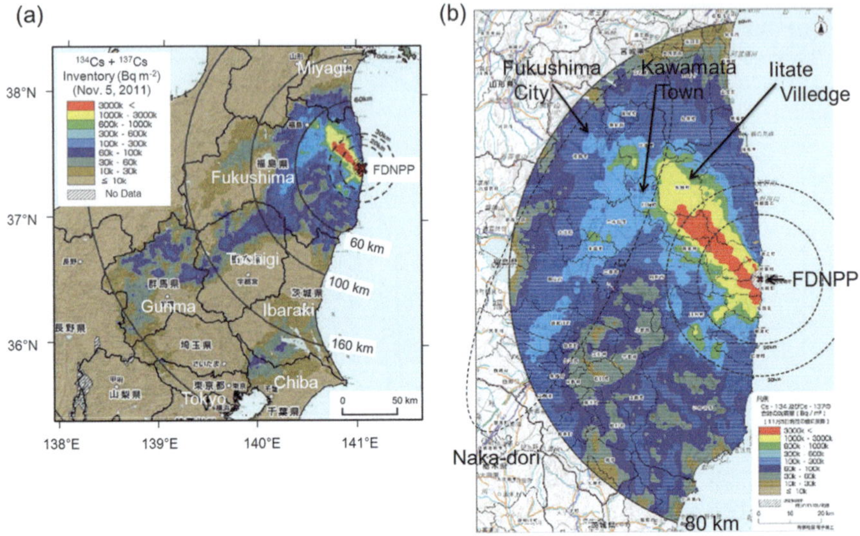

Fig. 3.7 Radiocesium (^{134}Cs and ^{137}Cs) deposition map for eastern Japan (a) and 80-km zone from the Fukushima Dai-ichi Nuclear Power Plant (FDNPP) (b) quoted from the fourth airborne monitoring survey by the Japanese Ministry of Education, Culture, Sports, Science, and Technology (MEXT 2011)

Figure 3.7 presents an inventory of the sum of ^{134}Cs and ^{137}Cs deposited on the ground surface. The surface deposition was estimated by the fourth airborne survey during 22 October–5 November in 2011 (MEXT 2011). The spatial distribution reveals marked external radioactivity in a northwest zone from the NPP, about 20 km wide and 50–70 km long. Very high deposition was recorded in Iitate Village. The highly contaminated area (>60 kBq m^{-2}) was located in the Naka-dori area. Total deposition of ^{134}Cs+^{137}Cs on the ground surface was estimated as 2.2 PBq using a chemical transport model (Morino et al. 2011).

Most of the material was released during 12–22 March with a maximum release phase of 14–17 March, 2011 (Chino et al. 2011). The radioactive gases and particles released during the accident were dispersed over the middle latitudes of the entire northern hemisphere and for the first time were also measured in the Southern Hemisphere (CTBTO 2011; Masson et al. 2011; Wetherbee et al. 2012). The radioactivity level and deposition distribution of radionuclides released from the Fukushima Dai-ichi NPP on the ground surface in Japan were ascertained from the release and weather conditions: wind direction and precipitation. The radionuclides released from the Fukushima Dai-ichi NPP on 12 March, 2011, in the afternoon moved eastward to the coast of Miyagi Prefecture and were then deposited on the ground surface. On the morning of 15 March, radionuclides were deposited southeastward to the coastal area of Fukushima and northeast Ibaraki area by dry deposition. However, on 15–16 March, wet deposition occurred at Gunma, Tochigi, and Fukushima area by rainfall. During 20–21 March, the radionuclides were

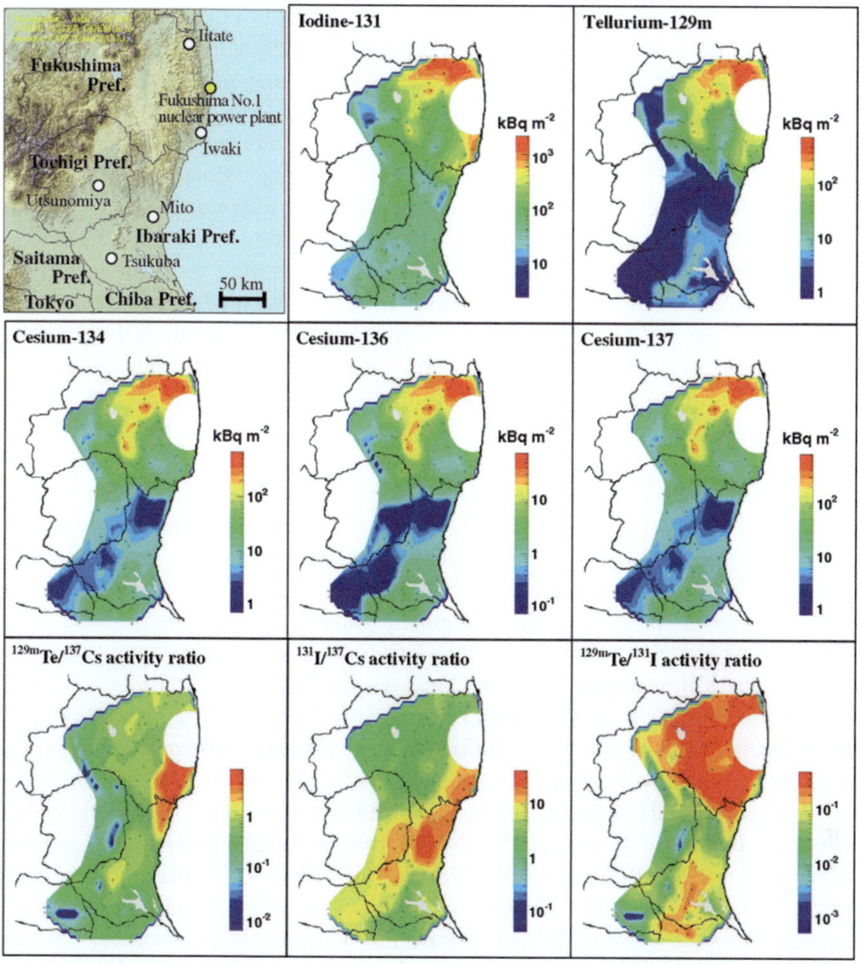

Fig. 3.8 Contamination maps of ^{129m}Te, ^{131}I, ^{134}Cs, ^{136}Cs, and ^{137}Cs on the soil surface during March–May 2011 from Kinoshita et al. (2011). Activities on 29 March, 2011, are shown. *Open circles* denote sampling point positions

deposited in Iwate, Miyagi, Ibaraki, and Chiba Prefectures by wet deposition (Nagai and Kurihara 2014).

For more detailed understanding of the transport and deposition of released radionuclides, Kinoshita et al. (2011) have reported contour maps of deposition for ^{131}I, ^{129m}Te, and ^{134,136,137}Cs in surface soils collected in Fukushima, Ibaraki, Chiba, Tochigi, and Saitama prefectures for March–May, 2011 (Fig. 3.8). Different spatial distributions of I, Te, and Cs were found as presented in Fig. 3.8. Very high accumulation was observed in the Iitate Village and Naka-dori regions for all the radionuclides listed above. ^{131}I was distributed rather homogeneously except for the contaminated region. The accumulation of ^{129m}Te in Fukushima region was higher than that of the other regions. The differences in the activity ratios reflect the existence forms of radionuclides in the atmosphere and geographical features. Gaseous materials were transported more easily than the particulate forms of radionuclides such as ^{129m}Te and ^{134,136,137}Cs, even though mountainous areas are located in the Abukuma Highland and the Ohu Mountains. Watanabe et al. (2012) reported the distribution of artificial radionuclides of ^{110m}Ag, ^{129m}Te, ^{134}Cs, and ^{137}Cs in surface soil (0–1 cm depth) collected from 60 to 190 km north (Miyagi Prefecture) of the NPP 1 month after the accident. In the south, close to the border with Fukushima Prefecture, the total activity concentrations of ^{110m}Ag, ^{129m}Te, ^{134}Cs, and ^{137}Cs during 16–29 April, 2011, reached approximately 27,600 Bq kg^{-1} dry-soil. High total concentrations (2600–6600 Bq kg^{-1} dry-soil) were also recorded in the northern part of Miyagi Prefecture, but low concentrations (400–1900 Bq kg^{-1} dry-soil) were found in Sendai City and other areas of central Miyagi Prefecture. The ^{129m}Te/^{137}Cs activity ratios varied: 0.36–1.19. Varying ratios were also observed in Fukushima Prefecture (Saito et al. 2014a). The variations in the ^{129m}Te/^{137}Cs ratio might be a consequence of ^{129m}Te migration in the soil following radionuclide deposition or a difference release attributable to the boiling point of Cs and Te.

Yamamoto et al. (2012, 2014) reported gamma-ray emitting radionuclides (^{129}Te, ^{129m}Te, ^{131}I, ^{132}I, ^{134}Cs, ^{136}Cs, ^{137}Cs, ^{140}La) and transuranium elements (Pu, Am, and Cm isotopes) within 80 km from the Fukushima Dai-ichi NPP. The surface soil samples at 0–5 cm depth exhibited a ^{238}Pu/^{239,240}Pu activity ratio of approximately 0.03 outside of the 20-km exclusion zone (20–40 km from the NPP), but 0.05–1.2 in Iitate Village. In Okuma Town, some surface soil samples were 0.059–0.103: slightly higher than the global fallout of ca. 0.03. TEPCO reported that trace amounts of Pu isotopes originated from the accident identified in several soils within the Fukushima Dai-ichi NPP site. The ^{238}Pu/^{239,240}Pu activity ratio was about 2.0 (TEPCO 2011b). Those samples over the activity ratio of 0.03 were considered to be affected by the Fukushima accident. The inventories of the Fukushima Dai-ichi NPP-derived ^{239,240}Pu were estimated as up to 2.8 Bq m^{-2} (Yamamoto et al. 2014).

3.3 Spatial Distribution of ^{134}Cs and ^{137}Cs in Surface Soil from Soil Survey Conducted After the Fukushima Nuclear Accident

A contamination map is necessary to estimate the radiation dose. The relevant soil survey was performed mainly within 80 km from the Fukushima Dai-ichi NPP during 2011–2012 by the Japanese government. The soil samples were collected from the upper 5 cm of surface soils at 2,168 locations in all during 4 June–8 July, 2011 (Saito et al. 2014a). Deposition maps were created for ^{134}Cs, ^{137}Cs, ^{131}I,

^{129m}Te, and ^{110m}Ag. The radioactivity ratios of ^{131}I and ^{129m}Te to ^{137}Cs were high in the regions south of the Fukushima NPP site. These regional features of radioactivity ratios were inferred from the different deposition pathways of radionuclides.

The NIAES (2012) reported spatial distributions of radionuclides in cultivated soils of 3420 locations in 15 prefectures from the Tohoku and Kanto areas. Soil samples were collected from the surface to 15 cm depth using a soil core sampler during November through December in 2011. Radiocesium activity of more than 10,000 Bq kg^{-1} dry-soil is distributed northwestward from the Fukushima Dai-ichi NPP. Naka-dori area from the south of Miyagi to the central to north Tochigi Prefecture shows radioactivity of 1000–5000 Bq kg^{-1} dry-soil. The spatial distribution of ^{134}Cs and ^{137}Cs in surface soil from forest, farm, and urban areas is similar to that estimated by airborne survey.

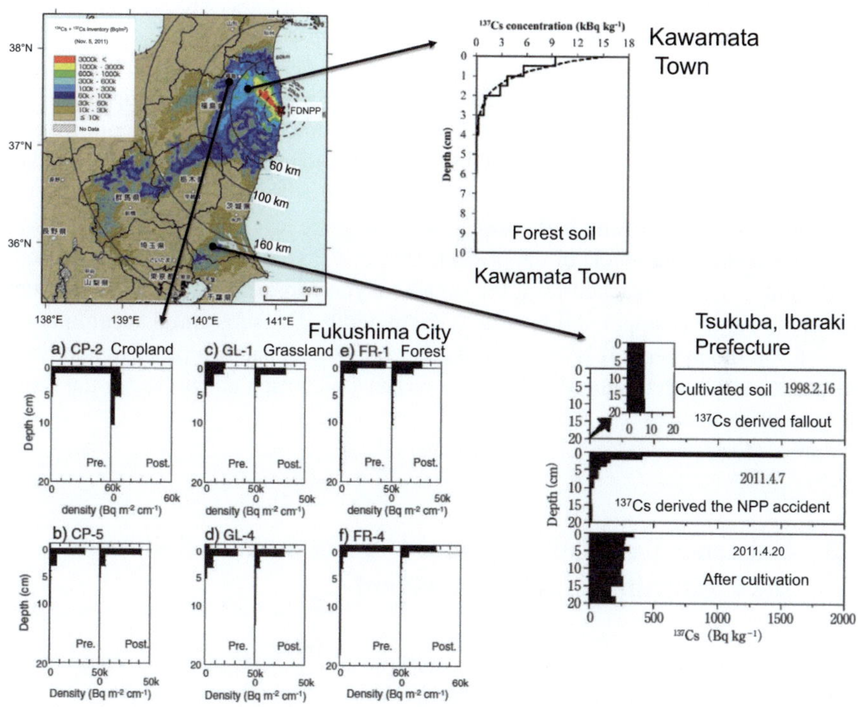

Fig. 3.9 Vertical distributions of ^{137}Cs activity in soil core samples from Fukushima City (Matsunaga et al. 2013) and Kawamata Town (Kato et al. 2012) in Fukushima Prefecture, and in Tsukuba from Ibaraki Prefecture, Japan (Yamaguchi et al. 2012). The values above the *dashed lines* in the soil profiles from Fukuhsima City indicate the inventories in the organic layer (*litter layer*). The unit of Bq kg^{-1} denotes a dry-soil weight basis

3.4 Soil Profiles of ^{134}Cs and ^{137}Cs Before and After the Fukushima Nuclear Accident

Surface soil profiles of radiocesium were investigated in Fukushima Prefecture, Japan, after the Fukushima Dai-ichi NPP accident. Various soils with several types were collected from cultivated and forest soils during April through June before the rainy season in 2011 (Fujiwara et al. 2012; Kato et al. 2012; Koarashi et al. 2012, 2013; Tanaka et al. 2012; Matsunaga et al. 2013). Vertical profiles show that ^{134}Cs and ^{137}Cs were distributed in the surface layer at depths of 0–5 cm. They accounted for more than 87 % of the total radioactivity in the core samples (Fig. 3.9). Similar results were reported for soil cores from forest areas in Tochigi Prefecture (Teramage et al. 2014) and cultivated areas in Ibaraki Prefecture, Japan (Yamaguchi et al. 2012) in 2011–2012.

The organic layer on the mineral soil layer is also important for storage in a soil environment. Teramage et al. (2014) reported the radioactivity in the surface organic layer such as litter (Ol) and semi-decomposed litter (Of) at 10 months after the accident. The raw organic layer holds 52 % of the Fukushima-derived ^{137}Cs and 25 % of the pre-Fukushima ^{137}Cs at the time of the soil sampling. Fujii et al. (2014) also investigated the role of the organic layer of soil collected in Fukushima Prefecture (Kawauchi, Ohtama, and Tadami Town) during August–September of 2011 and 2012. The vertical soil distribution of ^{134}Cs and ^{137}Cs suggested that most cesium was retained in the organic layer and upper mineral soil layer under different levels of deposition. Within 1.5 years after the accident, both ^{134}Cs and ^{137}Cs had leached from the organic layer. Most of these (59–73 %) were accumulated in the upper soil layer at 0–5 cm depth. The substantial accumulation of ^{137}Cs in the upper soil layer suggests that sorption capacities of clays and litter are sufficiently high to retain ^{137}Cs in the surface soil during at least the initial stage of contamination.

3.5 Existence forms of ^{134}Cs and ^{137}Cs in Surface Soil After the Fukushima Nuclear Accident

Many investigations have assessed the sorption of ^{137}Cs to smectite, vermiculite, illite, and mica in soils (e.g., Tamura and Jacobs 1960; Ohnuki and Kozai 1994; Iijima et al. 2010). Cesium sorbs to the clay minerals through a cation-exchange reaction (Sposio 2008). Figure 3.10 presents a schematic illustration of the sorption of radiocesium to clay minerals. The binding sites of clay minerals are (1) the exchangeable sites, and (2) the selective sorption of frayed edge sites. The Cs^+ ions in the interlayer of montmorillonite can exchange with other cations. The interlayer of bidellite is reduced according to Cs sorption and is then fixed to Cs^+ ion. Vermiculite and illite have frayed edge sites, which are fixed selectively and

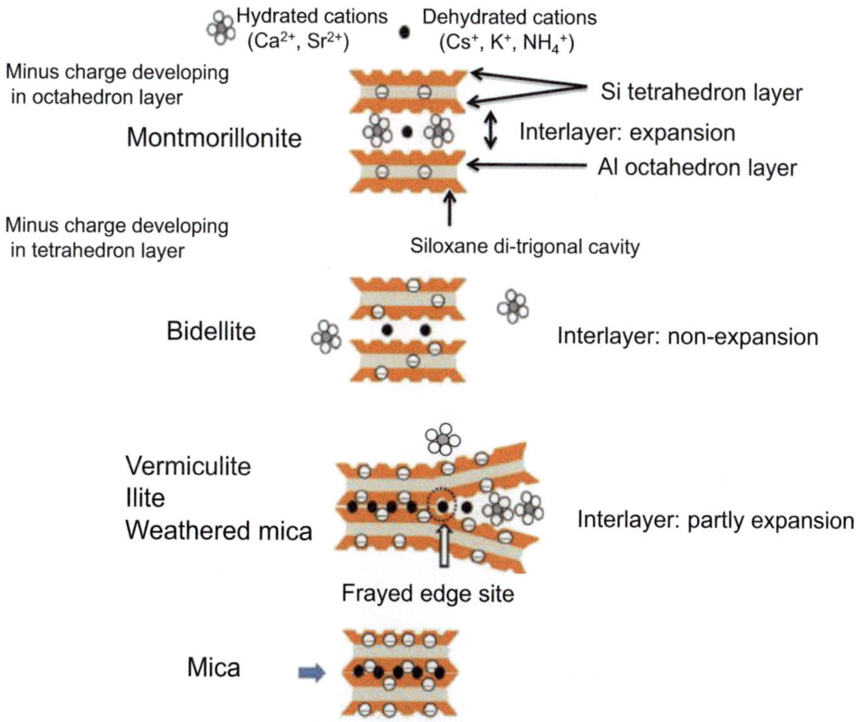

Fig. 3.10 Schematic illustration on radiocesium sorption to 2:1 silicate clay minerals (The illustration was modified from Yamaguchi et al. 2012)

strongly to Cs^+ ion. Mica has non-expansion of the interlayer because of the fixation of K^+ ion.

The most important parameters related to radiocesium sorption are clay mineralogy, stable Cs concentration, cation composition, and ionic strength in a soil solution (Sawhney 1972; Tsumura et al. 1984; Stauton and Roubaud 1997). Cesium is adsorbed specifically onto micaceous minerals with frayed-edge sites. Maes et al. (1999) demonstrated in south Belgium that the magnitude of Cs ion fixation is positively and strongly related to the soil vermiculite content. The proportion of ^{137}Cs fixed to clay minerals decreases concomitantly with increasing concentration of stable 3.1^{33}Cs because of dilution and exchange of ^{137}Cs (Tsumura et al. 1984). The sorption of ^{137}Cs depends on the cationic composition in the soil solution. The order of decreasing sorption affinity for ^{137}Cs is $Cs^+ \gg NH_4^+ > K^+ > Mg^{2+} \fallingdotseq Ca^{2+} \fallingdotseq Sr^{2+} \fallingdotseq Ba^{2+} \fallingdotseq Na^+$. This section presents a summary of the existing forms of radiocesium in surface soil from Fukushima area in Japan before and after the accident.

3.5.1 Size Distribution

Tanaka et al. (2013) reported the distribution of ^{137}Cs in surface soil of Kawamata Town in Fukushima Prefecture on 29 May, 2011. The ^{137}Cs activity of the elutriated sample with particles smaller than 2 μm, corresponding to the clay size fraction, is about 70 % higher than that of the bulk sample. This result is attributed to smaller particles with larger specific surface area giving more sorption sites for Cs (He and Walling 1996) and strongly sorbing onto clay minerals (Sawhney 1972; Maes and Cremers 1986; Cornel 1993). Tsukada et al. (2008) reported similar results with fallout-derived ^{137}Cs in agricultural soils in Aomori Prefecture, Japan in 1991, 1992, 1994, and 2004. The radioactivity of ^{137}Cs in the clay-size fraction after removing the oxidizable organic-bound fraction was highest among the particle-size fractions for agricultural soils. The ^{137}Cs in the clay fraction was 4–25 times higher than fine and coarse sand. This result shows agreement with those of previous studies demonstrating that ^{137}Cs was concentrated into the finer fraction in the soils, and where the concentration ratios of ^{137}Cs in clay against sand were 8–46 (Livens and Baxter 1988) and 3–43 (Spezzano 2005).

3.5.2 Chemical Extraction

Matsunaga et al. (2013) examined the water solubility of radiocesium for surface soil (croplands, grasslands, and forest) collected from the southwestern part of Fukushima City during the pre-rainy season (June 18–20, 2011). The radioactivity of ^{137}Cs was 1210–7672 Bq kg^{-1} dry-soil. The percentage of ^{137}Cs in the water soluble-fraction was quite low (0.4–1.6 %) at 0–1 cm depth and was undetected in the soils at 1–3 cm and 3–5 cm depth. Tanaka et al. (2012) reported that dissolved ^{137}Cs into the aqueous phase at pH 1–11 was less than 1 % for lowland, andosol, and brown forest soils. These results suggest that ^{137}Cs in the soil is present as less water-soluble forms.

Ammonium acetate (NH_4Ac) extraction has been used to examine the association of radiocesium with clay minerals (e.g., Tsumura et al. 1984; Roig et al. 1998; Takeda et al. 2006). The percentage of ^{137}Cs in the NH_4Ac-extractable fraction is shown in Table 3.1. The NH_4Ac-extractable percentage is 1–20 %. In the Fukushima samples, ^{137}Cs fractions extracted with 1 M NH_4Ac at pH 7 were 7.6–12.1 %, except for one location (1.4 %) (Matsunaga et al. 2013). The percentage of the NH_4Ac-extractable ^{137}Cs is higher in the soil with a higher soil pH (Matsunaga et al. 2013). Korobova et al. (2008) reported similar correlation between the percentage of extracted ^{137}Cs and soil pH for the Chernobyl soil samples. The results indicate that ion-exchangeable forms are important for NH_4Ac-extractable ^{137}Cs in studied soils. However, more than 81 % (81–99 %) of ^{137}Cs remained present in the residual phase in this experimental condition. Qin et al. (2012) conducted a sequential extraction experiment (modified BCR method) for the

Table 3.1 Percentage of ^{137}Cs extracted from surface mineral soil samples using 1 M ammonium acetate (NH_4Ac) at pH 7

Area	Depth (cm)	Soil type	^{137}Cs (Bq kg^{-1})	Extraction (%)
Fukushima[1]	0–1	——[a]	1210–7672	1.4–12.1(8.6 ± 3.9)[b]
Aomori[2]	0–5, 0–20	Andosol	1–37	(12 ± 5)[b]
Germany[3]	2–5	Alfisol	317	1–2
Sweden[4]	0–2	Forest	–	13
Fukushima[5]	0–5	Loam[c]	–	<20 (14 ± 7)[b]

Numbers in references are as follows: [1]Matsunaga et al. (2013), [2]Tsukada et al. (2008), [3]Bunzel et al. (1999), [4]Fawaris and Johanson (1995), [5]Saito et al. (2014b). The ratio of soil to the solution is 1:10 except for the Sweden soil (1:3). Extraction with 1 M NH_4Ac was conducted for 1–24 h

[a]Land-use type is cropland, grassland, and forest

[b]Numbers in the blanket of extraction indicate average ± standard deviation

[c]Loam samples are classified into sandy loam, loam and sandy clay loam

surface soil sample (0–5 cm depth) collected from Kawamata Town in the northern part of Fukushima Prefecture on 29 May 2011. Less than 0.1 % of the total ^{137}Cs in the natural sample was extracted in the first step (exchangeable fraction), which implies that extremely small amounts of ^{137}Cs can be found in the soluble and exchangeable species in the soil. More than 94 % of ^{137}Cs remained in the residue fraction after three-step extraction (exchangeable, Fe and Mn oxyhydroxides associated and organic matter and sulfide associated fractions). The clay mineral composition of soils in Fukushima area is dominated by smectite (Nakao et al. 2014). Vermiculite clays are also ubiquitous in the soil around Fukushima (Shimane 1968). Mukai et al. (2014) showed that ^{137}Cs is sorbed uniformly in the porous weathered biotite in forest soil near the boundary between Iitate village and Namie Town in Fukushima. These findings are explainable by the fact that most Cs occurs in the interlayer or frayed-edge sites (FES) of clay minerals (Choi et al. 2005).

Surface soil conditions are important for migration of radiocesium in the early stages after an accident. Matsunaga et al. (2013) reported that the aboveground litter layer retained more than half of the inventory in forest locations in the post rainy season. Bunzel et al. (1999) applied sequential chemical extraction to soil collected from northern Finland and southern Germany in 1997. Extracted fractions of ^{137}Cs derived fallout differed from the mineral soil and podozol (organic soil layer). The litter and semi-decomposed litter (Ol + Of) layer removed litter indicates an exchangeable fraction of 30–40 %, the oxide + organic fractions of 10 %, and the resident fraction of 40–50 %. However, the mineral soil showed that ^{137}Cs is found only in extremely small amounts (1–2 %). Tsukada et al. (2008) conducted similar extraction experiments for andosol in Aomori Prefecture. The extracted fractions of ^{137}Cs are the exchangeable of 10 %, the organic bound of 20 %, and the residue of 70 %. Downward movement of ^{137}Cs derived from the global fallout and the Chernobyl accident in soil is extremely low such that ^{137}Cs was retained in the surface layer (e.g., Almgren and Isaksson 2006; Yamaguchi et al. 2012). Rosèn et al. (1999) also reported that most (72–93 %) of the ^{137}Cs fallout remained present in the upper 5 cm of mineral soils 8 years after the Chernobyl accident. However, peat soil showed ^{137}Cs deeper than 50 cm from the surface (Rosèn et al. 1999) because of lower contents of clay minerals and weak sorption of organic matter to radiocesium (Rigol et al. 2002). Therefore, it is important to elucidate the existence forms of radiocesium in soils from the perspective of soil properties, clay mineral composition, amount of stable Cs, and time dependency after the nuclear accident.

3.5.3 EXAFS Analysis

Cesium adsorption on clay minerals such as vermiculite and montmorillonite is described as a function of surface coverage using extended X-ray adsorption fine structure spectroscopy (EXAFS). Bostick et al. (2002) determined the Cs–O distance between 3.2 and 4.3 Å so that adsorption recognizes both inner sphere and outer sphere adsorption complexes. Qin et al. (2012) applied EXAFS to soil, granite, and river sediment samples to elucidate the sorption of Cs soil components. The Cs species adsorbed onto the natural samples were similar to those adsorbed onto clay minerals and micas. This finding provided evidence related to the significant contribution of clay minerals or micas to Cs retention in soils from Fukushima Prefecture. Fan et al. (2014) investigated the cesium adsorption to clay minerals (illite, montmorillonite, and vermiculite) using sequential extraction, batch sorption, x-ray diffraction, and EXAFS analysis with molecular simulations. The inner sphere complex of dehydrated Cs^+ mainly formed at the frayed edge site and interlayer site on illite (non-expansion) without producing any illite structural changes. However, on vermiculite (intermediate expansion), the dehydrated Cs^+ can be adsorbed as an inner sphere complex associated with the siloxane group of the di-trigonal cavity in the tetrahedral SiO_4 sheet. However this is not a direct finding because these analyses were conducted for samples saturated with a 0.50 M CsCl solution.

3.6 Dynamics of ^{134}Cs and ^{137}Cs in Surface Soil

Elucidating the effects of intense rainfall on the distribution of Fukushima-derived ^{137}Cs in soil is important to assess transport in the early stages after an accident. Matsunaga et al. (2013) determined inventories and the vertical distribution of ^{137}Cs in soil before and after the rainy season in 2011 at 15 locations in Fukushima City. The ^{137}Cs inventory levels scarcely changed between points in time spanning the first rainy season after the accident, except for the CP-2 soil core (Fig. 3.9). The deeper penetration of ^{137}Cs after the rainy season results from tilling after the first sampling. The vertical profiles were almost unchanged at most locations. These results suggest that rainfall during the rainy season had a limited effect on the ^{137}Cs

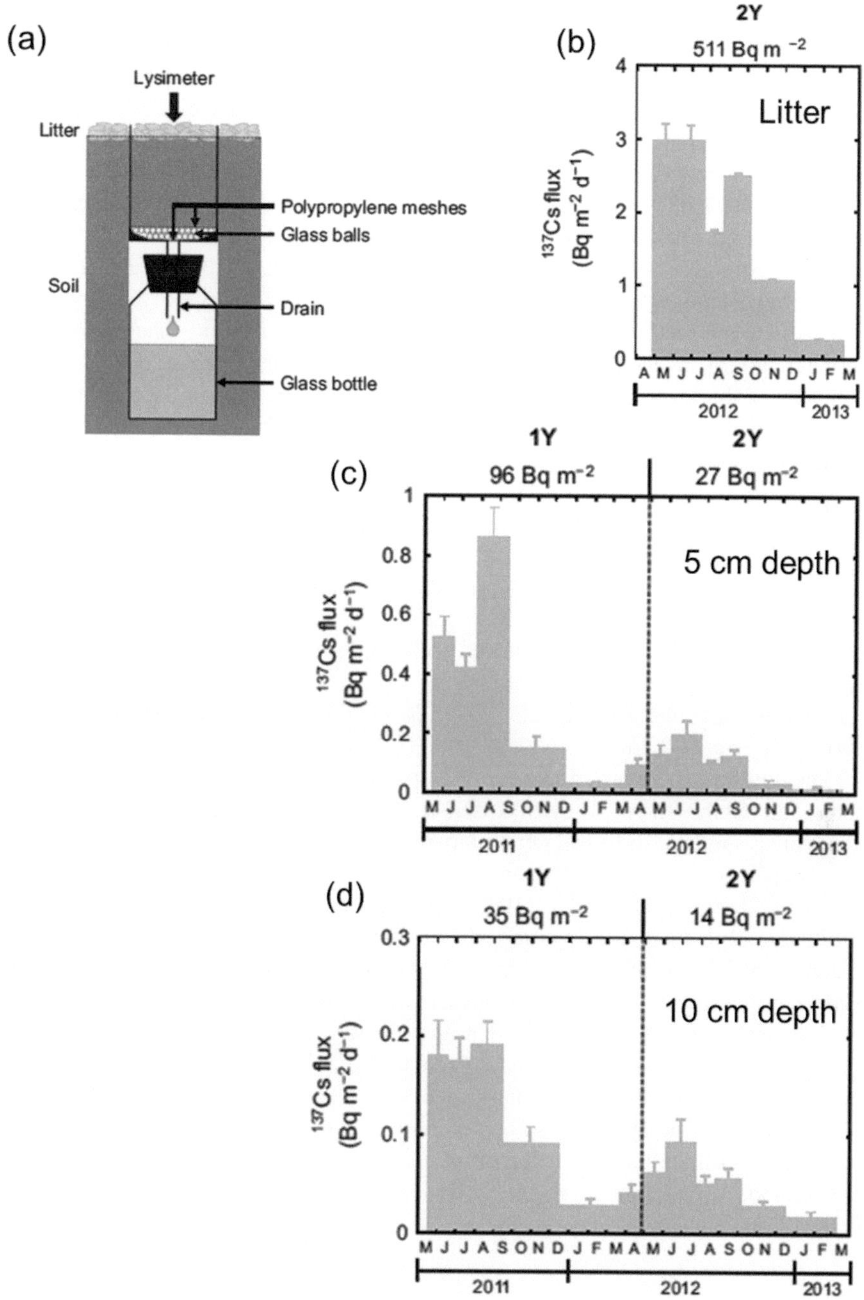

Fig. 3.11 *In situ* experimental system (**a**) and determined ^{137}Cs flux: in the litter leachate (**b**), at 5 cm depth (**c**) and at 10 cm depth (**d**) (modified from Nakanishi et al. 2014). *In situ* experiments were carried out in the Ogawa forest at the southern edge of the Abukuma Highland region

distribution in the soil. Uchida and Tagami (2014) reported soil profiles of ^{137}Cs activity at 0–10 cm depth in paddy fields and grassland in Kawamata Town in a highly contaminated area of Fukushima Prefecture. Data were referenced from the database of environmental monitoring on Fukushima Dai-ichi NPP from the Nuclear Regulatory Authority. The ^{137}Cs profiles are shown for the samples on the collection date (22 June 2011, 18 January, 28 August, and 11 December 2012). More than 80 % of ^{137}Cs was distributed at 0–3 cm depth. The results suggest very low downward movement of ^{137}Cs in soil environment for about 1 year after the accident. Similar results were observed for soil in Sweden 3–5 years after the Chernobyl accident (Rosèn et al. 1999).

In-situ experiments were conducted by the Japan Atomic Energy Agency research group in deciduous forest soil of the watershed of Yotudoki River, a tributary of the Same River, over a period spanning 2 months to 2 years after the accident (Nakanishi et al. 2014). Seepage water was collected from the study site using PVC zero-tension lysimeters at each of two depths (5 and 10 cm). Figure 3.11 presents results of the downward flux of ^{137}Cs at each depth layer. The major part of ^{137}Cs in the litter layer moved into the mineral soil within 1 year after the accident. The topsoil prevented migration of ^{137}Cs. Only 2 % of ^{137}Cs penetrated below 10 cm depth. These results are supported by a column experiment by Tsumura et al. (1984). The 370 kBq of ^{137}Cs solution was added on the top of a column (15 cm diameter, 9 cm height) packed with sandy soil and andosol. The columns were set up outside of the laboratory in Tokyo for 7 months with rainfall of 1,260 mm. After the experiments, 90–95 % of ^{137}Cs activity was retained in the surface layer at 0–2 cm.

The downward migration rate of ^{137}Cs in soil is 0.34–0.81 cm year^{-1} for brown forest soil (Chibowski et al. 1999; Rosèn et al. 1999; Hölgye and Malý 2000), which is a typical soil type in forest area in Japan. Mahara (1993) reported the downward rate of 0.1 cm year^{-1} for the unsaturated soil layer in the Nishiyama area of Nagasaki, Japan. Rosèn et al. (1999) reported that the migration rate decreases concomitantly with increasing time of the experiment: first year 0.5–1.0 cm year^{-1}; second year 0.2–0.6 cm year^{-1}. These results support the validity of the results of *in situ* experiments conducted by Nakanishi et al. (2014). However, peat soil with high organic matter content exhibited a migration rate of more than 1.0 cm year^{-1} (Rosèn et al. 1999; Chibowski and Zygmunt 2002). The soil type is related to the downward migration of ^{137}Cs in surface soil. Consequently, the elucidation of physicochemical properties of soil environments is extremely important to evaluate and to predict the migration behavior of radiocesium derived from the accident in Fukushima.

References

Almgren S, Isaksson M (2006) Vertical migration studies of Cs-137 from nuclear weapons fallout and the Chernobyl accident. J Environ Radioact 91:90–102

Aoyama M, Hirose K, Igarashi Y (2006) Re-construction and updating our understanding on the global weapons tests ^{137}Cs fallout. J Environ Monit 8:431–438

Bostick BC, Vairavamurthy MA, Karthikeyan KG, Chorover J (2002) Cesium adsorption on clay minerals: an EXAFS spectroscopic investigation. Environ Sci Technol 36:2670–2676

Bunzel KW, Schimmack P, Schramel P, Suomela M (1999) Effect of sample drying and storage time on the extraction of fallout Pu239 + 240, Cs-137 and natural Pb-210 as well as stable Cs, Pb and Mn from soils. Analyst 124:1383–1387

Chibowski S, Zygmunt J (2002) The influence of the sorptive properties of organic soils on the migration rate of Cs-137. J Environ Radioact 61:213–223

Chibowski S, Zygmunt J, Klimowicz Z (1999) Investigation of adsorption and vertical migration of ^{137}Cs in three kinds of soil at Lublin vicinity. J Radioanal Nucl Chem 242:287–295

Chino M, Nakayama H, Nagai H, Terada H, Katata G, Yamazawa H (2011) Preliminary estimation of release amounts of ^{131}I and ^{137}Cs accidentally discharged from the Fukushima Daiichi nuclear power plant into the atmosphere. J Nucl Sci Technol 48(7):1129–1134

Choi S, Crosson G, Mueller KT, Seraphin S, Chorover J (2005) Clay mineral weathering and contaminant dynamics in a caustic aqueous system II. Mineral transformation and microscale partitioning. Geochim Cosmochim Acta 69:4437–4451

Cornel RM (1993) Adsorption of cesium on minerals: a review. J Radioanal Nucl Chem 171:483–500

CTBTO (Comprehensive Nuclear-Test-Ban Treaty Organization Preparatory Commission) (2011) Fukushima-related measurements by the CTBTO http://www.ctbto.org/press-centre/highlights/2011/fukushima-related-mesurements-by-thectbto. Accessed 5 Jan 2015

de Cort M (1998) Atlas of caesium deposition on Europe after the Chernobyl accident. Rep. 16733, Office for Official Publications of the European Communities, Luxembourg

Endo S, Kimura S, Takatsuji T, Nanasawa K, Imanaka T, Shimizu K (2012) Measurement of soil contamination by radionuclides due to the Fukushima Dai-ichi Nuclear Power Plant accident and associated estimated cumulative external dose estimation. J Environ Radioact 111:18–27

Fan QH, Tanaka M, Tanaka K, Sakaguchi A, Takahashi Y (2014) An EXAFS study on the effect of natural organic matter and the expansibility of clay minerals on cesium adsorption and mobility. Geochim Cosmochim Acta 135:49–65

Fawaris BH, Johanson KJ (1995) Fractionation of cesium (Cs-137) in coniferous forest soil in Central Sweden. Sci Total Environ 170:221–228

Fujii K, Ikeda S, Akama A, Komatsu M, Takahashi M, Kaneko S (2014) Vertical migration of radiocesium and clay mineral composition in five forest soils contaminated by the Fukushima nuclear accident. Soil Sci Plant Nutr 60:751–764

Fujiwara T, Saito T, Muroya Y, Sawahata H, Yamashita Y, Nagasaki S, Okamoto K, Takahashi H, Uesaka M, Katsumura Y, Tanaka S (2012) Isotopic ratio and vertical distribution of radionuclides in soil affected by the accident of Fukushima Dai-ichi nuclear power plants. J Environ Radioact 113:37–44

He Q, Walling DE (1996) Interpreting particle size effects in the adsorption of ^{137}Cs and unsupported ^{210}Pb by mineral soils and sediments. J Environ Radioact 30:117–137

Hirano M, Yonomoto T, Ishigaki M, Watanabe N, Maruyama Y, Sibamoto Y, Watanabe T, Moriyama K (2012) Insight from review and analysis of the Fukushima Dai-ichi accident. J Nucl Sci Technol 49:1–17

Hölgye Z, Malý M (2000) Sources, vertical distribution, and migration rates of $^{239,240}Pu$, ^{238}Pu, and ^{137}Cs in grassland soil in three localities of central Bohemia. J Environ Radioact 47:135–147

IAEA (International Atomic Energy Agency) (2006a) Environmental consequences of the Chernobyl accident and their remediation: twenty years of experience/report of the Chernobyl Forum Expert Group ‘Environment’. IAEA, Vienna

IAEA (International Atomic Energy Agency) (2006b) Radiological conditions in the Dnieper River Basin. IAEA, Vienna

IAEA (International Atomic Energy Agency) (2014) The international nuclear and radiological event scale. http://www-ns.iaea.org/tech-areas/emergency.ines.asp. Accessed 09 Jan 2015

IDCRM (Inter-Departmental Commission on Radiation Monitoring) (2000) Map of contemporary (normalized to 1997) contamination of soils by ^{90}Sr of territories of the southern Urals region as a result of the 1957 and 1967 accidents at the Mayak PA, 1:200000. Institute of Global Climate and Ecology, Moscow

Iijima K, Tomura T, Shoji Y (2010) Reversibility and modeling of adsorption behavior of cesium ions on colloidal montmorillonite particles. Appl Clay Sci 49:262–268

International Advisory Committee (1991) The international Chernobyl project: technical report. IAEA, Vienna

JNREG (Joint Norwegian–Russian Expert Group) (1997) Sources contributing to radioactive contamination of the Techa River and the area surrounding the 'Mayak' production association, Urals, Russia. Joint Norwegian–Russian Expert Group for Investigation of Radioactive Contamination in the Northern Areas. NRPA, Østerås

Kato H, Onda Y, Teranaga M (2012) Depth distribution of ^{137}Cs, ^{134}Cs, and ^{131}I in soil profile after Fukushima Dai-ichi Nuclear Power Plant Accident. J Environ Radioact 111:59–64

Kinoshita N, Sueki K, Sasa K, Kitagawa U, Ikarashi S, Nishimura T, Wong YS, Satou Y, Handa K, Takahashi T, Sato M, Yamagata T (2011) Assessment of individual radionuclide distributions from the Fukushima nuclear accident covering central-eastern Japan. Proc Natl Acad Sci U S A 108:19526–19529

Knatko VA, Skomorokhov AG, Asimova VD, Strakh LI, Bogdanov AP, Mironov VP (1996) Characteristics of ^{90}Sr, ^{137}Cs and ^{239,240}Pu migration in undisturbed soils of southern Belerus after the Chernobyl Accident. J Environ Radioact 30:185–196

Koarashi J, Atarashi-Andoh M, Mtsunaga T, Sato T, Nagao S, Nagai H (2012) Factors affecting vertical distribution of Fukushima accident-derived radiocesium in soil under different land-use conditions. Sci Total Environ 431:392–401

Koarashi J, Moriya K, Atarashi-Andoh M, Matsunaga T, Fujita H, Nagaoka M (2013) Retention of potentially mobile radiocesium in forest surface soils affected by the Fukushima nuclear accident. Sci Rep 2:1005

Korobova E, Linnik V, Chizhikova N (2008) The history of the Chernobyl ^{137}Cs contamination of the flood plain soils and its relation to physical and chemical properties of the soil horaizons (a case study). J Geochem Explor 96:236–255

Kostyuchenko VV, Akleyev AV, Peremyslova LM, Popova IYa, Kazachonok NN, Melnikov VS (2012) Environmental migration of radionuclides (^{90}Sr, ^{137}Cs, ^{239}Pu) in accidentally contaminated areas of the southern Urals. In: Rahman RA (ed) Radioactive waste. http://intecopen.com/books/radioactive-waste/environmental-migration-of-long-lived-radionuclides-90sr-137cs-239pu-in-contaminated-areas-of-th. Accessed 20 Jan 2015

Livens FR, Baxter MS (1988) Particle-size and radionuclide levels in some west Cumbrian soils. Sci Total Environ 70:1–17

Maes A, Cremers A (1986) Highly selective ion exchange in clay minerals and zeolites. In: Davis JA, Hayes KF (eds) Geochemical processes at mineral surfaces. American Chemical Society, Washington, DC, pp 245–295

Maes E, Vielvoye L, Stone W, Delvaux B (1999) Fixation of radiocaesium tracers in a weathered sequence mica → vermiculite → hydroxyl interlayed vermiculite. Eur J Soil Sci 50:107–115

Mahara Y (1993) Storage and migration of fallout strontium-90 and cesium-137 for over 40 years in the surface soil of Nagasaki. J Environ Qual 22:722–730

Masson O, Baeza A, Bieringer J et al (2011) Tracking of airborne radionuclides from the damaged Fukushima Dai-ichi Nuclear Reactors by European Networks. Environ Sci Technol 45:7670–7677

Matsunaga T, Nagao S (2009) Environmental behavior of plutonium isotopes studied in the area affected by the Chernobyl accident. Humic Subst Res 5–6:19–33

Matsunaga T, Koarashi J, Atarashi-Andoh M, Nagao S, Sato T, Nagai H (2013) Comparison of the vertical distributions of Fukushima nuclear accident radiocesium in soil before and after the first rainy season, with physicochemical and mineralogical interpretations. Sci Total Environ 447:301–314

MEXT (Ministry of Education, Culture, Sports, Science and Technology of Japan) (2011) Result of fourth air-borne survey monitoring. http://radioactivity.nsr.go.jp/ja/contents/5000/4501/24/1910-1216.pdf. Accessed 16 Jan 2012

Morino Y, Ohara T, Nishizawa M (2011) Atmospheric behavior, deposition, and budget of radioactive materials from the Fukuhsima Daiichi nuclear power plant in March 2011. Geophys Res Lett 38:L00G11

Mukai H, Hatta T, Kitazawa H, Yamada H, Yaita T, Kogure T (2014) Speciation of radioactive soil particles in the Fukushima contaminated area by IP autoradiography and microanalyses. Environ Sci Technol 48:13053–13059

Nagai H, Kurihara O (2014) Outline of UNSCEAR 2013 Report (1): radionuclide release, dispersion and deposition. J At Energy Soc Jpn 56:791–795

Nakanishi T, Matsunaga T, Koarashi J, Atarashi-Andoh M (2014) ^{137}Cs vertical migration in a deciduous forest soil following the Fukushima Dai-ichi Nuclear Power Plant accident. J Environ Radioact 128:9–14

Nakao A, Ogasawara S, Sano O, Ito T, Yanai J (2014) Radiocesium sorption in relation to clay mineralogy of paddy soils in Fukushima, Japan. Sci Total Environ 468–469:523–529

National Institute for Agro-Environmental Sciences (NIAES) (2012) Study on spatial distribution of radionuclides in cultivated soils. http://www.niaes.affrc.go.jp/sinfo/result/result28/result28_02.html. Accessed 15 Jan 2015

NERH (Nuclear Emergency Response Headquarters, Government of Japan) (2011) Report of Japanese Government to the IAEA Ministerial Conference on Nuclear Safety—the accident at TEPCO's Fukushima Nuclear Power Stations—. http://www.kantei.go.jp/foreign/kan/topics/201106/iaea_houkokusho_e.html. Accessed 5 Jan 2015

Ohnuki T, Kozai N (1994) Sorption characteristics of radioactive cesium and strontium on smectite. Radiochim Acta 66–67:327–331

Pavlotskaya FI, Goryachenko TA, Myasoedov BF (1991) Behavior of transuranium elements in the environment. J Radioanal Nucl Chem 147:133–140

Peremylova LM, Tolstykh EI, Vorobiova MI, Degteva MO, Safronova NG, Shagina NB, Anspaugh LR, Napier BA (2004) Analytical review of data available for the reconstruction of dose due to residence on the East Ural radioactive trace and the territory of windblown contamination from Lake Karachay. Final report for milestone 10. http://biophys.urcrm.ru/publications/Milestone10.pdf. Accessed 20 Jan 2015

Qin H, Yokoyama Y, Fan Q, Iwatani H, Tanaka K, Sakaguchi A, Kanai Y, Zhu J, Onda Y, Takahashi Y (2012) Investigation of cesium adsorption on soil and sediment samples from Fukushima Prefecture by sequential extraction and EXAFS technique. Geochem J 46(4):297–302

Roig M, Vidal M, Rauret G (1998) Estimating the radionuclide available fraction in mineral soils using as extraction technique. Analyst 123:519–526

Rigol A, Vidal M, Rauret G (2002) An overview of the effect of organic matter on soil-radiozaesium interaction: implication in root uptake. J Environ Radioact 58:191–216. doi:10.1016/S0265-931X(01)00066-2

Rosèn K, Oborn I, Lonsjo H (1999) Migration of radiocaesium in Swedish soil properties after the Chernobyl accident. J Environ Radioact 46:45–66

Saito J, Tanihata I, Fujiwara M, Shinohara S, Otsuka T, Onda Y, Hoshi M, Ikeuchi Y, Takahashi F, Kinouchi N, Saegusa J, Seki A, Takemiya H, Shibata T (2014a) Detailed deposition density maps constructed by large-scale soil sampling for gamma-ray emitting radioactive nuclides from the Fukuhsima Dai-ichi Nuclear Power Plant accident. J Environ Radioact 139:308–319

Saito T, Makino H, Tanaka S (2014b) Geochemical and grain-size distribution of radioactive and stable cesium in Fukushima soils: implications for their long-term behavior. J Environ Radioact 138:11–18

Sawhney BL (1972) Selective sorption and fixation of cations by clay minerals. A review. Clay Clay Miner 20(2):93–100
Shimane H (1968) Vermiculite and allied minerals. In: Editorial Subcommitted for the Clays of Japan Organizing Committee International Clay Conference (ed) The clays of Japan. Geological Survey of Japan, Tsukuba, pp 139–142
Skipperud L, Salbu B, Oughton DH, Drozcho E, Mokrov Y, Strand P (2005) Plutonium contamination in soils and sediments at Mayak PA, Russia. Health Phys 89:255–266
Spezzano P (2005) Distribution of pre- and post-Chernobyl radiocaesium with particle size fractions of soils. J Environ Radioact 83:117–127
Sposito G (2008) The chemistry of soils, 2nd edn. Oxford University Press, New York
Stauton S, Roubaud M (1997) Adsorption of ^{137}Cs on montmorillonite and illite: effect of charge compensating cation, ionic strength, concentration of Cs, K and fulvic acid. Clay Clay Miner 45(2):251–260
Takeda A, Tsukada H, Takaku Y, Hisamatsu S, Inaba J, Nazyo M (2006) Extractability of major and trace elements from agricultural soils using chemical extraction methods: application for phytoavailablility assessment. Soil Sci Plant Nutr 52:406–417
Tamura T, Jacobs DG (1960) Structural implications in cesium sorption. Health Phys 2 (4):391–398
Tanaka K, Takahashi Y, Sakaguchi A, Umeo M, Hayakawa S, Tanida H, Saito T, Kanai Y (2012) Vertical profiles of Iodine-131 and Cesium-137 in soils in Fukushima Prefecture related to the Fukushima Daiichi Nuclear Power Station accident. Geochem J 46(1):73–76
Tanaka K, Sakaguchi A, Kanai Y, Tsuruta H, Shinohara A, Takahashi Y (2013) Heterogeneous distribution of radiocesium in aerosols, soil and particulate matters emitted by the Fukuhsima Daiichi Nuclear Power Plant accident: retention of microscale heterogeneity during the migration of radiocesium from the air into ground and river systems. J Radioanal Nucl Chem 295:1927–1937
TEPCO (Tokyo Electric Power Co.) (2011a) Estimation of reactor damage of the Fukuhsima Dai-ichi Nuclear Power Plant, Unit 1–3. http://www.tepco.co.jp/nu/fukushima-np/images/handouts_111130_07-j.pdf. Accessed 20 Jan 2015
TEPCO (Tokyo Electric Power Co.) (2011b) Result on radionuclides measured in soil within the site of Fukushima Dai-ichi nuclear power plant, Press Release March 28, 2011. http://www.tepco.co.jp/cc/press/11032806-j.html. 5 Jan 2015
Terada H, Furuno A, Chino M (2005) Improvement of worldwide version of system for prediction of environmental emergency dose information (WSPEEDI), (1); New combination of models, atmospheric dynamic model MM5 and particle random walk model GEARN-new. J Nucl Sci Technol 42(7):651–660
Teramage MT, Onda Y, Patin J, Kato H, Gomi T, Nam S (2014) Vertical distribution of radiocesium in coniferous forest soil after the Fukushima nuclear power plant accident. J Environ Radioact 137:37–45
Thakur P, Ballard S, Nelson R (2013) An overview of Fukushima radionuclides measured in the northern hemisphere. Sci Total Environ 458–460:577–613
Tsukada H, Takeda A, Hisamatsu S, Inaba J (2008) Concentration and specific activity of fallout Cs-137 and particle-size fractions of cultivated soils. J Environ Radioact 99:875–881
Tsumura A, Komamura M, Kobayashi H (1984) Behavior of radioactive Sr and Cs in soils and soil-plant systems. Bull Natl Inst Agric Sci Series B 36:57–113
Uchida S, Tagami K (2014) Environmental behavior of radiocesium. Bunseki 478:535–538
UNSCEAR (United Nations Scientific Committee on the Effects of Atomic Radiation) (2000) Sources and effects of ionizing radiation, UNSCEAR 2000 report to the general assembly with scientific annexes volume II scientific annex J. United Nations, New York
UNSCEAR (United Nations Scientific Committee on the Effects of Atomic Radiation) (2014) Sources, effects and risks of ionizing radiation, UNSCEAR 2013 report to the general assembly with scientific annexes volume I scientific annex A. United Nations, New York

Watanabe T, Tsuchiya N, Oura Y, Ebihara M, Inoue C, Hirano N, Yamada R, Yamasaki S, Okamoto A, Nara FW, Nunohara K (2012) Distribution of artificial radionuclides ($^{110\,m}$Ag, ^{129m}Te, ^{134}Cs, ^{137}Cs) in surface soils from Miyagi Prefecture, northeast Japan, following the 2011 Fukushima Dai-ichi nuclear power plant accident. Geochem J 46(4):279–285

Wetherbee GA, David A, Gay A, Debey TM, Lehmann CMB, Nilles MA (2012) Wet deposition of fission-product isotopes to north America from the Fukushima Dai-ichi incident. Environ Sci Technol 46:2574–2582

World Health Organization (WHO) (2006) Health effects of the Chernobyl accident and special health care programmes. Report of the UN Chernobyl Forum Expert Group "Health". WHO, Geneva

Yamaguchi N, Takada Y, Hayashi K, Ishikawa S, Kuramata M, Eguchi S, Yoshikawa S, Sakaguchi A, Asada K, Wagi R, Makino T, Akahane I, Hiradate S (2012) Behavior of radiocaesium in soil-plant systems and its controlling factors: a review. Bull Natl Inst Agro Environ Sci 31:75–121

Yamamoto M, Takada T, Nagao S, Koike T, Shimada K, Hoshi M, Zhumadilov K, Shima T, Fukuoka M, Imanaka T, Endo S, Sakaguchi A, Kimura S (2012) An early survey of the radioactive contamination of soil due to the Fukushima Dai-ichi Nuclear Power Plant accident, with emphasis on plutonium analysis. Geochem J 46(4):341–353

Yamamoto M, Sakaguchi A, Ochiai S, Takada T, Hamataka K, Murakami T, Nagao S (2014) Isotopic Pu, Am and Cm signatures in environmental samples contaminated by the Fukushima Dai-ichi Nuclear Power Plant accident. J Environ Radioact 132:31–46

Chapter 4
Test Methods for the Evaluation of Heavy Metals in Contaminated Soil

S. Mizutani, M. Ikegami, H. Sakanakura, and Y. Kanjo

Abstract Test methods for the evaluation of heavy metals contained in contaminated soil were summarized. The soil environmental standards, criteria for soils in each country, and setup backgrounds were studied. The methods for the extraction of heavy metals were classified into three categories: (1) digestion/decomposition, (2) extraction, and (3) leaching test. Regarding these three analytical methods, the scientific meaning of each method and concrete test procedures were described. Regarding the extraction method, the SCE (sequential chemical extraction) procedure for chemical forms of metals and extraction for bioavailability were discussed. In the section with regard to the leaching test, the relationship between operational factors and the leaching concentration was discussed and the availability test was also introduced. Furthermore, the movement of unifying the test methods in the ISO (International organization for standardization) was introduced. The environmental standards of soils and the criteria of 13 countries were compared. In addition, the setup backgrounds of the environmental standards and the environmental criteria for soils in Canada, Australia, the Netherlands, and Germany were investigated and discussed.

Keywords Digestion/decomposition • Extraction • Leaching • Bioavailability • Availability test • Environmental standard • Environmental criteria • ISO

S. Mizutani (✉) • Y. Kanjo
Department of Urban Engineering, Osaka City University, 3-3-138, Sugimoto, Sumiyoshi-Ku, Osaka 558-8585, Japan
e-mail: miz@urban.eng.osaka-cu.ac.jp

M. Ikegami
Research Reactor Institute, Kyoto University, 2, Asashiro-Nishi, Kumatori-cho, Sennan-gun, Osaka 590-0494, Japan

H. Sakanakura
Center for Material Cycles and Waste Management Research, National Institute for Environmental Studies, 16-2 Onogawa, Tsukuba, Ibaraki 305-8506, Japan

H. Hasegawa et al. (eds.), *Environmental Remediation Technologies for Metal-Contaminated Soils*, DOI 10.1007/978-4-431-55759-3_4

4.1 Introduction

"Soil contamination" is an extensive and often severe environmental problem that occurs globally. However, the term can be difficult to define: What exactly is contaminated soil? To answer that question, test methods and criteria for judging the extent of contamination in a soil are necessary. Beyond the identification of soil pollution, it is often required to treat or process the material to mitigate the contamination. Thereafter, it becomes necessary to confirm the satisfactory remediation of the soil after treatment. Test methods and criteria must also be established for confirmational purposes. Ultimately, the successful identification and control of soil pollutants depend on the definition of "contamination." The following questions must be considered:

- What should we treat as pollutants?
- How should we set the standard values or criteria?
- How are pollutant concentrations measured?

The answers to these questions vary widely among different countries and reflect societal concerns such as current or historical environmental problems or the background levels of natural contaminants (e.g., heavy metals). Another point of divergence between countries lies in whether the contaminated soil is handled as a solid waste. The soil may be regulated differently in this case, depending on the definition of hazardous waste in each locale. Therefore, societal perspectives complicate the issue of identifying and controlling contaminants.

It would be beneficial to develop uniform global definitions of "contaminated soil" to advance progress in addressing the issue. Such standardization would be analogous to the unification of the definition of hazardous waste that facilitated transboundary movement and global hazardous waste management under the Basel convention. However, such definitions are difficult to obtain due to the varying international sociohistorical backgrounds, as described earlier.

In the European Union (EU), the ongoing movement to harmonize various regulations has achieved the development of a unified test method for contaminated soil. After defining and ratifying a European standard, designated as "EN." However, the harmonization is just only for test method, and not for standard values. Even in the EU, each country uses its own standard values. Thus, arriving at a uniform, global definition of soil contamination will take some time.

In this chapter, the current knowledge on test methods for the digestion, extraction, and leaching of heavy metals is described, in terms of the criteria for the evaluation of contaminated soil. With respect to achieving a globally unified test method, standards established by the International Organization for Standardization (ISO) are important. In the Technical Committee No. 190 (ISO/TC190) in ISO, standardization such as extraction test, leaching test for the soil is discussed, and the standard are established. However, the ISO web site currently provides more than 150 standards when the database is searched using "soil quality" as a keyword. In those standards, especially those concerning the inorganic constituents of soils (including heavy metals), the fundamental characteristics of the soils (pH, moisture content, density, particle size distribution, etc.) are also summarized.

4.2 Mobility and Fraction of Metals in Soil – Content, Decomposition, Extraction, and Leaching

Heavy metals are present in contaminated soils as either easily or scarcely extractable fractions. The chart in Fig. 4.1 illustrates the distribution of the each fraction. The total metal content of a soil has a real value comprising all of the metal in the soil and represents a maximum value for the risk derived from soil contamination. Because it is difficult to determine this value accurately, the contents are generally analyzed after strong acid digestion.

To effect this chemical decomposition, alkali fusion or mixed-acid digestion methods using aqua regia, sulfuric acid, perchloric acid, hydrochloric acid, and hydrofluoric acid are widely employed. An important consideration is whether silicates, which constitute a major fraction of soils, can be decomposed. The disposition and distribution of the heavy metals is different for each soil, which will affect the characteristics of decomposition and extraction. Therefore, the most appropriate decomposition methods and reagents should be chosen after careful evaluation of the characteristics of both the soils and the metals.

4.2.1 Decomposition/Digestion

4.2.1.1 Alkaline Fusion

In alkaline fusion, reagents such as sodium carbonate, potassium carbonate, sodium perchlorate, and lithium metaborate are used. Hydrated silica, a major component of soil, can be dissolved by this method. Metal silicates or metals that are absorbed on silicates can be extracted. For example, bivalent metals in silicates are transformed into carbonates which are soluble in acids, as described by the following reactions.

$$Na_2CO_3 \rightarrow Na_2O + CO_2$$
$$MSiO_3 + Na_2CO_3 \rightarrow MCO_3 + Na_2SiO_3$$

In this method, a platinum crucible is usually used as the reaction container, because of its corrosion resistance.

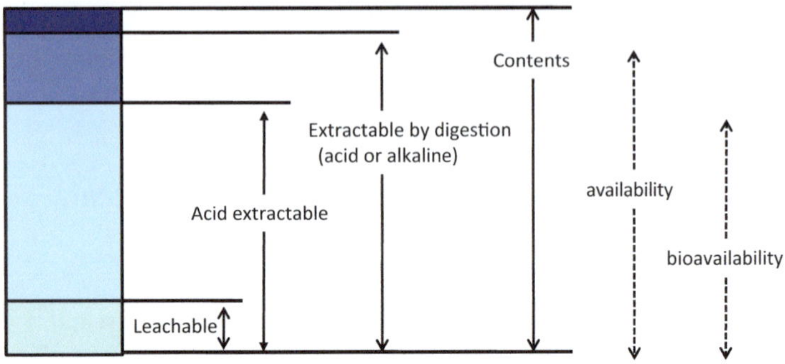

Fig. 4.1 Relationship between "contents," "acid extractable contents," and "leachable fraction"

4.2.1.2 Acid Digestion

Decomposition by Aqua Regia

This method is widely used not only for soils and sediment, but also for solid wastes such as incineration ash. Aqua regia is a 3:1 mixture of concentrated hydrochloric acid and concentrated nitric acid. The mixture is highly corrosive due to the formation of chlorine gas (Cl_2) and nitrosyl chloride (NOCl) by the following reaction:

$$3HCl + HNO_3 \rightarrow Cl_2 + NOCl + 2H_2O$$

The strong oxidizing power of these chemicals can decompose organic substances and dissolve very insoluble metals (e.g., Au, Pt) that do not easily ionize. However, As and Se can be underestimated as a result of volatilization, because of the low boiling points of their chlorides. Furthermore, silver chloride (AgCl) can be precipitated from aqua regia. Therefore, for these elements, other acids are recommended.

Nitric Acid/ Sulfuric Acid/ Perchloric Acid

This mixture of acids also exhibits very strong oxidizing power, and many kinds of metals can be extracted. However, the chromium extraction rate is reportedly sometimes lower than that in other acids. Furthermore, because of the poor solubility of lead sulfate ($PbSO_4$), lead can be underestimated. Calcium sulfate ($CaSO_4$) and barium sulfate ($BaSO_4$) also dissolve with difficulty. This system is also unfavorable for samples containing high levels of calcium.

Nitric Acid/ Hydrogen Chloride/ Hydrofluoric Acid

Silicate salts, as principal components of soils, can be dissolved in hydrofluoric acid, although they are insoluble in other strong acids. Therefore, hydrofluoric acid digestion is often used for the complete decomposition of soils along with alkaline fusion. However, special caution is required because hydrofluoric acid can dissolve experimental equipment made of borosilicate glass. Specifically, a platinum crucible or a polytetrafluoroethylene (PTFE) beaker must be used as the decomposition vessel. Furthermore, after hydrofluoric acid decomposition, any residual HF must be removed by the addition of sulfuric acid or perchloric acid and heating. Inadequate removal will damage scientific glassware or the quartz tube of an inductively coupled plasma (ICP) spectrometer.

4.2.1.3 ISO Standardized Digestion/Decomposition Methods

The ISO standard digestion/decomposition methods for soils are summarized in Table 4.1. Five methods pertain to the digestion/decomposition and extraction of heavy metals from soils: aqua regia (ISO 11466, ISO 12914), hydrofluoric acid/ perchloric acid (ISO 14869-1), alkaline fusion (ISO 14869-2), and nitric acid (ISO 16729). Each operational methodology is shown as a flow diagram in Fig. 4.2 (a)–(e).

4.2.2 Extraction Methods

4.2.2.1 Single Batch Extraction

The term "extraction" refers to the simple washing of soils by acid, rather than decomposition of the soil matrices. Generally, dilute rather than concentrated acid is used for this purpose. Furthermore, the extraction by other extractants such as the followings is also reported (Karstensen 1997; Rauret 1997).

- Chelating agent (EDTA (ethylenediaminetetraacetic acid), DTPA (diethylenetriamine pentaacetic acid) etc.)
- Buffered salt solution (NH_4-acetate + acetic acid buffer (pH 7 or pH 4.8))
- Unbuffered salt solution (0.01–0.1 mol/L of $CaCl_2$, $NaNO_3$, NH_4NO_3, $AlCl_3$, $BaCl_2$)

In Fig. 4.1, the fractions obtained by these methods are identified as the "acid extractable contents." These values are often used for evaluation of "bioavailability." It is described closely in Sect. 4.2.2.3.

Table 4.1 Standardized ISO decomposition/digestion methods for inorganic constituents

Designation	Name
ISO 11466: 1995	Extraction of trace elements soluble in aqua regia
ISO 12914: 2012	Microwave-assisted extraction of the aqua regia soluble fraction for the determination of elements
ISO 14869-1: 2001	Dissolution for the determination of total element content – Part 1: Dissolution with hydrofluoric and perchloric acids
ISO 14869-2: 2002	Dissolution for the determination of total element content – Part 2: Dissolution by alkaline fusion
ISO 16729: 2013	Digestion of nitric acid soluble fractions of elements

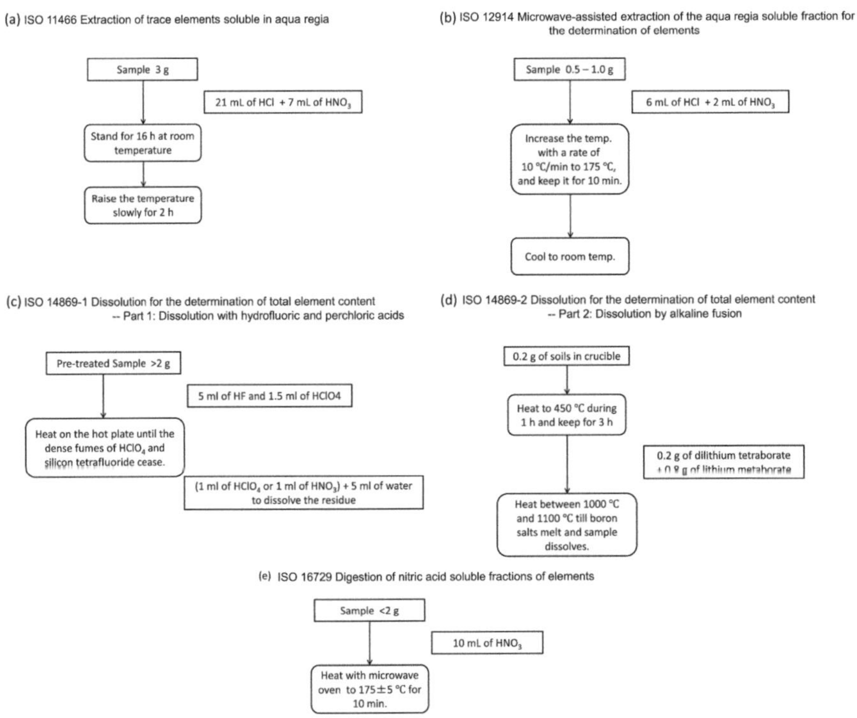

Fig. 4.2 (a)–(e) Operational flow diagrams for ISO digestion methods

4.2.2.2 Availability Test (Extraction for Availability)

Since the 1980s, a number of studies have reported that different leaching tests afforded inconsistent concentration results, even for the same solid waste. For example, van der Sloot et al. reported the results of standardized EU leaching tests performed on a single fly ash sample. The leached Pb concentrations were often different by a factor of 300, and those of Cu could be off by a factor of

100 (van der Sloot et al. 1991). These results were explained on the basis of the final pH of the leachate, which controlled the leached metal quantities. In light of this, a "best leaching test" that would be acceptable to all concerned would be desirable. The ongoing harmonization project in the European countries should incorporate the standardization of leaching test methods for waste management or contaminated soil management. In this social context, an availability test has been proposed.

When evaluating the leaching of heavy metals from soil or solid waste, the underestimation of the leaching risk must be avoided. However, overestimation is also undesirable because it results in excessive costs or inappropriate intervention. Thus, a prediction based on the total contents of soils or solid waste could overestimate the possibility of heavy metal leaching. van der Sloot reported that metals in silicate salts or insoluble minerals would not be leached within a hundred years (van der Sloot et al. 1991; van der Sloot 1991). Bishop also reported that 75 % of the Pb, 50 % of the Cd, and 85 % of the Cr were retained in soil residues after 15 serial batch extraction cycles using acid (Bishop 1988). The effective diffusivity coefficient in cement-solidified materials differs logarithmically according to each metal (van der Sloot et al. 1987).

In response to this situation, the International Ash Working Group (IAWG) proposed the concept of "availability" (i.e., the maximum leachable quantity); the test method for determining this quantity is known as the "availability test" (van der Sloot et al. 1994). This unique test is often described as a leaching test, but should be categorized as an extraction test.

Côté and Bridle defined "available for leaching" as "the fraction soluble in a leaching medium at infinite dilution" (Côté and Bridle 1987) for cement-based waste forms. Recently, this concept was expanded to solid waste fractions that become leachable under severe conditions, such as when, for example, "the particle size is physically decreased" or "the waste is kept under acidic conditions for a long term." Based on these concepts, the availability test has been standardized in the Netherlands (NEN 2004). The same concepts can be applied to soil, and availability in soil means the maximum value of the leachable quantity or the maximum leachable quantity of heavy metals in the soil. In Fig. 4.1, the concept of "availability" was also indicated. In general, this value is higher than the leachable quantity and lower than the total contents. However, the availability varies depending on the kind of soil or the chemical properties of the heavy metal.

4.2.2.3 Extraction for Bioavailability

Similar to the concept of "availability," this index focuses especially on "biological uptake" and is discussed as "bioavailability." It is important because the metal absorption degree varies according to chemical speciation of the metals widely. The index indicates the quantity of direct intake from the digestive system and is often used for human health risk assessment. To evaluate bioavailability, many extraction methods have been reported, and several review papers about bioavailability is published (Ruby et al. 1999; Nakamaru and Altansuvd 2014; Zia et al. 2011;

Peijnenburg et al. 2007; Hettiarachchi and Pierzynski 2004; Ehlers and Luuthy 2003; Traina and Laperche 1999).

In this area, however, several similar terms have been employed, leading to potential confusion. Ruby et al. provided definitions for the terms, which include bioavailability, relative bioavailability, relative absorption factor, and bioaccessibility (Ruby et al. 1999).

In order to evaluate quantity of extraction with digestive organs such as stomach or the small intestine, the extractants which simulated gastric juice and intestinal juice are often used for bioavailability. Since gastric juices contain hydrogen chloride with a pH value of approximately 1, hydrogen chloride is usually used as an extractant. But the concentration of the acid is varies, such as pH 1, pH 1.5, pH 2 or 1 mol/L. For example, the Japanese standard extraction test method for acid extractable contents of metals in soil, JLT-19, uses 1 mol/L of HCl.

Furthermore, a digestive enzyme (pepsine, pancreatin, and so on) and/or some digestive juice (porcine bile extract), and/or organic acid are often added in the extractant. In addition, the extraction operation is often performed at 37 °.

4.2.2.4 Sequential Chemical Extraction and Acid Extraction Methods

Some heavy metals in soils are easily leachable, and others are highly stable as insoluble minerals. To determine the disposition of the metals, a "sequential chemical extraction (SCE)" procedure is widely used. This comprises either serial or sequential batch tests. The leachants or extractants are various chemical reagents with different extraction strengths, ranging from weak (distilled water) to strong (aqua regia). Based on the extraction results, the distribution of the metal forms can be determined, such as the "exchangeable fraction" or the "fraction bound to carbonate."

These methods are well advanced in the fields of soil and sediment analysis, particularly the pioneering procedure proposed by Tessier et al. (Table 4.2) (Tessier et al. 1979). The methods are widely used for not only soil analysis but also incineration ash or dust. Furthermore, many modifications have been proposed. In 1987, the Community Bureau of Reference (BCR) Programme began to systematize the SCE procedure, and a unified method was proposed in 1992 (Table 4.3). Known as the "BCR sequential method," it has been applied in the analysis of soil, sediment, coal ash, waste incineration ash, dust, and so on (Ure et al. 1993; Fernández et al. 2004; Tokaliğlu et al. 2003). Many researchers have proposed modified procedures based on the characteristics of particular analytes (Pueyo et al. 2001; Rauret et al. 1999, 2000; Žemberyová et al. 2006; Silveira et al. 2006; Pueyo et al. 2008). He et al. compared four different SCE procedures for contaminated soil analysis (He et al. 2013). And some review papers about sequential chemical extraction procedures are published (Zimmerman and Weindorf 2010; Okoro et al. 2012).

The results of SCE can be affected by the soil matrix or other materials coexisting in the soil. Chemical characteristics, especially pH, must be considered

Table 4.2 Sequential chemical extraction procedure by Tessier et al. (1979)

Step	Fractions	Extractant	Time and agitation	Temp. °C
1	Exchangeable	1 M $MgCl_2$ (pH 7.0) 16 ml	1 h shaking	Room temp
2	Bound to carbonate	1 M NaOAc 16 ml (pH adjusted to 5 by HOAc)	15 h shaking	Room temp
3	Bound to iron and manganese oxide	25 % (v/v) HOAc in 0.04 M $NH_2OH{\cdot}HCl$ solution 40 ml	3 h agitation with double-boiling	96 ± 3
4	Bound to organic matter	0.02 M HNO_3 6 ml + 30 % H_2O_2 10 ml pH adjusted to 2 by HNO_3	2 h agitation with double-boiling	85 ± 2
		30 % H_2O_2 6 ml	3 h agitation with double-boiling	85 ± 2
		Up to 40 ml by 20 % HNO_3	30 min. shaking	Room temp
5	Residual	Calculation by substitution of contents (by aqua regia)	–	–

Table 4.3 Extraction procedure by BCR method

Step	Fraction	Extractant
1	Mobile, soluble in water or weak acids, or adsorbed to soil carbonates	Acetic acid
2	Bound to iron and manganese oxides	Hydroxylamine hydrochloride
3	Bound to organic matter and sulphides	Hydrogen peroxide + ammonium acetate

when evaluating the experimental results. It is important to note that, when reporting the results of a sequential extraction procedure, the heavy metals are often described as being in a particular state, such as "the metal was present in an oxidized form." However, as Tessier et al. emphasizes, the metal may be "bound to oxidized materials," which means it is adsorbed or chemically bound. It is not clear whether the metal is oxidized.

The advantages of the sequential extraction procedure are:

- It is easily performed in a laboratory.
- It is widely used and has a substantial knowledge base.
- We can obtain information about the heavy metal by the conditions necessary to wash or extract it from soil.

4.2.2.5 ISO Standardized Extraction Methods

For the extraction of heavy metals from soils, three methods are typically used: diethylenetriamine pentaacetic acid (DTPA) buffer (ISO 14870), ammonium nitrate solution (ISO 19730), and nitric acid (for thallium only) (ISO 20279).

Table 4.4 ISO standardized extraction methods for inorganic elements

Designation	Name
ISO 14870: 2001	Extraction of trace elements by buffered DTPA solution
ISO 19730: 2008	Extraction of trace elements from soil using ammonium nitrate solution
ISO 20279: 2005	Extraction of thallium and determination by electrothermal atomic absorption spectrometry

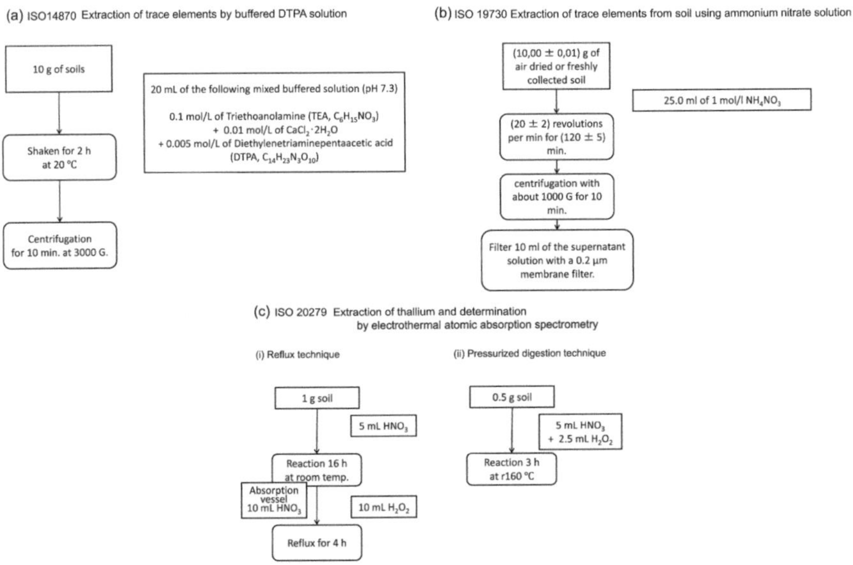

Fig. 4.3 (**a**)–(**c**) Operational flow diagrams for ISO extraction methods

These standard methods are summarized in Table 4.4. Each operational methodology is shown as a flow diagram in Fig. 4.3 (a)–(c).

4.2.3 *Leaching*

4.2.3.1 Single Leaching Test

The movement of heavy metals from the soil to groundwater constitutes one of the most important environmental impacts of contaminated soil after exposure to water sources such as rainfall or groundwater. The phenomenon of heavy metal transfer from the solid to the liquid phase is known as "leaching" or "leaching behavior." The test methods to evaluate "leaching behavior" or "leaching concentration" are termed "leaching tests."

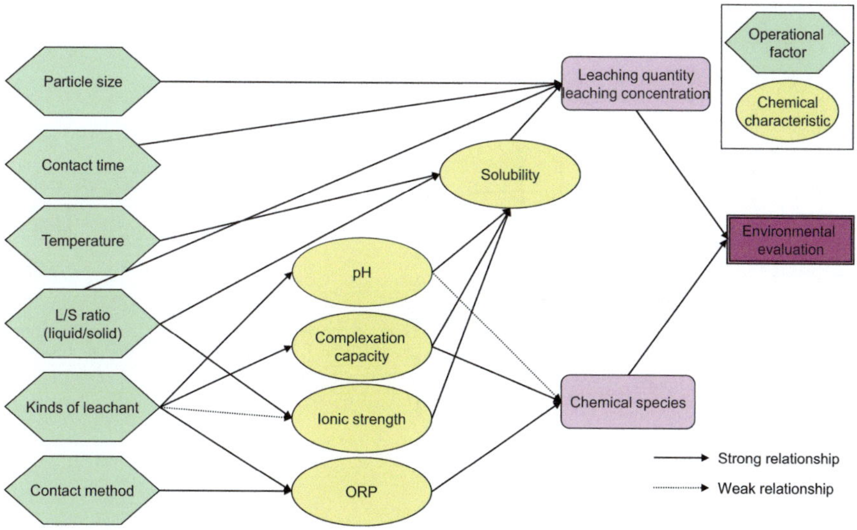

Fig. 4.4 Interrelationship between operating factors and chemical indexes for leaching tests

A wide variety of leaching tests have been applied to assess environmental impact risks before the final disposal of solid wastes or use of recyclable materials. Many operating factors can affect the assessment outcomes. The interrelationships between these operating factors and the affected chemical indexes are shown in Fig. 4.4. Among these factors, the pH of the leachant or leachate has the highest impact. Because each leaching test has advantages and disadvantages, it is impossible to designate any single test method as best.

4.2.3.2 pH-Dependent Tests

From the 1980s through the 1990s, standardized leaching tests for the same wastes were performed in several countries, although variable results were obtained. Therefore, the meaning of the data provided in the leaching tests was questioned. The data was later interpreted on the basis of the pH-dependent leaching test to afford a systematic explanation of the results from the various leaching tests. The pH-dependent test is a series of leaching tests which was developed to understand the leaching behavior of a metal as influenced by pH. A sample is leached using different concentration of an acid; after a defined contact time and agitation, the final pH of the leachate is examined and the metal concentration determined. The final pH of the leachate should be set by one between pH 2 and 12. A leaching concentration curve for pH is obtained when the leached quantity of metal or the leaching concentration of metals in the leachate is plotted against the pH of the final leachate. The results of a standardized leaching test from each country can be plotted on the leaching concentration curve as a function of the final pH of the

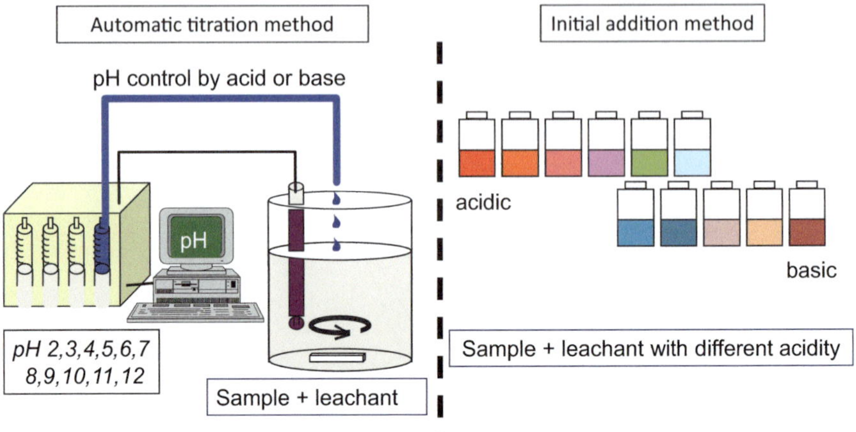

Fig. 4.5 Two types of pH-dependent tests

leachate, and various results can be explained without contradiction (IAWG 1997). In this way, the leaching test results from each country could be arranged on the basis of the pH-dependent test.

The pH-dependent testing protocol can be carried out in two ways. One is a titration-style examination in which acid is added under agitation and the pH is controlled to a target value. The other is a batch examination in which the sample is shaken with leachants of different acidities (Fig. 4.5). Various conditions have been reported for the leaching test, including variations of the acid used for pH adjustment and the pH control method, the liquid-to-solid ratio (L/S), stirring method, and leaching time. Generally, the L/S is set at a relatively low value of 10 or less to clarify the influence of the solubility on pH.

Although the pH-dependent test was developed as a leaching test for solid waste management, it is also very important and useful for contaminated soil management. The leaching of heavy metals and their adsorption on soil components are also controlled by pH. For example, Carter et al. used a combination of the pH-dependent leaching test and geochemical modeling to evaluate the partitioning of major, minor, and trace elements in relation to their phosphorus retention capability, the release of non-nutrient constituents, the reduction of soil acidity, and organic matter retention (Carter et al. 2009).

4.2.3.3 Column Test

Column test is a leaching test method to let leachant go through to the column which filled a sample continually. Small-scale test in the laboratory is called "column test" or "percolation test," and a large-scale experiment outdoors is called "lysimeter test." A purpose of the test is to grasp a time trend of substances concentration (heavy metal, anion, soluble salt, organic pollutant), or chemical characteristics (pH, ORP, EC, BOD, COD, TOC) in the leachate which flowed out. From the results of column test, basic data for seepage flow analysis can be derived. Leaching mechanism of the focusing metals can also be estimated.

Regarding the direction of leachant flow, there are two options of "up-flow" and "down flow." In the real environment, the water such as rainfall is basically down-flow. Generally, up-flow test is easier to control the leachant flow. In the case of up-flow of leachant, the filled soil or materials are water-saturated in the column, and the condition is similar to the lower condition of ground water level.

Column test method is considered that the leaching condition is more similar to the real environment than that of batch test. On the other hand, there are some disadvantages or concerns, including

- The long time required for the test
- The possible non-uniformity of leachant flow in the filling layer
- Concern over the plasticity of the method, meaning that the confinement conditions are different from those of the real environment

4.2.3.4 ISO Standardized Leaching Methods and Harmonization of Leaching Tests

Below is provided a list of ISO standardized leaching test methods (Table 4.5). ISO/TS21268-1 and ISO/TS21268-2 are compliance test. On the contrary, ISO/TS21268-3 and ISO/TS21268-4 are characterization test. These are established as technical specification (TS) that follows an official standard.

The ISO establishes international standard of test method and CEN determines standard of European common test method. In late years, they are harmonized including a test method of USEPA. Furthermore, this tendency is commonized for not only the commonization between standardization groups but also the different materials such as soil or the waste (Table 4.6) (van der Sloot et al. 2011). The accumulated data are compiled into a database and published on the web site (LeachXS Lite 2012).

4.3 Global Methods of Assessing Contaminated Soil

4.3.1 Analytical Methods of Metal Contents in Various Countries

The definition of "contaminated soil" varies for each country, as do the analytical methods for charactering it. Many countries use the contents of heavy metals as the criteria. As described above, the value of the "contents" can change according to the

Table 4.5 ISO standardized leaching test methods for inorganic elements

Designation	Name
ISO 18772:2008	Guidance on leaching procedures for subsequent chemical and ecotoxicological testing of soils and soil materials
ISO/TS 21268-1:2007	Leaching procedures for subsequent chemical and ecotoxicological testing of soil and soil materials – Part 1: Batch test using a liquid to solid ratio of 2 l/kg dry matter
ISO/TS 21268-2:2007	Leaching procedures for subsequent chemical and ecotoxicological testing of soil and soil materials – Part 2: Batch test using a liquid to solid ratio of 10 l/kg dry matter
ISO/TS 21268-3:2007	Leaching procedures for subsequent chemical and ecotoxicological testing of soil and soil materials – Part 3: Up-flow percolation test
ISO/TS 21268-4:2007	Leaching procedures for subsequent chemical and ecotoxicological testing of soil and soil materials – Part 4: Influence of pH on leaching with initial acid/base addition

Table 4.6 Harmonization approach of characterization test for various materials (van der Sloot et al. 2011)

Test	Soil, sediments, compost, sludge	Solid waste	Mining waste	Construction products
pH dependent test	ISO/TS 21268-4 EPA method 1313	CEN/TS 14429 CEN/TS 14997 EPA method 1313	CEN/TS 14429 CEN/TS 14997 EPA method 1313	CEN/TS 14429 EPA method 1313
Percolation test (Column test)	ISO/TS21268-3 EPA method 1314	CEN/TS 14405 EPA method 1314	CEN/TS 14405 EPA method 1314	CEN/TC351/TS-3 EPA method 1314
Monolith test	EPA method 1315	CEN/TS 15863 EPA method 1315	EPA method 1315	CEN/TC351/TS-2 EPA method 1315
Compacted granular test	EPA method 1315	NEN7347 EPA method 1315	EPA method 1315	CEN/TC351/TS-2 EPA method 1315
Redox capacity	–	NEN7348*		NEN7348*
Acid rock drainage	–		EN 15875	

*NEN: Standard in the Netherlands

determination method. On the contrary, many countries measure metal contents in the soil in various analytical methods, separately from a definition of "contaminated soil." In Table 4.7, the analytical methods for metal contents in soils published in several reports (Utermann et al. 2006; Langenkamp et al. 2001; USEPA 2002; CCME 1993a; CCME 1993b; Thailand 2004; Malaysia 2009; TaiwanEPA 2010, 2011) are shown for each country.

Table 4.7 Metal determination method for soil in many countries

Nation	Method (element) [Name of standard]
European countries (Utermann et al. 2006; Langenkamp et al. 2001)	
Austria	Aqua regia [Austrian Standard L 1085]
	($HNO_3 + HClO_4$ mixture) [Austrian Standard L 1085]
Estonia	$HF/HNO_3/HClO_4$ (Cd, Cu, Zn), XRF(Pb), gas analyzer (Hg)
Finland	Aqua regia
France	Aqua regia, $HF/HClO_4$ [NF X 31-147]
Germany	Aqua regia, XRF, HF acid
Ireland	$HF/HNO_3/HCl/H_2O$ + microwave
Italy	Aqua regia
Lithuania	Burning at a 450 °C for DC-Arc Emission Spectrometry
The Netherlands	Hot acid destruction (different for each element)
Norway	1 M NH_4NO_3 (for Zn)
Portugal	$HCl/HNO_3/H_2O_2$ (3:1:2)
Romania	$HNO_3/HClO_4/H_2SO_4$ (10:5:1)
Slovak Republic	HF/HNO_3(Cd, Cu, Ni, Pb, Zn), Na_2O_2 (Cr), AMA (Hg)
Sweden	7 M HNO_3 at 120 oC in autoclave [Swedish Standard SS 028311]
America and Canada (U.S.EPA 2002; CCME 1993a, b)	
United States	Aqua regia [SW-846 Method 3050B]
Canada	Aqua regia [SW-846 Method 3050B]
Asia (Thailand 2004; Malaysia 2009; Taiwan EPA 2010, 2011)	
Malaysia	Aqua regia
Thailand	Aqua regia
Korea	Aqua regia
Japan	1 N HCl for acid extractable contents
Taiwan	Aqua regia [NIEA S321.63B]

4.3.2 Analytical Method for Soil Criteria

Regarding analytical method of metal contents for soil criteria or for judgment of "contaminated soil," limited methods are used. In several countries or regions, the methods are being unified.

4.3.2.1 EU

In the EU, the "Harmonization Project" to establish common laws and standards across the European Union is ongoing, which includes the unification of evaluation methods for contaminated soils. These processes are under discussion by a technical committee No.190 (TC190) in ISO. On the contrary, TC 345 (characterization of the soil) in the Comite Europeen de Normalisation (CEN, European Committee for Standardization) discuss the EN standard. It publishes a standard established in ISO/TC190 as EN standard based on Vienna agreement (CEN/TC345 2013).

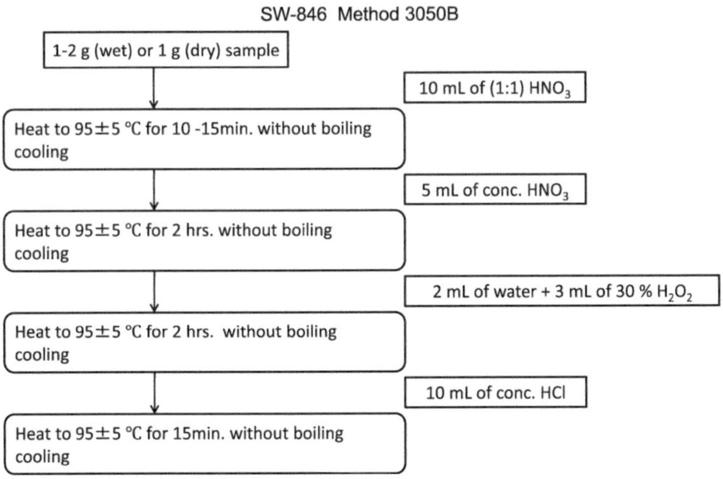

Fig. 4.6 Flow diagram for SW846 Method 3050B

For the determination of trace elements such as heavy metals, the standardized method ISO 11466 (aqua regia digestion) is employed.

4.3.2.2 United States and Canada

Both countries employ the same test methods for assessing contaminated soil and have the same evaluation scheme. They use SW846 Method 3050B, as mandated by the U.S. Environmental Protection Agency (USEPA) (Fig. 4.6).

4.3.2.3 Asian Countries

In the Asian countries, a harmonization project has not been established; each country uses its own judgment criteria and test methods. Each country also uses the standard methods of the EU, ISO, and USEPA as references for judging contaminated soil.

Malaysia and Thailand

Malaysia uses SW-846 for the determination of the total contents of metals, and thus judges whether the soil is contaminated. Thailand similarly determines heavy metal contents via SW-846 and judges the degree of soil contamination on this basis.

Korea

In Korea, the determination of contaminated soil is made on the basis of the "total contents of hazardous elements (heavy metals)." Until 2009, the criteria were based on the "acid extractable contents" and "total contents of hazardous elements (heavy metals)." In 2010, the method of judging whether a soil is contaminated has been changed based on the "total contents of hazardous elements."

Taiwan

In Chinese Taipei, the judgment is based on the total contents of heavy metals. For the analysis, an aqua regia digestion method (NIEA S321.63B) is used.

Japan

In Japan, the judgment of contaminated soil is based on two criteria: the acid extractable contents by 1 N hydrogen chloride and the leached quantity by distilled water. The former is an evaluation of the risk of bioavailability from the direct intake of soil and heavy metals. In contrast, the latter assesses the risk from drinking water by way of groundwater contamination.

4.4 Global Environmental Quality Standards and Criteria for Soil

The soil contamination becomes the problem all over the world, and most countries establish a standard for soil contamination. The standard value and the setting grounds are different every country. In this section, environmental standard for soil of many countries were compared, and the setting grounds of the standard value were studied.

4.4.1 Criteria of Soil Contamination in Various Countries

The soil contamination countermeasures law, which went into effect in Japan in February 2003, was enacted to prevent adverse effects on public health due to contamination of soils with toxic substances (Japan 2002b). In addition to Japan, many European countries as well as the United States (USA), Canada, and Australia have established their own criteria for soil contamination. Not surprisingly, the number of contaminants monitored as well as the criteria for soil contamination among different land usages differs between countries. Here, we therefore review and compare the soil contamination criteria of various countries.

As summarized in Table 4.8, the number of substances for which soil criteria have been established varies across countries (USEPA 2002; Canada 2007; Australia 1999a; VROM 2000; BBodSchV 1999; Carlon et al. 2007). While Italy maintains the largest number of criteria, with 234, Germany has established criteria for 15 substances, which is the lowest amount among the countries listed. Meanwhile, Japan has established criteria for 25 distinct substances. Many countries have also created soil criteria according to the intended use of the land. The majority of these countries have characterized land use as Residential or Commercial/Industrial. In contrast, Japan uses the criteria of Cd and As for agricultural land use (Tables 4.8 and 4.9) (Japan 2001). In addition, some countries have established distinct criteria based on different risk levels (USEPA 2002; Canada 2007; Australia 1999a; VROM 2000; BBodSchV 1999; Carlon et al. 2007). In general, these risk levels can be divided into three categories: negligible, intermediate, and potentially unacceptable risk. The negligible risk criteria are typically used for long-term environmental objectives. Meanwhile, in the case of a potentially unacceptable risk, decontamination of soil is required. Table 4.8 shows the phased criteria of several countries. Notably, however, Japan has not established phased criteria.

Table 4.9 summarizes the wide range of criteria for specific soil contaminants established by each country. As stated above, many criteria are established based on the purpose of the land. Soil ingestion by humans is one of the important factors that contribute to the generation of contamination criteria; however, there are marked variations in the assumed amounts of soil ingestion between countries. For example, the USA and Japan base their criteria on 200 mg of soil consumption per day for children and 100 mg/day for adults (USEPA 2002; Japan 2002a). By comparison, the Netherlands and Sweden assume 150 mg/day for children and 50 mg/day for adults (RIVM 1994; Sweden 1996), Canada has adopted the standard of 80 mg/day for toddlers (6 months to 4 years old) and 20 mg/day for individuals older than 4 years old (Canada 2006), and Germany has adopted 500 mg/day for children but no value for adults (Carlon et al. 2007). Furthermore, there are differences between countries in the method by which the levels of heavy metal contaminants are assessed. While several countries use the aqua regia extraction method for measuring the levels of heavy metals in soil (Carlon et al. 2007), the 1 N hydrochloric acid (HCl) extraction method was adopted as part of the soil contamination countermeasures law and is used for evaluating the risks associated with direct ingestion of soil in Japan (Japan 2002b). This method is also thought to be capable of measuring the approximate quantity of heavy metals that are absorbable by the human body.

Table 4.8 Summary of the setup of the soil criteria of various countries

Country	Number of substances with criteria	Categories of intended land use	Phased criteria
USA	109	Residential Commercial/industrial	Soil screening level Cleanup standard
Canada	32	Agricultural Residential/parkland Commercial Industrial	–
Australia	27	Residential with garden/accessible soil Residential with minimal opportunities for soil access Park, recreational open space and playing fields Commercial/industrial	Health-based investigation level Health-based response level
Italy	234	Residential/green areas Commercial/Industrial	–
Austria	16	Agricultural/gardening Residential, sport fields, playgrounds	Trigger value Intervention threshold value
Netherlands	78	Generic	Target value Intervention value
Sweden	36	Sensitive land use Less sensitive land use, groundwater protection Less sensitive land use	–
Denmark	34	Generic	–
Germany	15	Playgrounds Residential areas Parks and recreational facilities Industrial and commercial real properties Agriculture, vegetable garden grassland	Action value Trigger value Precautionary value
Finland	53	Residential area	Threshold value Lower guideline value Upper guideline value
Belgium (Flanders)	55	Nature Agricultural Residential Recreational Industrial	Background value Clean-up value

Table 4.9 Soil contamination criteria of various countries (mg/kg)

Country	Phased criteria	Purpose of land use	As	Cd	Cr	Hg	Ni	Pb
USA	Soil screening level	Residential	0.4	70	1.2×10^5(III) 230 (VI)	23	1600	–
		Commercial/industrial	2	900	1×10^6 (III) 3400 (VI)	340	23000	–
Canada	–	Agricultural	12	1.4	0.4 (VI)	6.6	50	70
		Residential/parkland	12	10	0.4 (VI)	6.6	50	140
		Commercial	12	22	1.4 (VI)	24	50	260
		Industrial	12	22	1.4 (VI)	50	50	600
Australia	Intervention value	Residential with garden/accessible soil	100	20	100 (VI)	15	600	300
		Residential with minimal opportunities for soil access	400	80	400 (VI)	60	2400	1200
		Park, recreational open space, and playing fields	200	40	200 (VI)	30	600	600
		Commercial/industrial	500	100	500 (VI)	75	3000	1500
Italy	–	Residential, green areas	20	2	2 (VI)	1	120	100
		Commercial/industrial	50	15	15 (VI)	5	500	1000
Austria	Trigger value	Agricultural/gardening	20	2	50	2	70	100
	Intervention value	Residential, sport fields, playground	50	10	250	10	140	500
Netherlands	Target value	–	29	0.8	100	0.3	35	85
	Intervention value	–	55	12	380	10	210	530
Sweden		Sensitive land use	15	0.4	120(III), 5(VI)	1	35	80
		Less sensitive land use, groundwater protection	15	1	250(III), 15(VI)	5	150	300
		Less sensitive land use	40	12	250(III), 20(VI)	7	200	300
Denmark	–	–	10	0.3	50 (III), 2 (VI)	0.1	10	50
Germany	Trigger value	Playground	25	10	200	10	70	200
		Residential	50	20	400	20	140	400

		Parks and recreational facilities	125	50	1000	50	350	1000
		Industrial	140	60	1000	80	900	2000
Finland	Threshold value	–	5	1	100	0.5	50	60
	Lower guideline value		50	10	200	2	100	200
	Upper guideline value		100	20	300	5	150	750
Belgium (Flanders)	Clean-up value	Nature	45	2	130 (III)	10	100	200
		Agricultural	45	2	130 (III)	10	100	200
		Residential	110	6	300 (III)	15	470	700
		Recreational	200	15	500 (III)	20	550	1500
		Industrial	300	30	800 (III)	30	700	2500
Japan	Threshold value	–	150	150	250 (VI)	15	–	150
Taiwan	Threshold value	–	60	20	250	20	200	2000
		Agricultural	60	5	250	5	200	500

4.4.2 Setup of the Criteria of Selected Countries

4.4.2.1 Canada (Canada 2006, 2007, 1996)

Canadian environmental quality guidelines were established in 1999. These guidelines include environmental soil quality guidelines (SQG_E) and human health soil quality guidelines (SQG_{HH}), which were derived specifically for protection of the ecological receptors in the environment and for the protection of the health of humans exposed to the various land categories, respectively. Furthermore, these guidelines were calculated based on certain exposure pathways. In Canada, land use is divided into four categories: agricultural, residential/parkland, commercial, and industrial areas. Both children and adults are considered as sensitive receptors at agricultural, residential/parkland, commercial areas, while only adults are considered sensitive receptors at industrial areas. Meanwhile, the exposure period is assumed to be 24 h/day, 365 days/year at agriculture and residential/parkland sites, and 10 h/day, 5 days/week, 48 weeks/year at commercial and industrial areas, respectively. Ingestion and inhalation of soil, and dermal contact with soil are considered direct exposure pathways at each of these land categories. In contrast, the indirect exposure pathways vary depending on land use. For example, groundwater, indoor air, and backyard produce are considered indirect exposure pathways in residential/parkland areas. The SQGHH were calculated using tolerable daily intakes (TDIs) or estimated daily intakes (EDIs). The default amounts of soil ingested by toddlers (6 month to 4 years), who comprise the most sensitive population, and by individuals older than 4 years have been set at 80 mg/day and 20 mg/day, respectively.

4.4.2.2 Australia (Australia 1999a, b; EPHC 2000)

The National Environment Protection Council in Australia chose to divide soil contamination into two categories: investigation levels and response levels. If investigation levels, which consist of health-based investigation levels (HILs) and ecologically based investigation levels (EILs), are exceeded, more detailed investigation is required to determine whether the soil is contaminated by a toxic substance. Meanwhile, response levels are defined at particular sites to protect both public health and the environment. Distinct HILs have been established for the following categories of land use: (i) residential with garden/accessible soil; (ii) residential with minimal opportunities for soil access; (iii) parks, recreational open space, and playing fields, and (iv) commercial/industrial. HILs are calculated using acceptable daily intakes (ADIs) or provisional tolerable weekly intakes (PTWIs). The default exposure duration is set as follows: 24 h/day, 365 days/year for 70 years at site (i), 2 h/day, 365 days/year, 70 years at site (iii), and 8 h/day, 5 days/week, 48 weeks/year, 30 years at site (iv). The direct exposure pathways are considered the ingestion and inhalation of soil, and dermal contact with soil at sites (i), (iii), and (iv), and the ingestion and inhalation of soil at site (ii). Additionally, indirect exposure, which is assumed to occur only at site (i), consists of consumption of crops grown in home gardens. When soil criteria were set, 13.2 kg was typically used as the body weight for children; however, 12 kg has been utilized for certain substances. In addition, 10 kg of body weight was used for calculating the standard for zinc (Zn). Children are assumed to ingest 100 mg of soil per day, but 80 mg/day was adopted for calculating the standard for lead (Pb).

4.4.2.3 The Netherlands (VROM 2000; Carlon et al. 2007; RIVM 1994)

The Soil Protection Act came into effect in the Netherlands in 1996. In this country, two soil screening values were created to assess soil quality: target values and intervention values. While target values indicate the levels necessary to recover the functional characteristics of the soil, intervention values indicate that serious soil contamination has occurred and that the functional characteristics of the soil have been heavily damaged. The potential exposure to contaminants in soil is calculated using the C-SOIL© exposure model. In this model, each of the potential exposure pathways is assumed to occur in residential areas. The direct exposure pathways consist of ingestion, inhalation, and dermal uptake of soil, and inhalation and dermal uptake of air. Furthermore, several indirect exposure pathways might exist, including the consumption of drinking water, dermal contact, inhalation when showering, and consumption of vegetables. Soil ingestion occurs through the consumption of soil particles, and the licking of fingers and hands, and the default amounts of soil ingested by children and adults have been set at 150 and 50 mg/day, respectively. In the case of inhalation of soil particles, this is especially true for particles smaller than 10 μm, 75 % of which are retained within the bodies. In contrast, no exposure can take place via the dermal contact pathway, as the adsorption factor of the skin is zero for inorganic substances. The mean lifelong ingestion and inhalation exposure is calculated by combining the exposure of children (during 6 years) and adults (during 64 years).

4.4.2.4 Germany (BBodSchV 1999; Carlon et al. 2007)

The Federal Soil Protection Act came into effect in Germany in March 1998. This act provides three categories of assessment criteria for soil contamination levels: action values, trigger values, and precaution values. When trigger values are exceeded, further detailed investigation is carried out to verify whether the soil is contaminated by hazardous substances. If soil levels exceeded the action values, however, it is considered necessary to decontaminate the soil. Meanwhile, precaution values indicate a possibility of future soil contamination. Action values and trigger values have been created for the "soil–human being" pathway. Furthermore, four categories of land usage have been distinguished for this pathway (playgrounds, residential areas, parks and recreational facilities, and industrial and commercial real properties), and distinct contamination criteria have been established for each category. Two pathways for human exposure are taken into consideration: the uptake of contaminated soil by children at play and the inhalation of soil particles. While both pathways are taken into consideration in playgrounds, residential areas, and parks and recreational facilities, only the inhalation route of exposure is assumed to occur in industrial and commercial real properties. The trigger levels at playgrounds are calculated using following values: 10 kg (body weight of children), 500 mg/day (soil ingestion), 0.625 m^3/h (respiration), and 2 h/day for 240 days/year (frequency). The daily ingestion and inhalation rates at residential areas and at parks and recreational facilities are considered 50 % and 20 % of the rates at playgrounds, respectively. In addition, the trigger values are calculated using toxicological data and exposure factors. Specifically, substance-specific toxicological data, which define a tolerable body dose (D_{tb}), are used for calculating these values. D_{tb} levels are equivalent to a "no observed adverse effect level (NOAEL)" in sensitive human subpopulations and are typically derived from the LOAEL (lowest observed adverse effect level) in animal studies by using specific uncertainty factors. Furthermore, it is assumed that 80 % of the tolerable ingestion of non-carcinogenic, hazardous substances occurs via other routes, including the ingestion of contaminated food and drinking water.

4.5 Other ISO Standard for Soil Quality

There are a lot of ISO standard for soil quality other than decomposition, extraction, and leaching test method. Those standards can be categorized into the following four groups:

- Definition, fundamental parameters or characteristics regarding soil
- Physical/chemical characteristics for soils
- Instrumental analytical method for inorganic constituent
- Inorganic constituents except for heavy metals

Here those standards are summarized in Tables 4.10, 4.11, 4.12 and 4.13.

Table 4.10 ISO standard for definition, fundamental parameters or characteristics regarding soil

Designation	Name of standard
ISO 11074:2005	Vocabulary
ISO 15903:2002	Format for recording soil and site information
ISO 16133:2004	Guidance on the establishment and maintenance of monitoring programmes
ISO 19258:2005	Guidance on the determination of background values
ISO 25177:2008	Field soil description
ISO 28258:2013	Digital exchange of soil-related data

Table 4.11 ISO standard for physical/chemical characteristics for soils

Designation	Name of standard
ISO 10390:2005	Determination of pH
ISO 10573:1995	Determination of water content in the unsaturated zone – Neutron depth probe method
ISO 10930:2012	Measurement of the stability of soil aggregates subjected to the action of water
ISO 11260:1994	Determination of effective cation exchange capacity and base saturation level using barium chloride solution
ISO 11265:1994	Determination of the specific electrical conductivity
ISO 11271:2002	Determination of redox potential – Field method
ISO 11272:1998	Determination of dry bulk density
ISO 11274:1998	Determination of the water-retention characteristic – Laboratory methods
ISO 11275:2004	Determination of unsaturated hydraulic conductivity and water-retention characteristic – Wind's evaporation method
ISO 11276:1995	Determination of pore water pressure – Tensiometer method
ISO 11277:2009	Determination of particle size distribution in mineral soil material – Method by sieving and sedimentation
ISO 11461:2001	Determination of soil water content as a volume fraction using coring sleeves – Gravimetric method
ISO 11465:1993	Determination of dry matter and water content on a mass basis – Gravimetric method
ISO 11508:1998	Determination of particle density
ISO 13536:1995	Determination of the potential cation exchange capacity and exchangeable cations using barium chloride solution buffered at pH = 8,1
ISO 14254:2001	Determination of exchangeable acidity in barium chloride extracts
ISO 15709:2002	Soil water and the unsaturated zone – Definitions, symbols, and theory
ISO 16586:2003	Determination of soil water content as a volume fraction on the basis of known dry bulk density – Gravimetric method
ISO 17312:2005	Determination of hydraulic conductivity of saturated porous materials using a rigid-wall permeameter
ISO 17313:2004	Determination of hydraulic conductivity of saturated porous materials using a flexible wall permeameter
ISO 23470:2007	Determination of effective cation exchange capacity (CEC) and exchangeable cations using a hexamminecobalt trichloride solution

Table 4.12 ISO standard for soil quality: Instrumental analytical method for inorganic constituent

Designation	Name
ISO 11047:1998	Determination of cadmium, chromium, cobalt, copper, lead, manganese, nickel, and zinc – Flame and electrothermal atomic absorption spectrometric methods
ISO 11262:2011	Determination of total cyanide
ISO 13196:2013	Screening soils for selected elements by energy-dispersive X-ray fluorescence spectrometry using a handheld or portable instrument
ISO 15192:2010	Determination of chromium(VI) in solid material by alkaline digestion and ion chromatography with spectrophotometric detection
ISO 16772:2004	Determination of mercury in aqua regia soil extracts with cold-vapor atomic spectrometry or cold-vapor atomic fluorescence spectrometry
ISO 17380:2013	Determination of total cyanide and easily liberatable cyanide – Continuous-flow analysis method
ISO 17402:2008	Requirements and guidance for the selection and application of methods for the assessment of bioavailability of contaminants in soil and soil materials
ISO 18227:2014	Determination of elemental composition by X-ray fluorescence
ISO 20280:2007	Determination of arsenic, antimony, and selenium in aqua regia soil extracts with electrothermal or hydride-generation atomic absorption spectrometry
ISO 22036:2008	Determination of trace elements in extracts of soil by inductively coupled plasma – atomic emission spectrometry (ICP – AES)
ISO 23161:2009	Determination of selected organotin compounds – Gas-chromatographic method
ISO/ TR18105:2014	Detection of water soluble chromium(VI) using a ready-to-use test-kit method
ISO/TS 16727:2013	Determination of mercury – Cold vapor atomic fluorescence spectrometry (CVAFS)
ISO/TS 16965:2013	Determination of trace elements using inductively coupled plasma mass spectrometry (ICP-MS)
ISO/TS 17073:2013	Determination of trace elements in aqua regia and nitric acid digests – Graphite furnace atomic absorption spectrometry method (GFAAS)
ISO/TS 17924:2007	Assessment of human exposure from ingestion of soil and soil material – Guidance on the application and selection of physiologically based extraction methods for the estimation of the human bioaccessibility/bioavailability of metals in soil

Table 4.13 ISO standard for determination of inorganic constituents except for heavy metals

Designation	Name
ISO 10693:1995	Determination of carbonate content – Volumetric method
ISO 11048:1995	Determination of water-soluble and acid-soluble sulfate
ISO 11261:1995	Determination of total nitrogen – Modified Kjeldahl method
ISO 11263:1994	Determination of phosphorus – Spectrometric determination of phosphorus soluble in sodium hydrogen carbonate solution
ISO 13878:1998	Determination of total nitrogen content by dry combustion ("elemental analysis")
ISO 14255:1998	Determination of nitrate nitrogen, ammonium nitrogen, and total soluble nitrogen in air-dry soils using calcium chloride solution as extractant
ISO 14256-2:2005	Determination of nitrate, nitrite, and ammonium in field-moist soils by extraction with potassium chloride solution – Part 2: Automated method with segmented flow analysis
ISO 15178:2000	Determination of total sulfur by dry combustion
ISO 17184:2014	Determination of carbon and nitrogen by near-infrared spectrometry (NIRS)
ISO/TS 14256-1:2003	Determination of nitrate, nitrite, and ammonium in field-moist soils by extraction with potassium chloride solution – Part 1: Manual method

4.6 Conclusions

The methods for the extraction of metals from contaminated soil, to determine its metal concentration, are classified into the following three categories: (1) digestion/decomposition, (2) extraction, and (3) leaching test.

Digestion/decomposition is used for the evaluation of the total metal content of the soil. Various methods were suggested, depending on the chemical properties of the soil and the metals. The digestion with strong acid (such as aqua regia) and the alkali fusion method are representative methods; with these treatments, the soil is normally treated with heat.

Extraction with acid is used to determine the fraction that affects the environment and natural life forms. Different acids result in different extraction fractions. The index called "bioavailability," which indicates the quantity taken in by life forms, as one of the extraction fractions, is used for health risk assessment. Another characteristic extraction operation is SCE (sequential chemical extraction). This is the test method for the determination of the chemical forms (or the chemical speciation) of metals in soil, and various methods are suggested. The procedure proposed by Tessier et al. and the BCR method are famous and popular.

The leaching test is a test method for the determination of the quantity of metal that transfers to the liquid phase when the soil comes into contact with water. It is common to extract with distilled water, $CaCl_2$ aqueous solution, or dilute acid solution. The leaching test is used for the evaluation of the possibility of groundwater contamination. In addition, with regard to the maximum amount that can be leached into the environment, a test method called the "availability test," which is a Dutch standard, is widely recognized.

Furthermore, the soil environmental standards and criteria of each country were compared, and the setup backgrounds were also compared. There are many countries that set the graded standard value every use of the land. Based on local social conditions, various soil environmental standards values are set. Various test methods for the evaluation of contaminated soil are standardized by ISO. However, it seems to still take some time to unify the standard values.

References

Australia (1999a) 7A guideline on health-based investigation levels, National Environmental Health Forum, Australia

Australia (1999b) 7B exposure scenarios and exposure settings, National Environmental Health Forum, Australia

BBodSchV (1999) Federal soil protection and contaminated sites ordinance. https://www.elaw.org/system/files/de.soil.ord.eng.doc. Accessed 8 Aug 2015

Bishop PL (1988) Leaching of inorganic hazardous constituents from stabilized/solidified hazardous wastes. Hazard Waste Hazard Mater 5:129–143

Canada (1996) Guidance manual for developing site-specific soil quality: remediation objectives for contaminated sites in Canada, Canadian Council of Ministers of the Environment, The national contaminated sites remediation program, March

Canada (2006) A protocol for the derivation of environmental and human health soil quality guidelines, Canadian Council of Ministers of the Environment

Canada (2007) Canadian soil quality guidelines for the protection of environmental and human health, excerpt from Publication No.1299, ISBN 1-896997-34-1, Canadian Council of Ministers of the Environment

Carlon C, D'Alessandro M, Swartjes F (2007) Derivation methods of soil screening values in Europe. A review and evaluation of national procedures towards harmonization. Joint Research Center, Ispra, European Commission

Carter CM, van der Sloot HA, Cooling D (2009) pH-dependent extraction of soil and soil amendments to understand the factors controlling element mobility. Eur J Soil Sci 60:622–637

CCME (1993a) Guidance manual on sampling, analysis, and data management for contaminated sites Vol.1

CCME (1993b) Guidance manual on sampling analysis, and data management for contaminated sites Vol. 2: Analytical methods summaries

CEN/TC345 (2013) Business plan CEN/TC 345 Characterization of soils

Côté PL, Bridle TR (1987) Long-term leaching scenarios for cement-based waste forms. Waste Manag Res 5:55–66

Ehlers LJ, Luuthy RG (2003) Contaminant bioavailability in soil and sediment. Environ Sci Technol 37:295A–302A

EPHC (2000) National chemical reference guide – Standards in the Australian Environment. Australia

Fernández E, Jiménez R, Lallena AM, Aguilar J (2004) Evaluation of the BCR sequential extraction procedure applied for two unpolluted Spanish soils. Environ Pollut 131:355–364

He Q, Ren Y, Mohamed I, Maha A, Waseem H, Fangui Z (2013) Assessment of trace and heavy metal distribution by four sequential extraction procedures in a contaminated soil. Soil Water Res 8:71–76

Hettiarachchi GM, Pierzynski GM (2004) Soil lead bioavailability and in situ remediation of lead-contaminated soils: a review. Environ Prog 23:78–93

IAWG (1997) Leaching tests. In: Chandler AJ, Eighmy TT, Hartlén J, Hjelmar O, Kosson DS, Sawell SE, van der Sloot HA, Vehlow J (eds) Municipal solid waste incinerator residues. Studies in environmental science, Vol. 67. Elsevier, Amsterdam, pp 579–606

Japan (2001) Risk assessment of direct soil intake. Ministry of the Environment, Government of Japan. https://www.env.go.jp/en/water/wq/wemj/soil.html. Accessed 8 Aug 2015

Japan (2002a) The soil contamination control measures. Ministry of the Environment, Government of Japan. https://www.env.go.jp/water/dojo/sesaku_kondan/06/mat02.pdf. Accessed 8 Aug 2015 (in Japanese)

Japan (2002b) The soil contamination countermeasures Act. Ministry of the Environment, Government of Japan. https://www.env.go.jp/en/laws/water/sccact.pdf. Accessed 8 Aug 2015

Karstensen KH (1997) Contaminated soil. In: van der Sloot HA, Heasman L, Quevauviller Ph (ed) Harmonization of leaching/extraction tests. Studies in environmental science. Elsevier, Amsterdam, pp 57–74

Langenkamp H, Düwel O, Utermann J (2001) Trace element and organic matter contents of European soils. First results of the second phase of the "Short term action". Progress Report, Joint Research Center, Ispra, European Commision

LeachXS Lite (2012) http://www.vanderbilt.edu/leaching/downloads/leachxs-lite/. Accessed 5 Mar 2015

Malaysia (2009) Contaminated land management and control guidelines no. 1: Malaysian recommended site screening levels for contaminated land. Putrajaya: Department of Environment, Ministry of Natural Resources and Environment

Nakamaru YM, Altansuvd J (2014) Speciation and bioavailability of selenium and antimony in non-flooded and wetland soils: a review. Chemosphere 111:366–371

NEN7371 (2004) Leaching characteristics – determination of the availability of inorganic components for leaching – solid earthy and stony materials, The Netherlands Standardization Institute, Delft

Okoro HK, Fatoki OS, Adekola FA, Ximba BJ, Snyman RG (2012) A review of sequential extraction procedures for heavy metals speciation in soil and sediments. 1:181. doi:10.4172/scientificreports

Peijnenburg WJ, Zablotskaja M, Vijver MG (2007) Monitoring metals in terrestrial environments within a bioavailability framework and a focus on soil extraction. Ecotoxicol Environ Saf 67:163–179

Pueyo M, Rauret G, Lück D, Yli-Halla M, Muntau H, Quevauviller P, López-Sánchez JF (2001) Certification of the extractable contents of Cd, Cr, Cu, Ni, Pb and Zn in a freshwater sediment following a collaboratively tested and optimised three-step sequential extraction procedure. J Environ Monit 3:243–250

Pueyo M, Mateu J, Rigol A, Vidal M, López-Sánchez JF, Rauret G (2008) Use of the modified BCR three-step sequential extraction procedure for the study of trace element dynamics in contaminated soils. Environ Pollut 152:330–341

Rauret G (1997) Soil. In: van der Sloot HA, Heasman L, Quevauviller Ph (eds) Harmonization of leaching/extraction tests. Studies in environmental science. Elsevier, Amsterdam, pp 41–56

Rauret G, López-Sánchez JF, Sahuquillo A, Rubio R, Davidson C, Ure A, Quevauviller P (1999) Improvement of the BCR three step sequential extraction procedure prior to the certification of new sediment and soil reference materials. J Environ Monit 1:57–61

Rauret G, Lopez-Sanchez JF, Sahuquillo A, Barahona E, Lachica M, Ure AM, Davidson CM, Gomez A, Luck D, Bacon J, Yli-Halla M, Muntau H, Quevauviller P (2000) Application of a modified BCR sequential extraction (three-step) procedure for the determination of extractable trace metal contents in a sewage sludge amended soil reference material (CRM 483), complemented by a three-year stability study of acetic acid and EDTA extractable metal content. J Environ Monit 2:228–233

RIVM (1994) Human exposure to soil contamination: a qualitative and quantitative analysis towards proposals for human toxicological intervention values. National Institute of Public Health and the Environment, Bilthoven

Ruby MV, Schoof R, Brattin W, Goldade M, Post G, Harnois M, Mosby DE, Casteel SW, Berti W, Carpenter M, Edwards D, Cragin D, Chappell W (1999) Advances in evaluating the oral bioavailability of inorganics in soil for use in human health risk assessment. Environ Sci Technol 33:3697–3705

Silveira ML, Alleoni LRF, O'Connor GA, Chang AC (2006) Heavy metal sequential extraction methods- a modification for tropical soils. Chemosphere 64:1929–1938

Sweden (1996) Development of generic guideline values, Swedish Environmental Protection Agency

TaiwanEPA (2010) Soil and groundwater pollution remediation act. http://sgw.epa.gov.tw/resag/Update_Data/Act2015495Taiwan_Pollution%20Remediation%20Act_v20100203.pdf. Accessed 8 Aug 2015

TaiwanEPA (2011) Soil pollution control standards. http://ivy5.epa.gov.tw/docfile/140120.pdf. Accessed 8 Aug 2015 (in Chinese)

Tessier A, Campbell PGC, Bisson M (1979) Sequential extraction procedure for the speciation of particulate trace metals. Anal Chem 51:844–851

Thailand (2004) Soil quality standard. http://www.pcd.go.th/info_serv/en_reg_std_soil01.html. Accessed 17 Mar 2015

Tokaliğlu Ş, Kartal Ş, Bịrol G (2003) Application of a three-stage sequential extraction procedure for the determination of extractable metal contents in Highway soils. Turk J Chem 27:333–346

Traina SJ, Laperche V (1999) Contaminant bioavailability in soils, sediments, and aquatic environments. In: National Academy of Sciences colloquium "Geology, Mineralogy, and Human Welfare", Irvine, 8–9 November, 1998, pp 3365–3371

Ure AM, Quevauviller P, Muntau H, Griepink B (1993) Speciation of heavy metals in soils and sediments. An account of the improvement and harmonization of extraction techniques undertaken under the auspices of the BCR of the Commission of the European Communities. Int J Environ Anal Chem 51:135–151

USEPA (2002) Supplemental guidance for developing soil screening levels for superfund sites. US Environmental Protection Agency, Washington, DC

Utermann J, Düwel O, Nagel I (2006) Part II contents of trace elements and organic matter in European soils, in Background values. In: European soils and sewage sludges, Results of a JRC-coordinated study on background values. Joint Research Center, Ispra, European Commision

van der Sloot HA (1991) Systematic leaching behavior of trace elements from construction materials and waste materials. In: Goumans JJJM, van der Sloot HA, Aalbers TG (eds) WASCON 1991: waste materials in construction. Elsevier, Amsterdam, pp 19–36

van der Sloot HA, de Groot GJ, Wijkstra J (1987) Leaching characteristics of construction materials and stabilization products containing waste materials. In: Côté P, Gilliam M (eds) Environmental aspects of stabilization and solidification of hazardous and radioactive wastes. American Society for Testing and Materials, Philadelphia, pp 125–149

van der Sloot HA, Hoede D, Bonouvrie P (1991) Comparison of different regulatory leaching test procedures for waste materials and construction materials. Netherlands Energy Research Foundation ECN, ECN-C-91-082, The Netherlands

van der Sloot HA, Kosson DS, Eighmy TT, Comans RNJ, Hjelmar O (1994) Approach towards international standardization: a concise scheme for testing of granular waste leachability. In: Goumans JJJM, van der Sloot HA, Aalbers TG (eds) WASCON 1994, environmental aspects of construction with materials. Elsevier, Amsterdam, pp 453–466

van der Sloot HA, Kosson DS, Hjelmar O, Comans RNJ (2011) Recent developments in testing, modeling and environmental impact assessment applicable to landfilled waste, treated waste and waste considered for beneficial use. In: Proceedings 13th international landfill and waste management symposium, Sardinia, Italy

VROM (2000) Circular on target values and intervention values for soil remediation.

Žemberyová M, Barteková J, Hagarová I (2006) The utilization of modified BCR three-step sequential extraction procedure for the fractionation of Cd, Cr, Cu, Ni, Pb and Zn in soil reference materials of different origins. Talanta 70:973–978

Zia MH, Codling EE, Scheckel KG, Chaney RL (2011) In vitro and in vivo approaches for the measurement of oral bioavailability of lead (Pb) in contaminated soils: a review. Environ Pollut 159:2320–2327

Zimmerman AJ, Weindorf DC (2010) Heavy metal and trace metal analysis in soil by sequential extraction: a review of procedures. Int J Anal Chem 2010:387803

Chapter 5
Soil Contamination and Remediation Measures: Revisiting the Relevant Laws and Institutions

M. Monirul Azam

Abstract This chapter analyzed existing international and regional legal developments to deal with soil contamination and remediation measures. It further examined characteristics of different national legal approaches in the field of soil contamination and remediation process. It concluded that there are little international initiatives for the adoption of an international soil regime to deal with soil protection including issues of contamination and possible remediation measures. Absence of international soil protection regime and hence lack of concerted global and national efforts on remediation action on contaminated sites could pose not only serious health risks but also long-term sustainable development challenge. However, a credible, comprehensive model law on soil contamination could make a considerable progress for remediation of contaminated sites globally and hence could reduce health and environmental risks for future generations.

Keywords Soil contamination • Remediation • Soil-framework directive • Command-control approach • Brown-field • Polluters-Pay Principle

5.1 Introduction

Soil contamination, also known as soil pollution, has emerged as a major environmental issue in most industrialized countries over the last 30 years (Brandon 2011). It is often caused by the presence of man-made chemicals in the natural soil environment in the form of industrial activity, agricultural chemicals or the

M.M. Azam (✉)
Graduate School of Decision Science and Technology, Tokyo Institute of Technology, 2-12-1 Ookayama, Meguro-ku, Tokyo 152-8550, Japan

Institute for the Advanced Study of Sustainability, United Nations University, 5-53-70 Jingumae, Shibuya-ku, Tokyo 150-8925, Japan

Department of Law, Stockholm University, 10 Universitetsvägen, 114 18 Stockholm, Sweden

Faculty of Law, University of Chittagong, Chittagong 4331, Bangladesh
e-mail: monirul.azam@gmail.com

H. Hasegawa et al. (eds.), *Environmental Remediation Technologies for Metal-Contaminated Soils*, DOI 10.1007/978-4-431-55759-3_5

improper disposal of waste, but also military bases, landfills, use of contaminated soils for residential development and petroleum storage (Fowler 2007, 2012). Mining, fertilizer application, oil and fuel dumping and a multitude of other environmental issues can also cause pollution of the soil. Again, in soil contamination can also be attributed to natural causes in many countries, for example, the wide-scale contamination of groundwater by naturally occurring arsenic in South and East Asia (World Bank 2012).

Some researcher use 'site contamination' rather than 'soil contamination' as the use of the word 'site' allows for a broader contamination issues without limiting the range of sub-issues to be addressed (Brandon 2011). 'Contamination often affects not only the soil surface, but the unsaturated zone "(the layer between the land surface and the water table, consisting of porous materials) and the saturated zone"(including the water table and any areas beneath it that also contain water) as well' (Brandon 2011).

This chapter used both soil and site contamination interchangeably while considering contamination from broader perspectives therefore it is non-specific in relation to soil, land or water, and embraces all physical features.

Although the full extent of site contamination problems largely unknown still in many regions, such as Eastern Europe and most developing countries – due to lack of coordinated surveys of potential sites, some statistics so far available regarding contaminated sites presents only partial picture of the magnitude of the problem, for example, European Environment Agency estimated that there are between 300,000 and 1.5 million contaminated sites in Europe (Prokop et al. 2000); it is estimated that there are more than 450,000 contaminated sites in the USA (Haninger et al. 2012) and Australia, estimated 80,000 contaminated sites (Fowler 2007).

The biggest concern associated with soil contamination is the possibility of potential harm to human health and surrounding environment. There are significant health risks involved with direct contact with contaminated soil for example, 'the vulnerability of the individual affected, and the level of exposure, may include an increased risk of cancer, respiratory illness (including asthma), reproductive problems, impairment of neurological functions and hormone disruption' (EEA 2007).

In addition to, health impacts, there are also serious socio-economic impacts of soil contamination. The confirmed or suspected existence of contamination on a site can affect its property value and marketability, as prospective purchasers or developers may not want to take on the financial and legal burden of dealing with contamination. This is particularly critical, when there is uncertainty as to the degree of contamination and required financial cost for dealing with contamination (Brandon 2011).

Considering the health risk, environmental issues and related economic and property rights issues trigger for finding possible solution to deal with contamination problem. A solution to the problem of soil contamination is soil remediation, which is a way of purifying and revitalizing the soil. It is the process of removing contaminants in order to minimize health risk and protect the environment.

The primary objective of the remediation process is to restore the soil to its natural, pollution-free state. It does not necessarily require removal of the contaminants themselves, although that was a common practice back in 1970s and till 1990s (Brandon 2011). However, remediation of all historically contaminated sites, commonly of industrial origin, to suitable levels for all uses often is not viewed as technically or economically feasible.

As a result, remediation strategies are increasingly designed to employ using functional approach specific to potential future use. Now countries are increasingly trying to encourage sustainable, long-term remediation process considering a risk-based approach to land management aimed at achieving 'fitness for use' appropriate to the location (US EPA 2009).

Remediation also includes post-remediation care and maintenance, which usually includes a combination of two broad types of measures: engineering (or physical) controls and institutional (legal and administrative) controls. Engineering controls comprise the physical barriers or structures designed to monitor and prevent or minimize exposure to contamination whereas institutional controls include administrative or legal instruments that minimize the potential for human exposure to contamination by limiting land or resource use (Brandon 2011).

Apart from the regulatory context and applicable soil standards, several factors affect the decision as to which remediation method should be used for a particular site. In general, excavation and removal of contaminated materials is a short-term remediation solution, which is used when remediation needs to be carried out quickly and completely. Where the excavated contaminated materials are not treated, but only removed to another location, the contamination is relocated but not resolved, and may present a problem in the future.

There are three main soil remediation technologies: soil washing, bioremediation and thermal desorption. 'Soil washing is a process that uses surfactants and water to remove contaminants from the soil. The process involves either dissolving or suspending pollutants in the wash solution and separates the soil by particle size. Bioremediation involves the use of living microorganisms, such as bacteria and fungi, to break down organic pollutants in the soil. In thermal desorption, heat is used to increase the volatility of contaminants, so that they can be separated from the solid material. The contaminants are then either collected or destroyed' (Cooper 2013).

On the other hand, considering the absence of specific international standard to address the remediation of contaminated soil, there are varied approach to impose obligation for remediation and associated costs.

Most of the European Union (EU) member countries apply the 'polluter pays' principle to varying degrees in remediation process in the absence of any mandatory specific EU guideline or specific law for doing so. However, public money or state aid also has been used in a number of EU States to finance remediation costs, when necessary.

Further, the effects of soil contamination may go beyond the border of a particular state and considering associated health risks, environmental and economic issues may have a wider global impact that is why issues of remediation of soil contamination require analysis of broader international, regional and national legal perspectives. This chapter tried to contribute in such direction with an analysis of existing international and regional legal developments and also overview on the national legal approaches to deal with soil contamination and remediation issues.

5.2 International Law: International Environmental Law and Soil Contamination

The debates over environmental concerns implicate core principles of international law, and therefore, global and regional environmental issues are increasingly debated as such as under the customary principles of international law and also utilizing numerous international agreements and declarations. However, the modern development of environmental law originated from national governments approach to address specific pollution issues based on growing community concern back to late 1960s and early 1970s (Bates 2006).

'It was not until the late 1970s that awareness first emerged among developed countries of a more specific environmental issue requiring regulation: the presence of potentially toxic pollutants in soil and groundwater' (Brandon 2011). One of the finest examples of early environmental legislation is the National Environmental Policy Act of 1969 in the United States (Pub L No 91-190, 42 USC 4321-4347).

However, environmental laws at that time maintained piecemeal approaches for distinct issues; the recognition and integration of linkages between them came later, in the 1990s (Lye and Manguiat 2003).

In the context of site contamination, it was stated in a study that 'when the problem was first noticed, the primary concern once again was public health, such as access to safe drinking water in communities that were heavily dependent on groundwater sources'(Brandon 2011).

But several major incidents of large-scale contamination in a handful of countries soon raised further concerns about the broader, longer-term impacts of such pollution and therefore created concern to deal with remediation of polluted sites.

One of the earliest international judicial decisions on international environmental law disputes concerning Trail Smelter Arbitration (United States vs Canada) considered as having some potential relevance to the subject of site contamination and obligation for compensation and remedial action by the responsible states in case of trans boundary pollution.

In this case (Trail Smelter Arbitration), a lead and zinc smelter situated at Trail, in British Columbia, Canada, released sulphur dioxide fumes that were carried by air across the border to the United States and hence caused pollution. The output of the Canadian smelter, and cross-border environmental damage to the United States, increased from 1925. To settle this issue, an Arbitral Tribunal was convened to examine among others, whether further cross-border pollution had occurred since 1932 (that year some monetary compensation was paid to the United States for damage caused), what the appropriate damages should be, whether the smelter operations should cease and what measures should be taken to this end.

In the Tribunal while analyzing relevant international laws, it observed that—'[...] no State has the right to use or permit the use of its territory in such a manner as to cause injury by fumes in or to the territory or the properties or persons therein, when the case is of serious consequence and the injury is established by clear and convincing evidence' (Trail Smelter Disputes) (Viñuales 2015).

On this basis, the Tribunal found that Canada was responsible in international law for the operations of the Trail Smelter and must refrain from causing further damage to the United States (Read 2006). The Trail Smelter decision has long been regarded as embodying the fundamental principle of state responsibility and is often cited in relation to pollution cases.

However, the trail smelter case have some relevance for the management of site contamination only, if the contaminant is sufficiently mobile to cross an international border, causes serious harm, and that harm could be proved with facts and evidence accordingly. In such cases, the national government in whose territory the contamination originates has an obligation to ensure that the contaminating activity either ceases or subsequent cross-border environmental harm is prevented by way of remediation process.

Some scholar argues that 'the scope of the decision is confined to situations where serious harm is caused by fumes and evidence of such harm is clear and convincing. These conditions may be difficult to satisfy in all but the most extreme cases of cross-border environmental harm, and perhaps only harm caused by air pollution. Therefore, Trail Smelter may offer little assistance for the consequences of gradual emissions of pollutants over time, even though these may be potentially devastating' (Ellis 2006).

Despite having greater national concern on the contamination and associated remediation issues, there is no specific international law to deal with contamination and remediation of contaminated sites. From a review of current international and regional legal initiatives having broader global significance, 'it is apparent that, with the exception of the United Nations Convention to Combat Desertification (CCD) and the EU Soil Protocol to the Alpine Convention, international soil protection law as such is non-existent' (Hannam and Boer 2002a). The study of Hannam and Boer (2002a) listed examples of non-binding instruments relevant to soil, which include the following:

- The Stockholm Declaration on the Human Environment (UN, 1972)
- The European Soil Charter (Council of Europe, 1972)
- The World Conservation Strategy (IUCN, 1980)
- The World Charter for Nature (UNEP, 1982)
- The World Soil Charter (FAO, 1982) and the World Soils Policy (UNEP, 1982)
- The Rio Declaration on Environment and Development (UN, 1992) and Agenda 21 (UN, 1992)

The Stockholm Declaration on human environment is considered as one of the oldest soft law instrument with some relevance to site contamination as it focused on the need for careful management of natural resources, including land and water, for present and future generations (principle 2 and 17). It also specifically linked pollution to ecosystem damage, calling for an end to the discharge of toxic substances or of other substances [. . .], in such quantities or concentrations as to exceed the capacity of the environment to render them harmless (principle 6).

However, there is no reference to the other impacts of pollution and the Stockholm Declaration as such provides no framework or guideline related to the soil contamination and remediation process to deal with contamination.

The *World Soil Charter* and the *World Soils Policy* adopted under the Food and Agricultural Organization (FAO) and United Nations Environmental Programme (UNEP) respectively were intended to promote international cooperation for rational use of soil resources. "Both the Charter and the Policy are over 25 years old. . .soil degradation is now much more serious and widespread than at the time of their creation" (Brandon 2011). Hannam and Boer commented that, "although many of the principles contained in the instruments are still relevant, they fall well short of the basic necessities of a modern day suitable non-binding-soft law instrument and it is time for them to be superseded" (Hannam and Boer 2002a).

In this regard, Rio declaration made some progress for overall environmental management from sustainable development perspectives, but has not been effectively used by the state parties for taking adequate further measures on soil contamination.

The Rio Declaration on Environment and Development recognizes the need for effective environmental legislation and asks States to develop national laws on liability and compensation for pollution victims with references to the 'polluter pays' principle.

The polluter pays principle, which is considered as one of the key issues for determining responsible parties for doing remediation, was mentioned for the first time in 1972 as an outcome of Stockholm conference on the human environment. Principle 22 stated that 'States shall cooperate to develop further the international law regarding liability and compensation for the victims of pollution and other environmental damage caused by activities within the jurisdiction or control of such States to areas beyond their jurisdiction.'

However, a widely recognized description of the polluter pays principle is given in Principle 16 of the Rio Declaration:

> National authorities should endeavor to promote the internalization of environmental costs and the use of economic instruments, taking into account the approach that the polluter should, in principle, bear the cost of pollution, with due regard to the public interest and without distorting international trade and investment.

It is yet to be tested by any specific international disputes to what extent polluters pay principle to be applied for imposing obligations for remediation of contaminated sites. On the other hand, from national perspectives, implementation of the polluter pays principle in developing countries, is a big challenge due to lack of requisite political and regulatory conditions. The successful implementation of the polluter pays principle very much depends and varied based on the quality and standard of national environmental legislations along with effective regulatory enforcement free from bureaucratic corruption, and strong public opinion regarding environmental issues.

Unfortunately most of the developing countries either lack standard environmental legislation or effective enforcement mechanism or even strong legislation could not deliver effective outcome due to lack of public awareness and bureaucratic hurdles. But polluters pay principle could provide effective remedy for remedial action for contamination in the absence of specific law on the relevant issues or could guide effective application of relevant law to determine responsibility for polluting parties. While considering importance of this principle for remediation issues, Justice Preston of Australia remarked 'the polluter pays principle operates in the remediation of contaminated sites by requiring the polluter to bear the primary liability for the remediation of sites they have contaminated' (Preston 2008).

Furthermore, the Rio Declaration on Environment and Development, 1992 under principle 15 stated that 'in order to protect the environment, the precautionary approach shall be widely applied by States according to their capabilities. Where there are threats of serious or irreversible damage, lack of full scientific certainty shall not be used as a reason for postponing cost-effective measures to prevent environmental degradation'. Principle 15 could be used to guide remediation of contamination considering potential threats of serious or irreversible damage due to contaminated sites. But till date, it has not been used by the state parties to impose obligation for remediation process or not utilized under national soil contamination law or has not been reflected in any international legal instrument.

On the other hand, Agenda 21: The United Nations Programme of Action from Rio, which is described by the United Nations as a comprehensive plan of action to be taken globally, nationally and locally by organizations of the United Nations System, Governments and Major Groups in every area in which human impacts on the environment refers to the need for an integrated and sustainable approach to the use of land resources, through the development of planning and management tools, has also relevance for addressing remediation of contaminated sites. Agenda 21 under chapter 19.2 mentioned that 'Gross chemical contamination, with grave damage to human health [...] and the environment, has in recent times been continuing within some of the world's most important industrial areas. Restoration will require major investment and development of new techniques'. Again it has some specific recommendations for the environmentally sound management of chemicals at the national, regional and global level such as it recommended for establishing risk reduction programme and strengthening national capacities for chemicals management; requested national governments to identify, assess, reduce and minimize, or eliminate as far as feasible by environmentally sound disposal practices, risks from storage of outdated chemicals. It also identifies the capacity for rehabilitation of contaminated sites as one of the basic elements for sound management of chemicals [Agenda 21, 19.56 (f)].

However, 'apart from these two brief references in Chapter 19, site contamination receives no further mention in the text of Agenda 21...it suggests that, at the time [of adoption of agenda 21], the international community did not perceive site contamination to be an appropriate issue for inclusion in Agenda 21" (Brandon 2011).

Although these soft law instruments could facilitate adoption of voluntary action on site contamination and remediation, till date little progress has been made at the international, regional and national levels to that end to address contamination issues and remediation process. That is why, it is important to have mandatory law to compel state parties to address these issues of serious environmental concern.

Among the binding international agreement/protocol/treaty, as the case may be, following instruments have some relevance to soil issues—

- The Convention to Combat Desertification (CCD) (UN, 1994)
- The Convention on Biological Diversity (CBD) (UN, 1995)
- The United Nations Framework Convention on Climate Change (UNFCCC) (UN, 1995)
- The Kyoto Protocol (UN, 1997)

It is considered that from the context of site contamination and remediation issues, only CCD has some direct relevance. The CCD focuses primarily on the prevention and mitigation of soil degradation in arid, semi-arid and dry sub-humid regions, which together comprise approximately one-third of the world's land area. These regions are particularly susceptible to desertification, a process which already affects millions of inhabitants as well as fragile ecosystems. The definition of land degradation in the Convention is expressly limited to these specific regions, thereby excluding any form of land degradation in most developed countries. The CCD promotes measures for the sustainable development of affected land areas, including activities aimed at the prevention or reduction of land degradation, rehabilitation of partly degraded land and reclamation of deserted land. This obligation could cover to some extent soil protection and remediation actions, but limited to only specific countries in the defined region affected by desertification issues.

On the other hand, CBD, UNFCCC and Kyoto protocol has some indirect references for the prevention of pollution and addressing mitigation and adaptation action only linked to biodiversity and climate change. Therefore, those provisions are of little use to compel state parties for action on contamination and obliging responsible parties for doing remediation.

There are a few other binding agreements at the international level that regulate the use, transport and disposal of particularly toxic or hazardous pollutants and therefore could have some relevance for the subject matter of contamination and obligation to take remediation measures. For example, the Stockholm Convention on Persistent Organic Pollutants (POPs) entered into force in 2004 has relevance as it regulates the disposal of certain toxic chemicals, which are primarily lies in contaminated sites. The convention obliges state parties to take action against 12 POPs or groups of POPs, i.e. aldrin, chlordane, DDT, dieldrin, endrin, heptachlor, mirex, toxaphene, polychlorinated biphenyls (PCBs), dioxins and furans (PCDD/PCDF) and hexachlorobenzene (HCB).

This Convention further deal on the management of industrial sites in particular, countries are obliged to identify where POPs are used, stockpiled or contained in waste. The POPs need not be causing contamination at a site to require action, although by their very nature they are likely to do so. Once POP sites are identified, they must be managed or cleaned up in a safe, efficient and environmentally sound manner, so as to minimize any leaks. Article 7 of the Convention requires each Party to develop a National Implementation Plan containing information on what it has done and intends to do to meet its obligations under the Convention. In fact while doing national implementation plan under article 7 each party could take measures against all contaminated sites including action against 12 specific POPs.

The Basel Convention on the Control of Trans boundary Movements of Hazardous Wastes and their Disposal entered into force in 1992 with the aim of minimizing the generation and movement of hazardous wastes (Hackett 1990; Kummer 1992; UNEP 2014). Under the Basel Convention, persons involved in the management of hazardous wastes must take the necessary steps to prevent pollution [...] arising from such management and, if such pollution occurs, to minimize the consequences thereof for human health and the environment, art. 4 (2)(c). This is perhaps more significant for the management of contaminated sites than other provisions in the Convention. It could be argued that site contamination may arise from hazardous waste operations, such as waste disposal facilities, or chemical manufacturing plants which generate hazardous wastes. Therefore, where this happens, operators have a positive duty to prevent the contamination and minimize its detrimental effects by way of remediation process.

Thus, there are a number of convention which partially addressed contamination issues and related remediation process at the international level, which are however could not be used to impose obligation on the state parties for remediation action as of positive duty or obligation.

There are several possible reasons behind the absence of specific site contamination law at the international level. One common reason or argument is that site contamination is widely considered as a domestic or local issue and which predominantly occurs within national borders and therefore is most appropriately regulated at the national or even local government level (Layard 2004). Another important concern is the lack of consensus among different countries for required standard and necessity of having an international agreement on soil contamination and also to address remediation actions. Further soil contamination often occurs at a slow pace and its impacts are largely remained invisible to the general public, which contributes to a lack of public awareness, and hence lack of political action on the issue. As the consequences of site contamination for environmental and health risks are largely invisible to general public and hence underestimated, so does contribute to lack of any concerted global efforts to deal with the soil contamination and remediation issues.

Hannam and Boer (2002a) considering inadequacy of existing international law to deal with site contamination and remediation issue remarked—

> The existing global binding instruments are insufficient as a framework and fall well short in including a sufficient range of legal elements that are needed to protect and manage land in a sustainable way. Although some current international non-binding instruments include general concepts relating to the control and prevention of soil degradation that are still relevant in the 21st century, they do not recognize soil as an important element of the terrestrial ecology. Finally, the existing international environmental law regime does not provide specific guidelines for States to approach the reform or development of national soil legislation (Hannam and Boer 2002b).

Thus, there are no international binding treaties or law dealing either solely or comprehensively with site contamination as of January, 2015, although a few touch briefly on the topic in an ad- hoc manner. The absence of any specific international legal instrument leaves countries without any clear guidance as to how to manage the many aspects of the soil contamination and remediation issues. Considering lack of technical and financial resources the need for guidance is particularly acute in developing countries as some developing countries are facing the problem for the first time due to late industrialization process and often lack the required experience and expertise to deal with it.

Furthermore, there is also clear absence of a suitable multilateral environmental agreement or treaty, which could be used as a parent agreement and thereby to attach an additional protocol to deal with soil contamination and remediation issues. However, there are regional initiatives particularly by the EU to deal with soil contamination and related remediation measures to some extent to fill the vacuum at the international level.

5.3 Regional Approaches: EU and ASEAN

There is currently no comprehensive regional agreement on soil protection anywhere in the world. However, the European Union has been the most active region in the area of soil protection. European efforts to improve soil protection date back to the 1972, when European Soil Charter was adopted, which has been re-confirmed 20 years later in the Towards a Thematic Strategy for Soil Protection (Communication from The Commission to The Council, The European Parliament, The Economic and Social Committee and The Committee of The Regions, COM 2002). Among the other regional organization only Association of South-East Asian Nations (ASEAN) intended to take measures on soil contamination prevention and remediation by way of an Agreement on the Conservation of Nature and Natural Resources, although it has not been entered into force as of January, 2015.

5.3.1 The EU

Although European countries were active for environmental protection and contamination issues since 1972, most of the countries took national measures during the periods between 1980 and 1995. There was little or no progress at the regional level until 2000. The EU adopted general environmental protection instruments, such as the 6th Community Environment Action Programme in 2002 and the Draft European Charter on the Environment and Sustainable Development in 2003, where it has prioritized the soil protection issue and identified specific soil threats. The European Commission released its soil protection strategy, including the draft Soil Directive, in 2006 after lengthy preparations.

However, none of the above instrument has any binding force to compel member states to prevent contamination or to take remediation action on the pre-existing contaminated sites. That is why; it is considered that the Soil Protocol to the 1991 Alpine Convention in the Field of Soil Conservation (1998), while expressly confined to the European alpine environment, is the only binding agreement on soil protection in the world apart from the CCD.

The Soil Protocol to the Convention on the Protection of the Alps (Soil Protocol) entered into force in 2005 for its nine signatories. The Protocol is limited to the alpine regions contained within the borders of its signatories, and is largely preventive in character. The protocol is very significant for soil protection as 'it is the only specific binding instrument for soil in the world, and contains many of the ecological concepts and principles advocated by this report as being "essential" for national and international soil instruments' (Hannam and Boer 2002b).

The soil protocol has included a number of specific obligations for the Parties, which are related to the issues of soil contamination and remediation measures such as legal and administrative measures to protect soil and applying the precautionary principle (Article 2); to consider the objectives of the protocol in other policies (such as forestry, agriculture, nature protection) (Article 3); coordination and cooperation between institutions and territorial authorities to develop synergies for soil protection (Article 4) and a commitment to support international cooperation among institutions on soil risk and soil research (Article 5).

It is also noting that the soil protocol acknowledged soil as part of the common European heritage (at art. 8). It recognized implied rights to an 'unpolluted environment' and to uncontaminated soil fit for many uses (at art. 11). Under Article 7(4) it stated that, 'where natural conditions allow it, disused or impaired soils, especially landfills, slag heaps, infrastructures or ski runs, shall be restored to their original state or shall be re-cultivated'. Further, Article 17(2) stated that 'to avoid soil contamination and to ensure the environmentally compatible pretreatment, treatment and disposal of waste and residual materials, waste management concepts shall be drawn up and implemented'.

Thus Soil protocol to Alpine Convention has so many important features which could be very useful to deal with contaminated sites and also for remediation measures. But it is limited to only few signatories and only for specific region that is why till date is of little use to promote soil protection and remediation measures within the EU countries.

However, in 2006, the European Commission presented a draft Framework Directive for the Protection of Soil (draft Soil Directive), in an attempt to fill an apparent gap in existing European environmental legislation. The proposal was based on the 'EU Soil Thematic Strategy' (the strategy), which tried to incorporate the opinion of hundreds of experts, stakeholders, NGOs and politicians throughout Europe.

The strategy explains why further action is needed to ensure a high level of soil protection and explained what kind of measures must be taken. It established a ten-year work program for the European Commission. In addition, the Impact Assessment (SEC (2006) 1165 and SEC (2006) 620) analyzed economic, social and environmental impacts of the different options that were considered in the preparatory phase of the strategy and also preferred measures as finally retained by the Commission. Again, soil contamination was identified as one of eight major threats to soils in the European Union and draft soil directive was a key step to deal with that threat.

The objective of draft directive was establishing a common strategy for the protection and sustainable use of soil based on the principles of integration of soil concerns into other policies, preservation of soil functions within the context of sustainable use, prevention of threat to soil and mitigation of their effects, as well as restoration of degraded soils to a functional level at least consistent with the present and potential future use. Contamination was referred to in the preamble to the draft Directive, and Chapter III of the Directive laid down specific measures that need to be taken by Member States.

A contaminated site has been defined in the draft Soil Directive as sites 'where there is a confirmed presence, caused by man, of dangerous substances of such a level that Member States consider they pose a significant risk to human health or the environment'. Article 13 of the draft directive stated that 'Member States shall ensure that the contaminated sites listed in their inventories are remediated'.

It further endorsed support for the polluter pays principle and also laid down that Member States shall set up appropriate mechanisms to fund the remediation of the contaminated sites, where the person responsible for the pollution cannot be identified or cannot be held liable under Community or national legislation or may not be made to bear the costs of remediation.

Again, article 14 stated that Member states should adopt national remediation strategy and article 15 mentioned about awareness building, reporting and exchange of information.

Following its initial proposal by the European Commission in September 2006, the draft Directive was adopted at the first reading stage by the European Parliament in November 2007. However, the draft Directive requires adoption by both the Parliament and the Council of Ministers, and from its earliest stages, the draft Directive has met with considerable opposition from a minority of countries in the Council of Ministers.

Austria, France, Germany, the Netherlands, the United Kingdom and Malta opposed the directive stating that it breaches the 'subsidiary principle' and have together blocked any further action on the draft Directive. The subsidiary principle, as mentioned in the Treaty of Europe, states that any action taken by the European Union must either fall within its exclusive competence or be more effective than action taken at the national, regional or local levels. The opposing countries argued that soil is a local or national issue and therefore should be dealt with exclusively at that level.

Apart from subsidiary principle, it was also stated that draft directive contains superfluous provisions to existing national and regional measures and it would place undue extra cost burden on countries which has already implemented specific soil policies or legislation at the national level. Again, another drawback of the directive was it unfairly places the burden of liability on land-users rather than those responsible for soil damage.

On the other hand, supporters of the draft Directive particularly environmental groups have argued that the wording of the proposed draft could be clarified and strengthened rather than abandoning it altogether.

Despite several attempts to minimize the contention between the parties, it was not possible to reach a consensus between the different Member states, that is why the EU Commission on 30 April 2014 took the decision to withdraw the proposal for a Soil Framework Directive (See, Withdrawal of the proposal for a Soil Framework Directive, available at http://ec.europa.eu/environment/soil/process_en.htm).

However, 'in taking its decision, the Commission stated that it remains committed to the objective of the protection of soil and will examine options on how to best achieve this'; (See, http://ec.europa.eu/environment/soil/index_en.htm, accessed on Oct 1, 2014). While making withdrawal of the directive, the Commission 'recognizes that soil degradation is a serious challenge. It provides that by 2020 land is managed sustainably in the Union, soil is adequately protected and the remediation of contaminated sites is well underway and commits the EU and its Member States to increasing efforts to reduce soil erosion and increase soil organic matter and to remediate contaminated sites' (See, http://ec.europa.eu/environment/soil/index_en.htm, accessed on Oct 1, 2014).

Thus, the attempt to adopt a uniform legislation for soil protection along with necessary remediation action across Europe postponed until further action in future. Considering absence of specific legislation some other existing legislation could be used at the European level to deal with the soil contamination and remediation obligation. For example, EU members could use United Nations Economic Commission for Europe (UNECE) Convention on Access to Information, Public Participation in Decision-Making and Access to Justice in Environmental Matters (1998), popularly known as Aarhus convention considering its objective is to safeguard the right of present and future generations to live in an environment adequate to their health and well-being, by guaranteeing the rights of access to information, public participation in decision-making and access to justice in environmental matters.

Although Aarhus convention has limited and indirect relevance for domestic management of site contamination, the provisions of the Convention may result in information on contaminated (and potentially contaminated) sites could become more readily accessible to the general public, particularly to the residential population near such sites. As per Aarhus Convention, local authorities must collate, update and disseminate environmental information including information on soil.

In relation to contaminated sites, Aarhus convention could have significant contribution considering that public participation could play a vital role during the site investigation and assessment stages, and also while deciding potential remediation decision. In addition, the Convention requires that access to justice be facilitated by allowing members of the public with a sufficient interest 'to challenge the legality of any decision, act or omission'. This could enable affected parties to oppose a potential remediation decision on a particular contaminated site on the basis that procedures for public consultation had not been maintained (Dellinger 2012).

In addition to, Aarhus convention, EU Directive 2004/35/CE on Environmental Liability with regard to the prevention and remedying of Environmental Damage, The Environmental Liability Directive (ELD) could be useful for remediation issues (it was entered into force on 30 April 2004). The ELD acknowledged that 'there are currently many contaminated sites in the Community, posing significant health risks, and the loss of biodiversity has dramatically accelerated over the last decades. Failure to act could result in increased site contamination and greater loss of biodiversity in the future'.

The objective of ELD was to establish a framework of environmental liability within the EU region based on the 'polluter pays principle' to prevent and remedy environmental damage. However, one of the limitations of this directive is, it is not retrospective, so it applies only to damage caused after April 2007 therefore could not consider vast majority of contaminated sites in EU which are linked to historical industrial revolution time.

There are three categories of environmental damage under the ELD: 'damage to protected species and natural habitats', 'water damage' and 'land damage'. The issues of land damage particularly related to issues of contamination and remediation, which is 'any land contamination that creates a significant risk of human health being adversely affected as a result of the direct or indirect introduction, in, on or under land, of substances, preparations, organisms or micro-organisms'. Activities which are recognized as causing damage are listed in Annex III to the ELD.

The ELD requires 'Operators' to take immediate steps to prevent damage when there is an imminent threat of it occurring, and to inform the competent authority immediately, if the preventive steps are unsuccessful. The ELD defines 'Operator' under Article 2, sub-article 6 as 'Operator means any natural or legal, private or public person who operates or controls the damaging occupational activity or, where this is provided for in national legislation, to whom decisive economic power over the technical functioning of such an activity has been delegated, including the holder of a permit or authorization for such an activity or the person registering or notifying such an activity'. The operator must take all practicable steps to immediately control, contain, remove or otherwise manage the relevant contaminants and/or any other damage factors, if damage has already been occurred. The operator should try to mitigate the damage and inform the competent authority immediately. Then considering the nature of contamination further appropriate remediation measures must be approved by the competent authority and to be carried out by the operator.

All costs of the preventive and remediation measures must be borne by the operator, unless it is proved that the damage was caused by a third party and despite appropriate safety precautions, or that it resulted from compliance with a compulsory order from a public authority. In such cases, the operators must still take the action required under the Directive, but do not have to bear the cost of doing so.

However, Member States may exempt the operator from financial costs for remedial actions, where he could demonstrate that he was not at fault or negligent and that the environmental damage was caused by:

(a) An emission or event expressly authorized by, and fully in accordance with the conditions of, an authorization conferred by or given under applicable national laws and regulations;
(b) An emission or activity or any manner of using a product in the course of an activity which the operator demonstrates was not considered likely to cause environmental damage according to the state of scientific and technical knowledge at the time when the emission was released or the activity took place (Environmental Liability Directive: A Short Overview (2004), available at http://ec.europa.eu/environment/legal/liability/pdf/Summary%20ELD.pdf).

The ELD further required 'as far as remedying of land damage is concerned, the necessary measures shall be taken to ensure, as a minimum, that the relevant contaminants are removed, controlled, contained or diminished so that the contaminated land, taking account of its current use or approved future use at the time of the damage, no longer poses any significant risk of adversely affecting human health' (Environmental Liability Directive: A Short Overview (2004), available at http://ec.europa.eu/environment/legal/liability/pdf/Summary%20ELD.pdf).

In the absence of operator, or where operator is unwilling or unable to take preventive or remedial action, the competent authority may do so and recover the costs from the operator. The ELD contains provisions on allocation of costs among multiple parties, limitation periods for costs and intervention by concerned individuals or non-governmental organizations. It specifically indicates that Member States may adopt more stringent provisions than those contained in the ELD, including additions to the list of harmful activities and potentially responsible parties.

Furthermore, the Industrial Emission Directive (IED) adopted end of December 2010 revises and merges 7 existing directives into one by way of integrated approach. It requires that any permits must take into account the whole environmental performance of the plant, covering e.g. emissions to air, water and land, generation of waste, use of raw materials, energy efficiency, noise, prevention of accidents and restoration of the site upon closure.

The objective of the IED is to ensure a high level of protection of the environment taken as a whole. The IED requires operators to prepare a baseline report before starting an operation of an installation or before a permit is updated having regard to the possibility of soil and groundwater contamination, ensuring the integrated approach. It is expected that this merged directive in future could lead to progress on EU wide harmonized action on contaminated sites and remediation plan.

Nevertheless, considering existing EU-wide legal framework, the conclusion should be that the regulation or remediation of contaminated brownfields is left to the Member States unless a new treaty or agreement reached by the member states. However, the ruling of the European Court of Justice in Ministère Public v. Paul Van de Walle could prove to be very important with regard to the European legal framework for brownfield redevelopment (Case C-1/03, Ministère Public v Paul van de Walle and Others, ECJ CURIA, 29 January 2004; Available at: http://curia.europa.eu/en/content/juris/index.htm).

The Court broadened the definition of waste and decided that soil contaminated by fuels leaking from underground tanks should be regarded as waste under the Waste Framework Directive. The result of the ruling is that the Waste Framework Directive may apply to soil contamination despite the Directive itself was not meant to cover soil contamination.

Again, it is problematic for brownfield developers to follow several provisions of the Directive such as most national soil legislation includes a system of soil pollution norms or a form of risk assessment for deciding whether a brownfield requires remediation or not. But the Waste Directive does not have such a system for the disposal or recovery of waste. Therefore, the Directive could not be used to cover soil contamination.

The European Commission being aware of the difficult situation and to avoid the problem, the Commission approved a revised proposal for Waste directive (Proposal for a Directive of the European Parliament and of the Council on Waste, COM (2005) 667 final (Dec. 21, 2005), available at http://eur-lex.europa.eu/LexUriServ/site/en/com/2005/ com2005_0667en01.pdf). The proposal excludes unexcavated contaminated soil from the scope of the directive.

Thus, till date the European initiatives that do relate to site contamination have their limitations. The 1998 Soil Protocol to the Alpine Convention is confined to the European alpine region, and the ELD applies only to liability for significant damage caused after April 2007.

The draft EU Soil Framework Directive has targeted contamination only as a sub-issue, but still considered as controversial and finally failed to be adopted by the member countries. Nevertheless, despite EU adopted a piecemeal approach in its coverage of site contamination, still probably the most advanced of any regional legal system in dealing with certain aspects of the contamination and remediation issue.

It is noting that EU commission acknowledge that extensive and consistent regulatory measures on site contamination are needed, and continuously adopting numerous initiatives on environmental liability, water pollution, soil protection and chemical use, which are an attempt to at least partially fill the gap in dealing with contaminated sites and encourage potential remediation process.

Apart from the EU, only ASEAN attempted to take concrete action on contamination and remediation issues.

5.3.2 ASEAN

ASEAN Agreement on the Conservation of Nature and Natural Resources, concluded in 1985, is the only regional instrument outside of Europe that promotes measures for soil protection generally and the improvement and rehabilitation of damaged soils in particular. However, site contamination measures are not specific objectives of the Agreement. The agreement was concluded and signed by foreign ministers of all six ASEAN countries in 1985. Of the six ASEAN member states at the time (now ten member states), only Indonesia, the Philippines and Thailand ratified it in 1986, while no other member states have ratified it in last thirty years. So, the agreement has not entered into force yet.

There is currently no comprehensive regional agreement on site contamination. Every region has a general environmental agreement, containing broad objectives which are mostly preventive in character. Although there are also some specialized agreements on pollution and hazardous waste having references to contamination and remediation, these are not enough to specifically address site contamination as an issue in its own right.

5.4 National Approaches for Soil Contamination and Remediation Measures

Site contamination is a problem experienced worldwide, although to different extents. The absence of site contamination on the international law-making agenda is in clear contrast to the increasing regulation of site contamination at the domestic level, notably in Europe but also in North America.

The issue of site contamination has now progressed through at least two generations of law in most developed countries. The first generation is identified as having a 'command and control' approach, which primarily dealt with issues of remediation and liability. Relevant legislation commonly addressed matters such as the investigation and identification of contaminated sites, potentially responsible parties, development of remediation plans, the imposition of specific types of liability and cost recovery. The second generation of site contamination law is characterized by measures to address 'brown-fields' and encourage 'voluntary remediation'. These measures range from tax incentives, liability relief and transfer of liability, to loans, grants and a variety of other financial tools.

It is important to note that the term brown fields generally relates to any land that has been previously developed, and that contamination may not necessarily be present on such sites. The term should be distinguished from 'green fields', concerning land which has not already been developed.

Therefore, the most significant trend in the field of addressing site contamination and remediation process is the emergence of voluntary remediation, popularly known as 'brownfields redevelopment'. This approach has been particularly widespread in the United States of America (USA) since the mid-1990s and has also been vigorously pursued in Canada and the United Kingdom. 'Brownfields' sites have two basic characteristics: they generally constitute areas of unused land and they are most likely to be contaminated by prior use.

In some highly industrial Western countries, such as the United Kingdom and the Netherlands, land scarcity is a pressing issue and therefore continuous need to utilize brown-fields for housing, recreational commercial and other forms of land-use. Considering demand for more residential and recreational facilities most of the western countries now developing contaminated sites through a market- driven approach. Voluntary remediation of such sites is also undertaken by developers due to growing governmental support for remediation through a combination of measures, such as liability relief and financial incentives. This has resulted in significant reductions in the number of sites awaiting remediation. In this regard, the market-driven approach has so far been viewed as more effective than the command and control approach. Considering the changed approach for remediation Nathanail and Bardos (2004) stated:

> It is now widely recognized that drastic hazard or contaminant control, e.g. cleaning up all sites to background concentrations or to levels suitable for the most sensitive land use, is neither technically or economically feasible nor is such control compatible with sustainable development (Nathanail and Bardos 2004).

However, there is an apprehension that in this way more sites are now being remediated to a lower standard and with minimal regulatory supervision and could pose environmental and health risks in future again.

Another avenue to remediation has been the use of specific funds established by site contamination legislation to address 'orphan sites' where responsible parties are unable or unwilling to assume responsibility for assessment and remediation.

The Federal Superfund scheme in USA has been the outstanding example of this approach, but it has not been complemented by additional schemes in more than a few of the American States. In Europe, there also appears a greater willingness to apply public funds to site cleanup. The 2007 EEA report states that 'approximately 35% of total expenditure in the (16) surveyed countries costs on remediation of contaminated sites derives from public budgets'. But neither Canada nor Australia or any other jurisdictions, which are active in remediation actions, have chosen to establish specific funds for cleanup purposes or to otherwise commit public funds to cleanups.

Thus site contamination law evolved over the years based on the remediation approach and considering mostly social-economic and regulatory approaches.

In one study, it has been observed that three types of national policies were successively generated [to deal with soil contamination and remediation issues]—

- a systematic approach (inventories, protocols) with a drastic control of soil contamination, in the early 1980s,
- around 1990, a contaminated land and risk assessment approach, with a real focus on land use as the main criteria for assessing and decision-making,
- Since 2000, a risk based land management (RBLM) and solution design, which integrates spatial planning, soil and water management and socio-economy issues (Darmendrail 2014).

The Risk-Based Land Management (RBLM) concept developed by CLARINET (Vegter et al. 2002) is considered as the third generation of national approaches in soil contamination and remediation action. It emphasized on sustainable solutions, for recovering the usability and economic value of land and integrating protection of environment quality. According to CLARINET, the third generation of remediation process account for three core elements:

(a) Suitability for use – the fitness for use principle is used to ensure the safe use or reuse of contaminated sites considering potential risks for the Environment and human health and;
(b) Protection of the environment on the basis of the stand-still principle (no further degradation and also possibility of improvement of the soil and groundwater quality) and
(c) Long term post remediation care (Darmendrail 2014); emphasis added.

It is further stated that a fourth generation of site contamination and remediation law evolving based on, 'a risk-informed and sustainable land management', which should integrate three key principles: being risk-informed, managing adaptively and taking a participatory approach, which could be called sustainable remediation. It is stated that 'Sustainable Remediation of soil, sediment and groundwater involves the assessment and management of significant risks to human health and the environment, in a manner that identifies the environmental, social and economic benefits and impacts of remedial strategies and options, and which seeks to maximize the overall benefit through a balanced, evidence-based and transparent decision-making process' (Darmendrail 2014).

5.4.1 *Key Trends in National Remediation Measures and Good Practices*

Based on the nature of soil contamination law addressing remediation measures, countries could be divided into three categories: First, this category includes countries with extensive specific provisions on site contamination. Second, countries having specific provisions on site contamination but no extensive law, due to the fact that these legislations address only some aspects of site contamination management, such as legislation may introduce liability for site remediation but may not specify any site management procedures for identification or assessment. Third, countries having neither extensive law nor provisions that specifically refer to site contamination (Brandon 2011).

Among the first category of countries, some countries developed site contamination law either in response to large national contamination incident took place in the 1970s or 1980s (such as United States and The Netherlands) or due to rising adverse public concern over contamination, such as Germany and and Denmark, or due to gradual increase in urban pressures for remediation of contaminated sites, for example, United Kingdom (Brandon 2011). However, some countries having no extensive law at the federal level, but states or provinces have adopted extensive specific legislation in the absence of federal legislation such as Australia, Canada and Belgium.

Countries having no extensive site contamination law mostly include provisions on contaminated sites into a general environmental protection law. New Zealand, Sweden and Finland adopted legislations falls under this category. Some researchers considers that several countries, particularly in Southern Europe, Eastern Europe and Russia, have legislation which ostensibly deals with aspects of site contamination, therefore could also fall under this category (Brandon 2011).

On the other hand, the absence of site contamination legislation may suggest that the issue is being ignored, particularly in countries with a highly urbanized population or lack of democratic government or absence of strong public opinion. In countries without those urban and industrial pressures or strong public opinion, it may also suggest a lack of awareness of the issue or lack of democratic and accountable governance or is not a priority for concern considering so many ongoing developmental challenges. No countries in Africa, Central America, South America or the South Pacific Islands appear to have specific site contamination law (Brandon 2011).

Among the countries having extensive site contamination law, the first specific site contamination and remediation legislation was adopted in the USA. The discovery of large-scale contamination at Love Canal in New York State in 1978 led to the passage of the *Comprehensive Environmental Response, Compensation and Liability Act* (CERCLA) in 1980, which deals primarily with abandoned or orphan ‘contaminated sites.

It was largely addresses ‘historic’ sites that are not subject to any current use. Contamination on sites that are subject to ongoing Federal licensing is dealt with separately under another Federal Act, the *Resource Conservation and Recovery Act* (RCRA), which has been enacted as back in 1976 and enables site contamination on licensed, operating industrial facilities to be regulated.

Both of these are considered as landmark statutes in the area of site remediation, ensuring that the polluter pays principle is applied wherever possible and that orphan sites are remediated. Many other countries have adopted the CERCLA model in their own domestic approaches to site contamination. These Federal laws have been complemented subsequently by State-based site contamination laws in most States in USA.

On the other hand, Western European countries having extensive laws proceed to adopt first site contamination legislations in the 1980s such as Denmark and the Netherlands in 1983; Switzerland in 1987 and Austria in 1989. The United Kingdom adopted comprehensive law in 1995 and Germany in 1999. The Netherlands is considered pioneer in site contamination and remediation management process in Europe.

In 1983, the Dutch Government introduced the multi-functionality approach for remediation of contaminated sites 'in response to a high-profile contamination incident at Lekkerkerk in 1980' (Brandon 2011). The multi-functionality approach intended to maximise soil quality and risk control by setting very high standard for remediation, that is why, the costs of such an approach proved prohibitive. To minimise cost and encourage more remediation, the Dutch government moved to a site-based risk assessment approach in 1997 and promoted other risk-based sustainable strategies for contaminated site management considering intended use only.

In compare to countries having federal government and hence federal law on site contamination such as in USA and Germany, Canada and Australia have no federal law rather most of the Provinces and States respectively have adopted specific site contamination legislations.

In Asia, several jurisdictions have adopted specific site contamination and remediation measures legislations such as Japan, South Korea, Taiwan, Singapore and the special region of Hong Kong. While some others are under process for the development of such legislation, such as China and Malaysia reviewing possible legal framework for soil contamination and remediation process to deal with rapid industrialization in this two countries in the last twenty to thirty years.

Among the Asian countries, Japan is the pioneer to take environmental protection measures. But in compare to earlier intervention for environmental protection in other sectors (such as laws related to water and air enacted in the 1960s and 1970s), Japan responded at a later stage to deal with soil contamination and remediation issue. The Soil Contamination Countermeasures Law was enacted in 2002 to prevent adversary health effects from soil contamination, which came into force in February 2003. This law more or less followed the American approach of soil remediation and mostly relied on risk based approach means remediation for minimizing health risks and fit for particular use. But it has been criticized as 'the law solely focuses on human health issues and does not take into account social or economic impacts of soil contamination problems' (Mitsunari 2014).

Remediation of contaminated sites has become an urgent issue in Japan considering concern that in preparation for the Tokyo Olympics Games in 2020 while doing construction work to renew old facilities, buildings and infrastructure, remedial proceedings could be imposed on some of the proposed facility sites under the soil contamination countermeasures law.

On the other hand, China has made rapid industrialization in last 30 years and therefore resulted in huge contaminated sites. But till date it has taken too little action to deal with contaminated sites and rate of remedial measures are very slow. In late 2013, the Ministry of Land and Resources in China revealed that approximately 3.3 million hectares of farmland was 'moderately' polluted. Considering huge number of polluted sites, the Ministry of Environmental Protection (MEP) of China proposed an 'Action Plan for Prevention of Soil Pollution' in March 2014. Although there are some voluntary measures for remediation of contamination sites based on risk-based approach, there are no specific laws or regulation on soil pollution and remediation process.

However, Office of the State Council of China issued a document on short term arrangement for protection and remediation of soil and MEP issued a circular to deal with remediation of contaminated sites (Chun 2014). There is assumption that the revision of the Environmental Protection Law will include soil pollution, but it might not fulfill the purpose of specific law on soil contamination. A specific law would be ideal to add more detail treatment of contamination and remediation issues from Chinese perspectives.

Considering importance of specific law in China, it has been stated 'China is huge, with big differences in soil, hydrology and geology. China could categorize pollution differently across regions, to allow for a differentiated approach'. It further mentioned urgency of action stating 'China needs better monitoring of soil pollution, and to take action at any time... If you wait a decade the pollution has spread. That means it does more harm, and is much more costly to clean up' (Chun 2014).

It is expected that specific site Contamination law and remediation policy will be recognized as an important concern in the developing countries in the coming decade considering global concern for sustainable development and also greater public awareness of the effects of contamination on the surrounding environment and potential health risks.

5.5 Conclusion

There are little international initiatives for the adoption of an international soil regime to deal with soil protection including issues of contamination and possible remediation measures. The absence of action for soil specifically perhaps due to growing concern that too many environmental treaties addressing specific areas, resulting in gaps, duplication and an overwhelming body of legal obligations for States and resulting in dysfunctional and isolated environmental measures.

That is why, in the absence of specific soil protection regime, some argue that efforts would be better spent on strengthening existing soil provisions in the climate change, biodiversity and desertification treaties, and thereby to promote sustainable use of soils and also remediation of contaminated sites (Fowler 2012). In 2004, the secretariats of the UN Convention to Combat Desertification (UNCCD), UN Convention on Biological Diversity (CBD) and UN Framework Convention on Climate Change (UNFCCC) prepared a joint paper on improving synergies between the three treaties. Since then, the Secretariats and Conferences of the Parties for each of the Conventions have been working towards this goal. But till date the issues of soil has not been discussed in a way to progress further actions on the sustainable soil uses and integrated approach on soil for protecting biodiversity and also dealing with climate change actions and combat desertification. It is challenging to adopt new soils measures therefore, it would be less disputed, if any new measures are introduced by way of amendment to existing treaties, rather than in the form of a new treaty.

However, the 2005 Selfoss Declaration on the Conservation and Sustainable Use of Global Soil Resources, hosted by the International Union of Soil Sciences, called for a new international instrument on sustainable soil use, but also recognized the potential for synergy between the three main conventions.

In the context of site contamination, 'there is clear scientific evidence of environmental damage and risks to health caused by contaminated sites, and this would provide a sound basis for any action at the international level' (Brandon 2011). Indeed, some sections of the international community, particularly networks of soil scientists, have already been actively promoting global action on some aspects of site contamination relating to soil. For example, the International Union of Soil Sciences (IUSS) have specific expert group dedicated to the study of soil degradation control, remediation and reclamation. It is noting that the 68th United Nations(UN) General Assembly declared 2015 as the International Year of Soils(IYS) and the UN Food and Agriculture Organization was nominated to implement the IYS 2015 in cooperation with the Governments and the secretariat of the UN Convention to Combat Desertification. The prime objectives of the IYS includes, raising full awareness among civil society and decision makers about the importance of soil; support effective policies and actions for the sustainable management and protection of soil resources; promote investment in sustainable soil management activities and soil uses for sustainable development and also improve monitoring at local, regional and global levels (see, http://www.fao.org/soils-2015/about/en/).

In addition, the IUCN's Commission on Environmental Law is working for long time on the issues related to an international framework on soil protection and model national soil legislation. Several other international organizations prepared guidelines on the national management of contaminated land, such as, UNEP guidelines, FAO reports on various site contamination issues, remediation technologies and prevention of site contamination.

However, the lack of widespread public awareness and understanding and lack of financial and technical expertise and resources regarding contamination particularly in the developing countries may hamper the progress for adopting any universal law on site contamination and remediation of contamination measures. In the absence of strong public opinion, local political pressure, governments need to be persuaded of the need for collaborative, international measures against site contamination. In this regard, three important issues need particular attention while considering a globally applicable site contamination and remediation law—

First, there is a dilemma as to 'whether it is appropriate to link site contamination and soil protection together at the international level', given the major differences between the two issues. They have varied causes and effects, and therefore require particular regulatory approaches. For instance, soil management does not address ground-water contamination, which is an important part of the site contamination issue. That is why, a pragmatic view might be that linking site contamination with soil protection is the only option with any prospect of succeeding in terms of a binding treaty (Fowler 2012).

Second, consideration of possibility to adopt either hard law (binding nature) or soft law (with voluntary obligation). There are both advantages and disadvantages for adopting either a hard law or soft law approach. A binding treaty on specific soil issues could be a successful role model, contrary to the general assumption that treaties are too lengthy or costly to negotiate. Again, a soft law proposal does not unavoidably could assurance of prompt action or consensus, although its function as a short to medium-term a step towards future binding measures cannot be overlooked.

Third, considering the absence of a credible international organization willing to promote idea of a separate instrument on soil contamination some researcher consider that a realistic proposition in the current circumstances would be to employ a ‘bottom up’ approach based on harmonization via a model law such as IUCN guidelines on national soils legislation (Brandon 2011; Fowler 2012).

It is stated in one study that ‘A credible, comprehensive model law on site contamination could provide relatively immediate benefits if it is widely promoted by the appropriate institutions. It would draw on the combined experience of developed countries over the past 30 years, and could be made available for use by any country needing to introduce or improve its own regulatory approach. A key feature of the model law would be its versatility in terms of reflecting the domestic conditions of individual countries’ (Brandon 2011).

However, in the absence of a model law, countries have three basic options: either to take no action on the issue, leaving later generations to deal with its accumulated effects or to develop their own approach, possibly repeating the same mistakes made in early industrialization approaches to site contamination; or simply borrowing from other countries ‘laws, which may end up in a bad design or ill-suited to local circumstances’ (Brandon 2011). Therefore, none of these options could achieve a favorable outcome for the remediation of contamination. Nevertheless, model law at least could provide guidance to countries on some relevant issues, such as the management of hazardous waste, remediation of contamination sites, functional and market approach, post-remediation maintenance and sustainable use of soils etc. (Fowler 2012, Brandon 2011).

Furthermore, it is important to encourage different national governments having no site contamination and remediation related law to adopt a national law as a first step towards taking effective measures for prevention of site contamination and also to take remediation measures to prevent health risk and environmental pollution.

To conclude absence of international soil protection regime and hence lack of concerted global efforts on remediation action on contaminated sites could pose not only serious health risks but also long term sustainable development challenge. On the other hand, most of the national measures till date address contamination and remediation process from health risk perspective only rather than considering broader social and environmental concerns. If we would like to promote sustainable development for our future generations, sustainable uses of soil and remediation of existing contaminated sites would be first step in the right direction towards sustainable development path.

References

Bates GM (2006) Environmental law in Australia. LexisNexis ButterWorths, Chatswood

Brandon E (2011) The development and harmonization of domestic site contamination law: the role of international law and other mechanisms. Ph.D. thesis, University of South Australia, Australia

Chun Z (2014) China "lacks experience" to clean up its polluted soil. China Dialogue, Beijing

Cooper S (2013) Soil remediation and the environment. Environmental Protection, Dallas

Darmendrail D (2014) EU legislation applying to contaminated sites management and future needs. Paper presented at the 5th joint Nordic Meeting on Remediation of Contaminated Sites (NORDROCS 2014), Stockholm, Sweden, September 15–18

Dellinger M (2012) Ten years of the Aarhus Convention: how procedural democracy is paving the way for substantive change in national and international environmental law. Colo J Int Environ Law Pol 23:311–366

EEA (2007) Europe's environment – the fourth assessment. European Environment Agency, Copenhagen

Ellis J (2006) Has international law outgrown trail smelter? In: Bratspies RM, Miller RA (eds) Transboundary harm in international law: lessons from the Trail Smelter arbitration. Cambridge University Press, Cambridge, pp 56–61

Fowler R (2007) Site contamination law and policy in Europe, North America and Australia – trends and challenges. Paper presented at the 8th meeting of the International Committee on Contaminated Land, Stockholm, Sweden, September 10–11

Fowler R (2012) Other international environmental law aspects of soils: soil contamination and soil carbon sequestration. Paper presented at the IUCN World Conservation Congress, Jeju, Korea, September 6–15

Hackett DP (1990) An assessment of the Basel convention on the control of transboundary movements of hazardous wastes and their disposal. Am Univ Int Law Rev 5:291–323

Haninger K, Ma L, Timmins C (2012) Estimating the impacts of brownfield remediation on housing property values, Working paper EE 12-08. Duke Nicholas Institute for Environmental Policy Solutions, Durham

Hannam I, Boer B (2002a) Legal and institutional frameworks for sustainable soils – a preliminary report, IUCN Environmental Policy and Law Paper No. 45. International Union for Conservation of Nature (IUCN), Cambridge

Hannam I, Boer B (2002b) International and national legal and institutional frameworks for the sustainable use of soil. Paper presented at the 12th ISCO conference, Beijing, China, May 26–31

Kummer K (1992) The international regulation of transboundary traffic in Hazardous wastes: The 1989 Basel convention. Int Comp Law Q 41:530–562

Layard A (2004) The Europeanisation of contaminated land. Environmental Law Review 6:97–110

Lye LH, Manguiat MSZ (eds) (2003) Towards a "Second Generation" in environmental laws in the Asian and Pacific region: select trends, IUCN Environmental Policy and Law Paper No. 48. International Union for Conservation of Nature (IUCN), Cambridge

Mitsunari M (2014) Toward the next generation soil protection framework in Japan (Part 1 and 2). ENFOS, San Mateo
Nathanail CP, Bardos P (2004) Reclamation of contaminated land. Wiley, West Sussex
Preston B (2008) Ecologically sustainable development in the context of contaminated land. Paper presented at the Australian Land and Groundwater Association 1st annual conference, EcoForum 2008, Gold Coast, Queensland, February 28
Prokop G, Schamann M, Edelgaard I (2000) Management of contaminated sites in Western Europe (Topic report No 13/1999). European Environment Agency, Copenhagen
Read JE (2006) The trail smelter dispute [Abridged]. In: Bratspies RM, Miller RA (eds) Transboundary harm in international law: lessons from the trail smelter arbitration. Cambridge University Press, Cambridge, pp 27–31
UNEP (2014) Basel convention on the control of transboundary movements of hazardous wastes and their disposal: protocol on liability and compensation for damage resulting from transboundary movements of hazardous wastes and their disposal (texts and annexes). United Nations Environment Programme (UNEP), Geneva/Switzerland
US EPA (2009) Soils policy: soil contamination in Europe. United States Environmental Protection Agency, Washington, DC
Vegter J, Lowe J, Kasamas H (eds) (2002) Sustainable management of contaminated land: an overview. Austrian Federal Environment Agency (on behalf of CLARINET), Wien
Viñuales JE (ed) (2015) The Rio declaration on environment and development: a commentary. Oxford University Press, Oxford
World Bank (2012) Towards a more effective operational response: arsenic contamination of groundwater in South and East Asian countries. World Bank, Washington, DC

Chapter 6
Solidification/Stabilization: A Remedial Option for Metal-Contaminated Soils

Ismail M.M. Rahman, Zinnat A. Begum, and Hikaru Sawai

Abstract Decontamination of hazardous discards by immobilization of toxic components is a longstanding approach for managing waste, while it gained much attention in recent years due to the increasing number of statutes and regulations favouring the technology. The solidification/stabilization (S/S) technique is the commonly adopted immobilization option to treat the contaminated soils, which employ additives to convert the hazardous waste to non-hazardous mass in accordance with the legitimate landfill provisions. The discussion is further extended to the stabilization of toxic elements in contaminated soils using chemical amendments. The current paper presents a summarized overview on the application of S/S technique in managing metal-contaminated soil, including information about the frequently used additives for the purpose, and the steps involved in the implementation of S/S remediation.

Keywords Soil • Metal • Solidification • Stabilization • Binders • Chemical amendments

I.M.M. Rahman (✉)
Department of Applied and Environmental Chemistry, Faculty of Science, University of Chittagong, Chittagong 4331, Bangladesh

Graduate School of Natural Science and Technology, Kanazawa University, Kakuma, Kanazawa 920-1192, Japan
e-mail: I.M.M.Rahman@gmail.com

Z.A. Begum (✉)
Department of Civil Engineering, Southern University, 739/A Mehedibag Road, Chittagong 4000, Bangladesh
e-mail: Zinnat.Ara@gmail.com

H. Sawai
Graduate School of Natural Science and Technology, Kanazawa University, Kakuma, Kanazawa 920-1192, Japan
e-mail: hikaru.sawai@gmail.com

H. Hasegawa et al. (eds.), *Environmental Remediation Technologies for Metal-Contaminated Soils*, DOI 10.1007/978-4-431-55759-3_6

6.1 Introduction

Soil is a vital constituent of the rural and urban environments, and as constituent metals and metalloids occur naturally but rarely at toxic levels (Zhao and Kaluarachchi 2002; Pierzynski et al. 2005). Due to the anthropogenic activities, natural geochemical cycle is interrupted, and accumulation of metals occurs (D'Amore et al. 2005). Metals from anthropogenic sources possess increased mobility character than those resulting from pedogenic or lithogenic ones (Chlopecka et al. 1996) and cause risks to human health, plants, animals and related ecosystems (Samsøe-Petersen et al. 2002; D'Amore et al. 2005; Rahman et al. 2008, 2014). The mass balance of the accumulation of potentially toxic metals in soil from various sources can be expressed as follows (Alloway 1995; Lombi and Gerzabek 1998):

$$M_{\text{total}} = \left(M_p + M_a + M_f + M_{ag} + M_{ow} + M_{ip}\right) - \left(M_{cr} + M_l\right) \quad (6.1)$$

where *M*, *p*, *a*, *f*, *ag*, *ow*, *ip* stand for the 'heavy metal', 'parent material', 'atmospheric deposition', 'fertilizer sources', 'agrichemical sources', 'organic waste sources', 'inorganic pollutants', while *cr* and *l* denote the 'crop removal' and 'losses by leaching, volatilization, etc. The fate of hazardous contaminants in soil, as adapted from Sims et al. (1989), is illustrated in Fig. 6.1.

A simulated estimate shows that the anthropogenic emission into the atmosphere, for many of the toxic metals, is the one-to-three order of magnitude higher than the natural fluxes (Sposito and Page 1984), and the metal-bearing solids at the contaminated sites can originate from a wide-variety of human-made sources

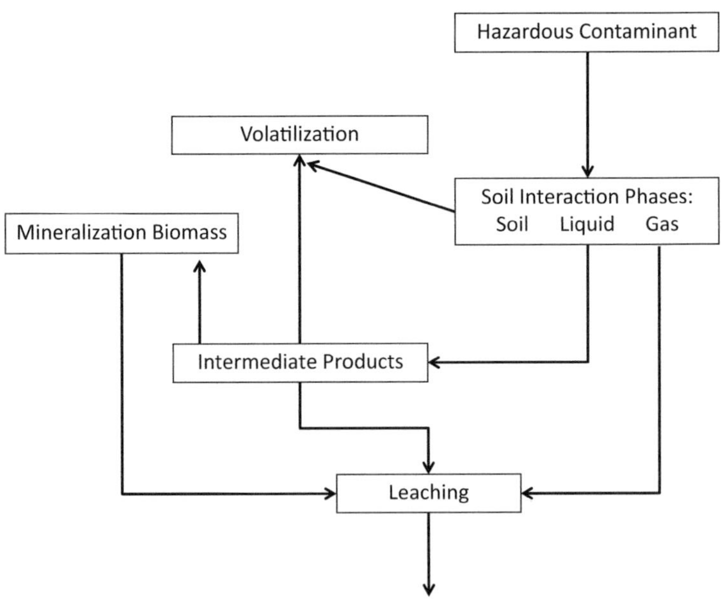

Fig. 6.1 Fate of hazardous contaminants in soil (Sims et al. 1989)

(Basta et al. 2005; Khan et al. 2008; Rahman et al. 2011; Begum et al. 2012, 2013). In accordance with the 'maximal permissible supplement' values of metals/metalloids as a pollutant, the following sequence in the soil has been assumed: Se > Tl > Sb > Cd > V > Hg > Ni > Cu > Cr > As > Ba (Vodyanitskii 2013). However, the aforesaid sequence differs from the popular series of the hazardous elements, e.g., US-EPA priority pollutant list, in which the relative toxicity from Pb and Zn is described as high while that from the V, Sb and Ba, is shown as the minimum (Evanko and Dzombak 1997; US EPA 2012; Vodyanitskii 2013).

It is important to limit the intrusion of toxic metal ions in the soils not only due to the ecological concern but also to avoid the expensive soil cleaning process. Therefore, regulatory limits are proposed to indicate the maximum allowable content in the disposable waste and so forth. The limits proposed by the U.S. Environmental Protection Agency (EPA) is shown in Table 6.1 (US EPA 1993) as an example, which include the information about the permissible annual loading rates (kg ha^{-1} $year^{-1}$) and the cumulative loading rates (kg ha^{-1}) of the toxic metals alongside the total acceptable contents (mg kg^{-1} or ppm). In spite of all the regulatory barriers, the distribution of contaminated sites around the world is growing, e.g., (a) in China, 46,700 ha of the land area associated with mining activities supposed to be degraded each year with metal (Wong 2003; Li 2006); (b) in USA, 50 million m^3 of soil are estimated to be contaminated with metals (US EPA 2004); (c) in Europe, several million ha of agricultural lands are reported to be polluted with metal (Flathman and Lanza 1998).

The main threat from the contaminated soil is the leaching of metals to the surroundings and consequential eco-toxicity. Hence, the soil remediation approaches are designed either to decrease the total cumulative metal content or to minimize the possibility of metal leaching. The remediation approaches based on the immobilization technologies modify the physical or leaching characteristics of the contaminated matrix. The target is either to restrict the physical contact between the contaminant and the surrounding water phase, or chemical alternation of the contaminant to make it more stable with respect to further dissolution (Evanko and Dzombak 1997). 'Solidification and/or stabilization' is the most commonly adopted immobilization option to treat the contaminated soils, due to its better viability in terms of affectivity and economic competence in handling heterogeneity in soil and

Table 6.1 Regulatory limits for the disposal of toxic metal ion containing waste to soils[a]

		Loading rates	
	Maximum allowable concentration	Annual	Cumulative
Metals	mg kg^{-1}	kg ha^{-1} $year^{-1}$	kg ha^{-1}
Cadmium	85	1.9	39
Chromium	3000	150	3000
Copper	4300	75	1500
Lead	420	21	420
Nickel	75	0.9	18
Zinc	7500	140	2800

[a]Adapted from US EPA (1993)

contaminant conditions as compared to the other remediation options. Moreover, the option offers an immediate solution to the soil contamination problem because it considers restricting the contaminants in soil instead of removal, and thus the sites become accessible for further development in no time (Smith et al. 1995; Evanko and Dzombak 1997; Al-Tabbaa and Perera 2006). In this chapter, we have discussed the facts and features of the solidification and stabilization approach used for the remediation of metal-contaminated soils.

6.2 Solidification and Stabilization: Definition and Objective

In spite of being two distinct technologies, the term's solidification and stabilization are frequently mentioned together as 'solidification/stabilization (S/S)', and are widely adopted to restrict the migration and exposure of toxicants from the polluted soil, sludge or sediments (US EPA 1997, 2000). The waste material is converted into solid forms via entrapping within a granular or monolithic matrix by adding appropriate reagents or using a mechanical process during the process of solidification. A chemical reaction is usually involved in the stabilization process to limit the mobility or solubility of the hazardous components in the wastes, and not necessarily alter the physical nature of the waste (US EPA 2000). The combined application of the solidification and stabilization process ensures the mixing of the contaminated waste materials with the treatment agents, and consequently, both the physical and chemical immobilization of the hazardous components occurs. The S/S technique is a non-destructive approach to eradicate or inhibit the mobility of contaminants in the waste materials (US EPA 1999).

The ultimate objective of the S/S approach is to complete transformation of the toxic components in wastes into a non-toxic form. However, the objective of S/S technology not only includes limiting the solubility of the contaminant when exposed to leaching fluid, but also improvement of the handling characteristics of the wastes and decreasing the surface area across which contaminant transport might occur is expected through solidifying of the waste materials. The objectives are most likely expected to be achieved via chemical transformation, which result in the formation of new compounds, although chemical changes are seldom occurred during the application of existing S/S approaches (Wiles 1987).

6.3 Development of Solidification/Stabilization Technique: Historical Timeline

The S/S concept was introduced, according to the documented evidence, in 3000 B. C. primarily for road construction, and continued until then for similar purposes, including the adoption of the technique for harbour protection starting from the second century B.C. (Barth 1990). The application of S/S technique for waste treatment was started in the 1950s by processing the liquid low-level radioactive waste into solid forms using urea formaldehyde and asphalt systems to accomplish easy transportation and better disposal (Conner 1990). However, the practices for the management of hazardous wastes were generally mandated after the 1970s, and the regulations for the S/S techniques were available in the 1980s as amendments to the Solid Waste Disposal Act (SWDA), Resource Conservation and Recovery Act (RCRA) and the Hazardous and Solid Waste Amendments (HSWA) (US EPA 1986a; Barth 1990).

Several guidance documents for S/S processes have been published by the US EPA until the date (US EPA 1982, 1986b, 1989, 1999). Historically, S/S has been one of the top five source control treatment technologies used for the remediation of the abandoned hazardous waste sites in USA. It has been evaluated that the average operational time for the S/S projects was shorter (~1.1 months) than many other approaches (e.g., soil vapor extraction, land treatment and composting), and cost-effective with an average cost per cubic yard between USD 194 to 264 (US EPA 2000).

6.4 Implementation of Solidification/Stabilization

The implementation of the S/S process starts with the characterization of the contaminated mass, followed by the treatability studies to determine the appropriate mix design of reagent and/or additive. The next step is the mobilization of the field equipment to execute the field-scale S/S mix design in accordance with the pre-determined construction specifications. The final step is the long-term monitoring of the S/S mass after the process is completed (Bone et al. 2004).

6.4.1 Characterization of the Contaminated Mass

The applicability of the S/S remediation to treat metal-contaminated soil is often determined based on the soil properties (physical and/or chemical) and nature of the contaminant, which controls the interaction between the soils and the contaminants to a considerable extent. The typical soil properties measured during the S/S design, as adapted from Harris et al. (1995), is listed in Table 6.2. However, several other factors might be required to be considered depending on the waste characteristics and have been discussed in detail elsewhere (Conner 1990; LaGrega et al. 2001; Bone et al. 2004). A brief list of site characterization considerations in the development of performance goals and specifications for the S/S application, as adopted from ITRC (2011), is listed in Table 6.3.

Table 6.2 The typical soil properties that are required to be measured before S/S application[a]

Parameter	Purpose
Physical characteristics	
Particle size	To determine pre-treatment and materials handling requirements, mixing characteristics, potential environmental impacts
Moisture content	To determine pre-treatment materials handling requirements, reagent formulation
Temperature	To determine impact on process performance, curing characteristics
Chemical characteristics	
Contaminants (type, concentration, variability)	To determine formulation of reagents, potential environmental impacts through emissions
Leaching behavior	To provide measures against which to judge results of treatability studies and full-scale application
Inhibitory species	To determine compatibility with formulation
pH	To predict reaction conditions, impact on leaching characteristics

[a]Adapted from Harris et al. (1995)

6.4.2 *Selection of the Reagents and/or Additives*

The reagent and/or additive, termed as 'binder' hereafter, to be used for the S/S treatment of contaminated soil is selected based on the end-use goal of the processed material, such as land filling, a resource of aggregate in construction or redevelopment of the sites. The common criteria used for the selection of binder to meet site-specific requirements with the corresponding typical target values (Al-Tabbaa and Perera 2005a, b, c) are listed in Table 6.4.

'Inorganic' and 'organic' are two broad categories that are used to differentiate the binder systems used for the S/S treatment. The inorganic binder systems include cement, fly ash or blast furnace slag, while the asphalt/bitumen is the common organic binder in use (Al-Tabbaa and Perera 2006). Besides, there are instances of using mixed binder systems, e.g., diatomaceous earth with cement and polystyrene, polyurethane and cement, and polymer gels with silicate and lime cement (Wiles 1987).

The metal-contaminated wastes are frequently processed through cement-based S/S treatment (LaGrega et al. 2001), because the high pH of the cement facilitates the retention of metal cations as insoluble hydroxide salts within the solidified structure. Portland cement (PC) is the most commonly used cement variant for S/S, and the process is executed via mixing of the contaminated mass and PC with or without water depending on the nature of waste (Conner 1990; LaGrega et al. 2001). Both the chemical fixation and immobilization via physical encapsulation is expected during the PC-based S/S processing (Harris et al. 1995; Al-Tabbaa and Perera 2006). The binders that have been used as a partial

Table 6.3 Site characterization considerations in the development of performance goals and specifications for the S/S application[a]

S/S evaluation factor	Analyses/observations	Significance to technology performance and monitoring
Groundwater	pH	Controlling variable for inorganic solubility and S/S material durability (ITRC 2011)
Geochemistry	Organic content	Key variable for organic concentrations due to complexation with dissolved organic carbon, which is soluble at high pH (Roskam and Comans 2009)
	Contaminant levels	High concentrations of some contaminants may affect S/S cure, requiring additives to overcome interference (Conner 1997)
	Sulphate content	Sulphate attack of Portland cement blends may lead to aggressive degradation through delayed ettringite formation (Little et al. 2005)
Contaminant characterization	Leaching behaviour of untreated material	Defines baseline against which treatability studies and full-scale application may be compared (ITRC 2011)
	Class(es) of contaminants	Defines list of contaminant of concerns, defines detection limits for analysis (ITRC 2011)
	Presence/distribution of non-aqueous-phase liquids	Defines phases/location of source and expected outcomes (ITRC 2011)
Hydrogeology	Hydraulic conductivity	Controlling value in comparison to hydraulic conductivity of S/S material for mode of water contact (e.g., infiltration vs. flow-around) (ITRC 2011)
	Water table depth and seasonal variability	Defines division between vadose zone, capillary fringe and saturated zone; non-aqueous-phase liquid impacts at water table (ITRC 2011)
	Geologic strata (including geometry of geology units)	Location of contaminant distribution/accumulation zones (ITRC 2011)
	Groundwater flow direction and gradients	Hydraulic head on S/S mass, evaluate fate and transport with respect to point of compliance (ITRC 2011)

[a]Adapted from ITRC (2011)

alternative to PC include blast furnace slag and pulverized fuel ash (Al-Tabbaa and Evans 1999; Al-Tabbaa and Perera 2006).

The pulverized fuel ash (PFA), which is a siliceous and aluminous material, is used as an alternative binder to PC in the S/S process in conjunction with lime, and the corresponding reactivity depend on the relative fineness of the content and the glass phase ratio (Dhir 1986; Harris et al. 1995; Neville and Brooks 2010). The metals in the contaminated soil, if treated with PFA-lime binder, might get chemically immobilized as hydrate complexes (Barth 1990; LaGrega et al. 2001). However, the hardening process of PFA is slower (Taylor 1992) and exhibit reduced sustainability during both leaching and durability tests than that of the PC (Harris et al. 1995).

Table 6.4 Commonly used design criteria and typical target values for the selection of the reagents and/or additives for the S/S application[a]

Design criteria	Typical target values
Unconfined compressive strength	>350kPa (soaked) at 28 days (US EPA 1986a)
Leachate pH	7–11 (Conner 1990; Harris et al. 1995)
Leachability	Usually, 100 times of the drinking water standards; in accordance to the standard leaching test, such as TCLP (toxicity characteristic leaching procedure; US EPA Method 1311), etc. (Conner 1990; US EPA 2007)
Permeability	$<10^{-9}$ m s^{-1} (as for cut-off walls) (Al-Tabbaa and Perera 2005b)
Freeze-thaw and wet-dry durability	In accordance with the recommended ASTM tests (ASTM 2001, 2009)
Acid neutralization capacity	In accordance with the Environment Canada test method; to measure the stability of the chemical environment in the contaminated mass in terms of the immobilization of metal contaminants (Stegemann and Cote 1990)
Microstructural analyses	In accordance with the standard test methods; to examine the development of the hydration products and their interaction with contaminants (Al-Tabbaa and Perera 2005b)

[a]Adapted from Al-Tabbaa and Perera (2005b)

The other material that have been used as a partial replacement of PC in the S/S process is the blast furnace slag (BFS), which is a by-product of pig iron manufacturing process and contain silica, alumina and lime as ingredients (Neville and Brooks 2010). The composition of the most common BFS variant corresponds to the latent hydraulic cement, and broadly remains between the pozzolanic substances and PC (Taylor 1992; Harris et al. 1995). The lime in the BFS initiates the process of hydration, while the several other factors, e.g., bulk composition, glass content ratio, the grinding fineness, etc. controls the overall reactivity pattern (Taylor 1992). The use of BFS in replacement to PC is preferred due to the less cost involvement (Al-Tabbaa and Perera 2006), and/or possibility to use as a standalone binder (Allan and Kukacka 1995).

Bentonite, a naturally-occurring clay as derived from the volcanic ash and having low hydraulic conductivity and high sorption capacity for cations (Matthes et al. 1999) is another suitable binder for the S/S process. The sorption property is attributable to the presence of mineral montmorillonite, which is a di-octahedral smectite and is chemically characterized as a hydro-alumino-silicate (Andini et al. 2006). Although it is considered as an advantageous material for treating metal-contaminated wastes, the presence of other organic and inorganic chemicals might alter the sorption ability of bentonite (Spooner et al. 1984). Hence, it is generally used as a co-additive with other binder, such as PC, etc (Al-Tabbaa and Perera 2006).

The calcium oxide and calcium hydroxide are the common lime variants, which are used as the S/S binder, and another frequently-used inorganic-type binder (Sherwood 1993). If the waste material components include alumina and silicates, lime can be used as a standalone binder because those materials can react with lime to generate cementitious solids (LaGrega et al. 2001). However, lime is mostly used as pH-controller additive with the other binders, such as, PC, PFA and clay (Conner 1990).

Bitumen, classified as an organic-type binder and consists mainly of hydrocarbons, does not react chemically with contaminants in the waste; rather a physical encapsulation mechanism comes into the act during the process of solidification of bitumen itself (Bone et al. 2004). Bitumen generally is an extremely viscous material in its natural state, and for S/S application it is either converted into solution with paraffin/diesel or changed to emulsion in water (Sherwood 1993). Besides bitumen, polyethylene and sulfur polymer are other examples of organic binders (US EPA 1997). The organic phase is hydrophobic in nature that reduces water permeability into the solidified product, while the presence of certain organic compounds, such as, oil and grease, can decrease the durability of the organic binder-encapsulated solids (AEPI 1998).

In addition to above-mentioned common binders, there are other binders that have been explored for S/S applications, e.g., activated carbon, phosphates, rubber particulates, chemical gellants, saw dust, straw, etc. (Kershaw and Pamukcu 1997; Ajmal et al. 1998; ITRC 2011). More detail discussion about the binders used in the S/S processes can be found in the works of Conner (1990), Trussell and Spence (1994), Glasser (1997), AEPI (1998), Bone et al. (2004), Paria and Yuet (2006), Al-Tabbaa and Perera (2006), Kumpiene et al. (2008), Chen et al. (2009), Ahmad et al. (2012), and references therein.

6.4.3 Process Technologies for the Field-Scale Application

There are several factors, which includes waste characteristics, purpose of the processing, handling plans, regulatory limitations and economics, are taken into consideration during the selection of process technologies for the field-scale S/S application (Wiles 1987; Wiles et al. 1988). The available S/S application technologies are categorized into *ex situ* and *in situ* operations. The *ex situ* approach requires excavation for the pre-confinement of the contaminated waste fraction before S/S processing, while the *in situ* methodology is executed in on-site locations (Conner 1990; Evans et al. 2001; LaGrega et al. 2001).

The *ex situ* S/S processing schemes available for consideration includes in-drum processing, in-plant processing and direct-mixing (Wiles 1987; Evans et al. 2001). The S/S binders and the excavated waste are mixed in a drum or other container during the in-drum processing, and the waste-binder matrix is usually disposed of in the drum after the setting process is completed. During the in-plant processing, the bulk waste material and the selected S/S binder are treated together in a specifically designed mixing plant that is either fixed or can easily be relocated within the contaminated sites. The contaminated solid is unearthed, transported to the disposal location and mechanically treated with the pre-selected S/S binder during the

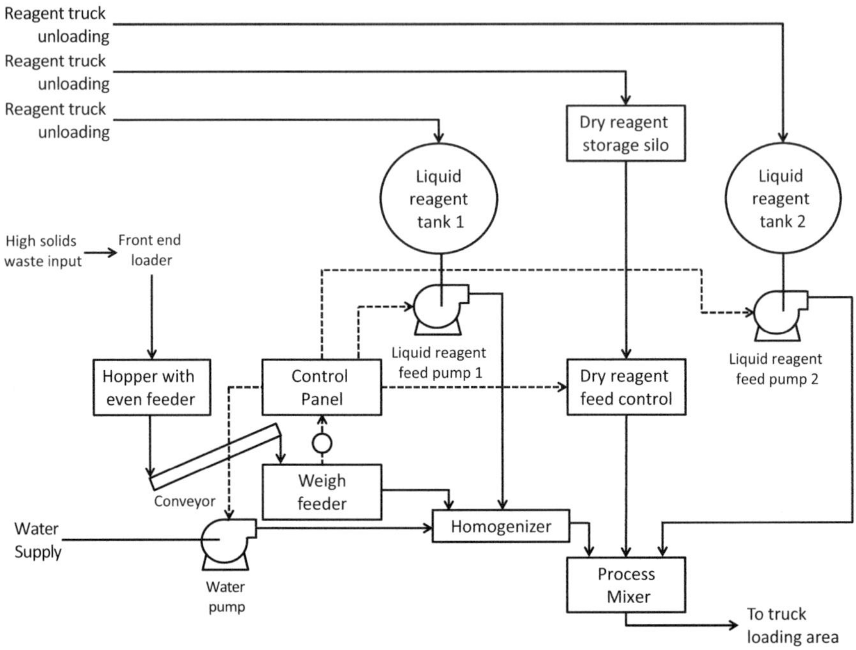

Fig. 6.2 Schematic diagram of a typical *ex situ* S/S system (Harris et al. 1995)

direct-mixing process, and left in-place (Wiles 1987; Harris et al. 1995; Al-Tabbaa and Perera 2006). A schematic diagram of the typical *ex situ* S/S processing scheme, as adapted from Harris et al. (1995), is shown in Fig. 6.2.

The *in situ* S/S approach includes three basic approaches for mixing binder with the matrix: (a) vertical auger mixing, (b) shallow in-place mixing, and (c) injection grouting (US EPA 2006). When the location of contaminated material is within the depth of 6 m, the site is usually divided into grid cells and subjected to shallow in-place mixing of binder with waste using conventional earth-moving equipment, e.g., draglines, backhoes or clamshell buckets (US EPA 2006; ITRC 2011). On the contrary, *in situ* treatment through deep soil mixing using auger tool or injection grouting is employed when the contaminated zone remains in greater than 6 m depths (ITRC 2011). A conceptual illustration of a typical *in situ* S/S system, as adopted from Palaia (2007), is shown in Fig. 6.3.

For *ex situ* or *in situ* S/S processing, the equipment used for binder mixing is determined based on contaminant characteristics and site conditions, and the proximity of surface water bodies. The overviews of the available equipment have been discussed in detail in several literatures, which includes the works of Al-Tabbaa and Perera (2005b), Bone et al. (2004) and ITRC (2011).

The pros and cons of the *ex situ* or *in situ* process technologies for the field-scale S/S application for the treatment of contaminated soils, as reported by Evans et al. (2001), are given in Table 6.5.

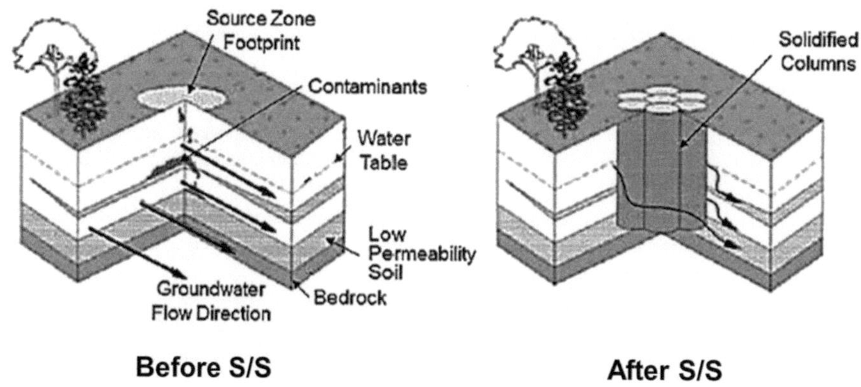

Fig. 6.3 Conceptual diagram of a typical *in situ* S/S system (Palaia 2007)

6.4.4 Monitoring of S/S Mass After Processing

The S/S mass after processing has been suggested to be monitored in a long-term basis, because the concept suggested the immobilization of the contaminants rather than removal (Barth 1990). It has been assumed that the S/S mass might degrade over time due to the individual or combined influence from several of the following factors: (a) Internal chemical reactions; (b) Geochemical and/or biological impacts of the surroundings; and (c) physical mechanisms, e.g., settling, wet-dry cycling or freeze-thaw cycling. Therefore, the regulatory provisions have been designed based on the stability of the S/S-treated materials both after the immediate period of the implementation and longer-term aspects of material durability (ITRC 2011). The case studies and predictive modeling approach have been used to describe the long-term performance, durability and reliability of the S/S application for the remediation of contaminated solids (Bone et al. 2004; Perera et al. 2005a; ITRC 2011).

The long-term durability of the S/S treated mass has been influenced by several physical and environmental factors both from internal and external sources, which are typically site-specific (Perera et al. 2005a). A graphical illustration of the internal and external stresses influencing the performance of S/S-treated materials, as adapted from Garrabrants and Kosson (2005), is shown in Fig. 6.4. The effects of internal and external factors on the long-term stability of the S/S-treated mass, as adapted from the ITRC (2011) is listed in Table 6.6. However, it should be noted that the relative stability of the S/S treated mass are seldom catastrophically hindered due to the minute variances in the internal or external stresses.

The impacts of physical and chemical environment of the surrounding on the S/S treated mass are assessed via measuring the basic material properties, which are often named as index tests, and the performance tests of the S/S treated mass itself. The objectives of the assessment tests include the verification of regulatory compliances, explanation of the reaction mechanisms, on-site evaluation of applied material, etc. (Conner 1990; Stegemann and Cote 1990; LaGrega et al. 2001;

Table 6.5 Advantages and disadvantages of *ex situ* and *in situ* process technologies for the field-scale S/S application in terms of the remediation treatment of contaminated soils[a]

	In situ S/S processing	*Ex situ* S/S processing
Advantages	The costs are typically lower for large and deep remediation projects.	A single plant at a central location can treat material from many sources thus minimizing plant mobilization costs.
	Recently developed *in situ* equipment allows controlled reagent injection and mixing, as well as effective authority on both volatile and particulate emissions.	It provides better control of reagent addition and mixing than *in situ*.
	Secondary spoil generation is little or zero.	Quality control sampling is easier than *in situ*.
	Noise and vibration remain at low levels.	It is suitable for site remediation at shallow depths, i.e., where groundwater or support of adjacent land is not an issue.
	It allows treatment close to structures where excavation could cause damage.	It is tolerant of unstable soil surface or soils with low-bearing capacity.
	There are no requirements for excavation or ground control.	It may be included as an additional component of treatment processes such as soil washing.
Disadvantages	The costs for small remediation projects may be strongly influenced by plant mobilization costs.	The cost may be higher than for *in situ* mixing for large remediation projects at substantial depths.
	The small sites may not accommodate *in situ* mixing equipment and associated batching plants.	The material may have to be excavated and transported to the treatment plant.
	The bearing capacity of the ground must be sufficient to support the *in situ* mixing equipment.	The practical considerations regarding excavation may preclude treatment where depth of contamination is significant.
	The presence of underground services may complicate operations.	Disposal arrangements must be made for cemented waste.
	The physical obstructions and clays, oily sands and cohesive soils may reduce auger penetration rate and depth of operation.	
	The ground may have to be excavated in advance of mixing.	

[a]Adapted from Evans et al. (2001)

Al-Tabbaa and Perera 2006). A list of selected physical tests that are conducted on S/S materials with the corresponding limiting values, as adapted from Stegemann et al. (2001), is given in Table 6.7. The detail of the test methods used for the

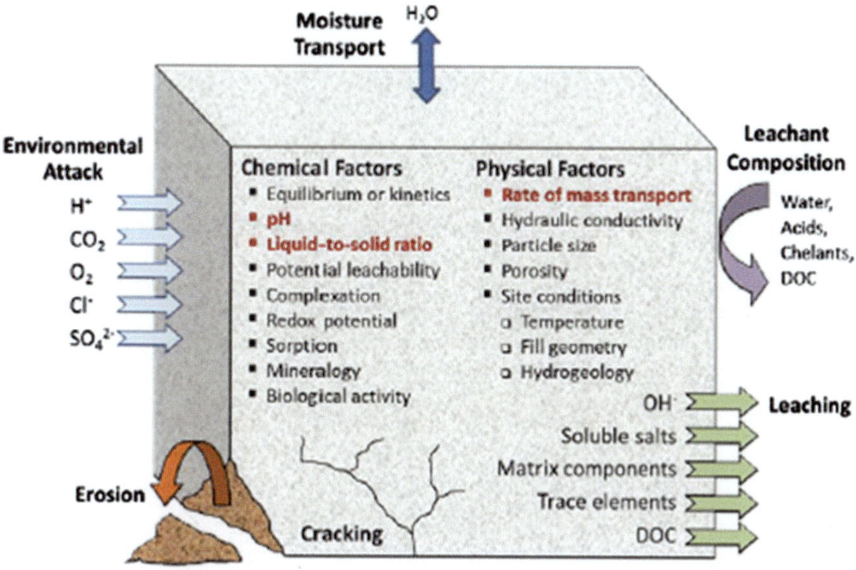

Fig. 6.4 Internal and external stresses influencing the performance of S/S-treated materials (Garrabrants and Kosson 2005)

monitoring of S/S mass after processing is available from Perera et al. (2004), Perera et al. (2005b), Al-Tabbaa and Perera (2006) and ITRC (2011).

6.5 Chemical Stabilization of Toxic Elements in Contaminated Soils

The S/S treatments' physico-chemically stabilize the contaminants in the soil matrices mostly using cement-based binders followed by the supplement of solid covers, barriers or so forth. Although the stabilization process is a common part of the S/S, the waste treatment method that involves the technique of reducing the contaminant mobility only by chemical means is usually called *chemical stabilization* and received a standalone entity (Kumpiene et al. 2008). The added chemical substance, mentioned as 'stabilizing amendments' hereafter, immobilize the metal contaminants in the waste by adsorption, complexation or coprecipitation, and the movements of the stabilized metal-species are sometime further restricted using ambient plant species (Vangronsveld et al. 1995; Gorman et al. 2000; Tordoff et al. 2000; Bleeker et al. 2002; Wong 2003; Adriano et al. 2004; Mench et al. 2006, 2009). However, several limiting factors, such as pH, redox potential, soil constituents, cation exchange capacity etc., can hinder the metal-immobilization process after the application of stabilizing amendments. The

Table 6.6 The impact of internal and external factors on the durability of the S/S-treated contaminated materials[a]

Factor		Impact on performance
Chemical factors	Equilibrium vs. kinetics	Equilibrium-controlled (e.g., steady-state) concentrations are generally higher than kinetic-controlled (e.g., time-based) concentrations such as some hydration and degradation reactions.
	pH	Solubility of inorganic species and organic carbon can be a strong function of pH.
	Liquid-to-solid ratio (L/S)	At low L/S, ionic strength increases, which can increase the solubility of some species.
	Maximum leachability	Fraction of total content that is leachable (i.e., availability) provides driving force for leaching.
	Complexation	Some contaminants form soluble complexes (e.g., $CdCl_2$, Pb-acetate, dissolved organic carbon [DOC]– PAHs), which increase aqueous concentrations and leaching rates.
	Redox potential	S/S mix designs may result in reducing conditions. Oxidation of reduced contaminant speciation, such as Cr(III) → Cr(IV), can increase concentrations, toxicity, and leaching rates.
	Sorption	Surface interactions with mineral phases (e.g., iron, aluminum and manganese oxides; calcium silicates) can reduce pore-water concentrations of some contaminants.
	Biological activity	Acids produced by biological activity can alter pore chemistry and locally degrade minerals.
Physical factors	Particle size	Unit particle size dictates whether material is monolithic or granular. Mean particle size of contaminated material (e.g., finely grained or gravely) may influence selection of reagents.
	Hydraulic conductivity	Water contact mode (e.g., flow through or flow around) is dictated by relative hydraulic conductivity of S/S material and surrounding soil.
	Pore structure	Materials with large connected pores generally have higher hydraulic conductivity, whereas lower hydraulic conductivity may be seen in materials with smaller or disconnected pores.
Site Conditions	Groundwater flow rate	Fast-moving groundwater limits contact time with the surface of S/S materials but may result in sufficient hydraulic head to force groundwater through the material pore structure.
	Fill geometry	Flux-based release of contaminants is proportional to the bulk surface area of the S/S fill.
	Temperature	Higher temperatures increase the rate of chemical reactions (e.g., mineral dissolution).
	Hydrogeological conditions	Determine water contact mode, liquid-to-solid ratios, infiltration rates, active surface area for leaching.
Moisture transport		Leaching from an S/S material is discontinued during drying; gas-phase reactions (e.g., oxidation, carbonation) require a pore vapor space for transport (i.e., partially dried material).
Leachant composition		Acids, chelants and organic carbon may alter solubility of surface/near-surface minerals and contaminants.

(continued)

Table 6.6 (continued)

Factor	Impact on performance
Environmental attack	Several species in the surrounding environment may accelerate leaching or degradation of the mineral structure through pH or redox changes and expansive precipitation reactions.
Leaching	Release of mineral phases increases pore diameters and connectivity, potentially leading to increase in hydraulic conductivity and increased release rates.
Cracking	All S/S materials have cracks on the micro and macro scales; formation of larger-aperture through-cracks may increase hydraulic conductivity but does not equate to catastrophic failure as complete through-cracks simply result in two monoliths of the same performance characteristics.

[a]Adapted from ITRC (2011)

Table 6.7 Physical tests applied to treated S/S materials for monitoring purpose[a]

Physical property	Minimum	Maximum
Bound water (%)	6.8	19.6
Chloride permeability (mg kg^{-1} wet wt)	2540	21,110
Density (bulk) (g cm^{-3})	0.466	2.86
Density (dry) (g cm^{-3})	0.145	1.18
Density (saturated) (g cm^{-3})	1.6	1.97
Flow table spread diameter (cm)	10.5	13.6
Permeability (m s^{-1})	4×10^{-18}	3.66×10^{-6}
Intrinsic permeability (m^2)	2.2×10^{-17}	1.74×10^{-16}
Modulus of elasticity (MPa)	10.2	21,000
Moisture content (% wet wt)	0.263	98
Oxygen permeability (m s^{-1})	4.0×10^{-16}	5.3×10^{-15}
Penetration resistance (Mpa)	16	52.4
Porosity (%)	2	75
Setting time – initial (minutes)	25	2400
Setting time – final (minutes)	65	12,000
Shrinkage/expansion (%)	-9.3×10^{-5}	7
Slump (mm)	180	220
Soundness (cm)	0.09	4.12
Specific gravity	0.905	5.189
Tensile strength (kPa)	3.4	10,270
Unconfined compressive strength (MPa)	0	395
Water absorption at 80 °C (%)	12.5	19.4

[a]Adapted from Stegemann et al. (2001)

selection of stabilizing amendments also requires the pre-characterization of the waste to determine the type and distribution of the contaminant (Kumpiene et al. 2008).

Metal oxides, typically of Fe and, to a letter extent, Al and Mn have been studied as the potential stabilizing amendment due to their extraordinary sorption property (Komárek et al. 2013). The Fe, Mn or Al oxides relevant for chemical stabilization of contaminated soils along with their basic properties, as adopted from the (Komárek et al. 2013), is given in Table 6.8. The metal oxides are generally good sinks of several toxic elements, such as arsenic, copper, lead, etc. (Kabata-Pendias and Pendias 1991; Kumpiene et al. 2008; Komárek et al. 2013). Although the immobilization of metals is generally attributable to the sorption ability of the oxides, the corresponding efficiency is varied with the elements. For example, Mn-oxides are an efficient stabilizing amendment to treat Pb-contaminated soils (Hettiarachchi et al. 2000), while it enhances the toxicity extent of chromium in soils (Guha et al. 2001; Kim and Dixon 2002).

The reaction between a chelating agent such as dithiocarbamate compound and metals, which produces a water-insoluble precipitate of metal–dithiocarbamate complexes, have also been employed for the chemical stabilization of toxic components in various metal-contaminated solid wastes, including soil (Osborne-Lee et al. 1999; Sukandar et al. 2009).

Table 6.8 The Fe, Mn or Al oxides relevant for chemical stabilization of contaminated soils and their basic properties[a]

Oxide	Formula	Specific surface ($m^2\ g^{-1}$)	pH_{zpc}	Solubility product log K_{sp}
Iron oxides				
Ferrihydrite	$Fe_5HO_8 \cdot 4H_2O$ (simplified as $Fe(OH)_3$)	100–700	7.8–8.8	−37 to −39
Goethite	α-FeOOH	8–200	7.5–9.4	−44
Lepidocrocite	γ-FeOOH	15–260	6.7–7.5	−38.7 to −40.6
Haematite	α-Fe_2O_3	2–115	7.5–9.5	−43.9 ± 0.2
Manganese oxides				
Birnessite	$(Na,Ca,K)_xMn_2O_4 \cdot 1.5H_2O$ (simplified as δ-MnO_2)	35.4	1.8–2.2	−85.5463
Manganite	γ-MnOOH	8.9	5.4	−0.1646
Pyrolusite	β-MnO_2	0.15	7.2	−17.6439
Aluminum oxides				
Gibbsite	γ-$Al(OH)_3$	120	9.8	7.7560
Boehmite	γ-AlOOH	224	8.6	7.5642
Diaspore	α-AlOOH	11	2.0–7.5	7.1603

[a]Adapted from Komárek et al. (2013)

The role of organic matters (e.g., bio solids, compost) and alkaline materials (e.g., fly ash, hydroxyapatite, $CaCO_3$) have also been evaluated as the stabilizing amendments to treat metal contaminated soils (Pantsar-Kallio et al. 2001; Seaman et al. 2001; Brown et al. 2004, 2005; Cao and Ma 2004; Sánchez-Monedero et al. 2004; Su and Wong 2004; Raicevic et al. 2005). Some inexpensive amendments, e.g., limestone, red-mud, and furnace slag, have also been studied to reduce contaminant availability in polluted soils (Lee et al. 2009, 2011a).

The inevitability for a long-term assessment of the immobilized elements using stabilizing amendments has been suggested using various approaches, such as chemical extractions, leaching tests, TCLP (Toxicity Characteristics Leaching Procedure), SPLP (Simulated Precipitation Leaching Procedure), eco-toxicity evaluation, biodiversity monitoring, etc. (Geebelen et al. 2003; Lombi et al. 2003; Sukandar et al. 2009; Lee et al. 2011b; Komárek et al. 2013).

6.6 Conclusion

The solidification/stabilization (S/S) technique is a cost-effective solution for the treatment of metal-contaminated soils, which immobilize the hazardous components in wastes rather than separating it from the wastes, and also satisfy the legislative guidelines for landfills. This paper presented a summary of the components involved in the execution of S/S for the treatment of metal-contaminated soils. A number of binders are available both from the inorganic and organic origin, and the S/S processing can be carried out either *ex situ* or *in situ*. The selection of the binder and S/S processing technique is decided based on the site characteristics, nature of the contaminants in the waste, after-processing purpose, regulatory limitations and economics of operation. The stabilization of toxic contaminants using chemical amendments only has also been discussed. The stability of the S/S treated material in a greater time-span, particularly the leaching of contaminants from the immobilized mass, is the major issue of concern regarding the application of S/S. Hence, the implementation of S/S technique for metal-contaminated soil remediation also requires a long-term monitoring plan along with.

Acknowledgement The research has partially been supported by the Grants-in-Aid for Scientific Research (15H05118 and 25 · 5863) from the Japan Society for the Promotion of Science. One of the authors, Ismail M.M. Rahman, acknowledges the financial grant from the 'The Public Foundation of Chubu Science and Technology Center, Japan' to support his research in the Kanazawa University, Japan.

References

Adriano DC, Wenzel WW, Vangronsveld J, Bolan NS (2004) Role of assisted natural remediation in environmental cleanup. Geoderma 122:121–142

AEPI (1998) Solidification technologies for restoration of sites contaminated with hazardous wastes. Army Environmental Policy Institute, Atlanta

Ahmad M, Hashimoto Y, Moon DH, Lee SS, Ok YS (2012) Immobilization of lead in a Korean military shooting range soil using eggshell waste: an integrated mechanistic approach. J Hazard Mater 209–210:392–401

Ajmal M, Hussain Khan A, Ahmad S, Ahmad A (1998) Role of sawdust in the removal of copper (II) from industrial wastes. Water Res 32:3085–3091

Allan ML, Kukacka LE (1995) Blast furnace slag-modified grouts for in situ stabilization of chromium-contaminated soil. Waste Manag 15:193–202

Alloway BJ (1995) Heavy metals in soils. Blackie, London

Al-Tabbaa A, Evans CW (1999) Laboratory-scale soil mixing of a contaminated site. Ground Improv 3:119–134

Al-Tabbaa A, Perera ASR (2005a) UK stabilisation/solidification treatment and remediation, Part II: Binders & technologies – research. In: Tabbaa AA, Stegemann JA (eds) Stabilisation/solidification treatment and remediation: advances in S/S for waste and contaminated land. Balkema, London, pp 387–398

Al-Tabbaa A, Perera ASR (2005b) UK stabilisation/solidification treatment and remediation, Part III: binders & technologies – applications. In: Tabbaa AA, Stegemann JA (eds) Stabilisation/solidification treatment and remediation: advances in S/S for waste and contaminated land. Balkema, London, pp 399–414

Al-Tabbaa A, Perera ASR (2005c) UK stabilisation/solidification treatment and remediation, Part I: binders & technologies – basic principles. In: Tabbaa AA, Stegemann JA (eds) Stabilisation/solidification treatment and remediation: advances in S/S for waste and contaminated land. Balkema, London, pp 365–386

Al-Tabbaa A, Perera ASR (2006) UK stabilization/solidification treatment and remediation – Part I: binders, technologies, testing and research. Land Contam Reclam 14:703

Andini S, Cioffi R, Montagnaro F, Pisciotta F, Santoro L (2006) Simultaneous adsorption of chlorophenol and heavy metal ions on organophilic bentonite. Appl Clay Sci 31:126–133

ASTM (2001) Standard test method for determining the resistance of solid wastes to freezing and thawing; D4842-90(2001). ASTM International, West Conshohocken. doi:10.1520/D4842-90R01

ASTM (2009) Standard test method for wetting and drying test of solid wastes; D4843-88(2009). ASTM International, West Conshohocken. doi:10.1520/D4843-88R09

Barth EF (1990) An overview of the history, present status, and future direction of solidification/stabilization technologies for hazardous waste treatment. J Hazard Mater 24:103–109

Basta NT, Ryan JA, Chaney RL (2005) Trace element chemistry in residual-treated soil: key concepts and metal bioavailability. J Environ Qual 34:49–63

Begum ZA, Rahman IMM, Tate Y, Sawai H, Maki T, Hasegawa H (2012) Remediation of toxic metal contaminated soil by washing with biodegradable aminopolycarboxylate chelants. Chemosphere 87:1161–1170

Begum ZA, Rahman IMM, Sawai H, Mizutani S, Maki T, Hasegawa H (2013) Effect of extraction variables on the biodegradable chelant-assisted removal of toxic metals from artificially contaminated European reference soils. Water Air Soil Pollut 224:1381

Bleeker PM, Assunção AGL, Teiga PM, de Koe T, Verkleij JAC (2002) Revegetation of the acidic, As contaminated Jales mine spoil tips using a combination of spoil amendments and tolerant grasses. Sci Total Environ 300:1–13

Bone BD, Barnard LH, Boardman DI, Carey PJ, Hills CD, Jones HM, MacLeod CL, Tyrer M (2004) Review of scientific literature on the use of stabilisation/solidification for the treatment of contaminated soil, solid waste and sludges (SC980003/SR2). The Environment Agency, Bristol

Brown S, Chaney R, Hallfrisch J, Ryan JA, Berti WR (2004) In situ soil treatments to reduce the phyto- and bioavailability of lead, zinc, and cadmium. J Environ Qual 33:522–531

Brown S, Christensen B, Lombi E, McLaughlin M, McGrath S, Colpaert J, Vangronsveld J (2005) An inter-laboratory study to test the ability of amendments to reduce the availability of Cd, Pb, and Zn in situ. Environ Pollut 138:34–45

Cao X, Ma LQ (2004) Effects of compost and phosphate on plant arsenic accumulation from soils near pressure-treated wood. Environ Pollut 132:435–442

Chen QY, Tyrer M, Hills CD, Yang XM, Carey P (2009) Immobilisation of heavy metal in cement-based solidification/stabilisation: a review. Waste Manag 29:390–403

Chlopecka A, Bacon JR, Wilson MJ, Kay J (1996) Forms of cadmium, lead, and zinc in contaminated soils from Southwest Poland. J Environ Qual 25:69–79

Conner JR (1990) Chemical fixation and solidification of hazardous wastes. Van Nostrand Reinhold, New York

Conner JR (1997) Guide to improving the effectiveness of cement-based stabilization/solidification. Portland Cement Association, Skokie

D'Amore JJ, Al-Abed SR, Scheckel KG, Ryan JA (2005) Methods for speciation of metals in soils. J Environ Qual 34:1707–1745

Dhir RK (1986) Pulverised fuel ash. In: Swamy RN (ed) Cement replacement materials. Surrey University Press, Bishopbriggs, pp 197–255

Evanko CR, Dzombak DA (1997) Remediation of metals-contaminated soils and groundwater (TE-97-01). Ground-Water Remediation Technologies Analysis Center (GWRTAC), Pittsburgh

Evans D, Jefferis SA, Thomas AO, Cui S (2001) Remedial processes for contaminated land – Principles and practice, CIRIA Report C549. Construction Industry Research Information Association (CIRIA), London

Flathman PE, Lanza GR (1998) Phytoremediation: current views on an emerging green technology. J Soil Contam 7:415–432

Garrabrants AC, Kosson DS (2005) Leaching processes and evaluation tests for inorganic constituent release from cement-based matrices. In: Spence RD, Shi C (eds) Stabilization and solidification of hazardous, radioactive, and mixed wastes. CRC Press, Boca Raton, pp 229–279

Geebelen W, Adriano DC, van der Lelie D, Mench M, Carleer R, Clijsters H, Vangronsveld J (2003) Selected bioavailability assays to test the efficacy of amendment-induced immobilization of lead in soils. Plant Soil 249:217–228

Glasser FP (1997) Fundamental aspects of cement solidification and stabilisation. J Hazard Mater 52:151–170

Gorman JM, Sencindiver JC, Horvath DJ, Singh RN, Keefer RF (2000) Erodibility of fly ash used as a topsoil substitute in mineland reclamation. J Environ Qual 29:805–811

Guha H, Saiers JE, Brooks S, Jardine P, Jayachandran K (2001) Chromium transport, oxidation, and adsorption in manganese-coated sand. J Contam Hydrol 49:311–334

Harris MR, Herbert SM, Smith MA (1995) Remedial treatment for contaminated land, vol VII, Ex-situ remedial methods for soils, sludges and sediments (SP 107). Construction Industry Research and Information Association (CIRIA), London

Hettiarachchi GM, Pierzynski GM, Ransom MD (2000) In situ stabilization of soil lead using phosphorus and manganese oxide. Environ Sci Technol 34:4614–4619

ITRC (2011) Development of performance specifications for solidification/stabilization. Interstate Technology & Regulatory Council, Washington, DC

Kabata-Pendias A, Pendias H (1991) Trace elements in soils and plants, vol V, 2nd edn. CRC Press, Boca Raton

Kershaw DS, Pamukcu S (1997) Ground rubber: reactive permeable barrier sorption media. In: Conference on in situ remediation of the geoenvironment (Geotechnical special publication no. 71), American Society of Civil Engineers, pp 26–40

Khan S, Cao Q, Zheng YM, Huang YZ, Zhu YG (2008) Health risks of heavy metals in contaminated soils and food crops irrigated with wastewater in Beijing, China. Environ Pollut 152:686–692

Kim JG, Dixon JB (2002) Oxidation and fate of chromium in soils. Soil Sci Plant Nutr 48:483–490

Komárek M, Vaněk A, Ettler V (2013) Chemical stabilization of metals and arsenic in contaminated soils using oxides – A review. Environ Pollut 172:9–22

Kumpiene J, Lagerkvist A, Maurice C (2008) Stabilization of As, Cr, Cu, Pb and Zn in soil using amendments – a review. Waste Manag 28:215–225

LaGrega MD, Buckingham PL, Evans JC (2001) Hazardous waste management. McGraw-Hill, Boston

Lee S-H, Lee J-S, Jeong Choi Y, Kim J-G (2009) In situ stabilization of cadmium-, lead-, and zinc-contaminated soil using various amendments. Chemosphere 77:1069–1075

Lee S-H, Kim EY, Park H, Yun J, Kim J-G (2011a) In situ stabilization of arsenic and metal-contaminated agricultural soil using industrial by-products. Geoderma 161:1–7

Lee S-H, Park H, Koo N, Hyun S, Hwang A (2011b) Evaluation of the effectiveness of various amendments on trace metals stabilization by chemical and biological methods. J Hazard Mater 188:44–51

Li MS (2006) Ecological restoration of mineland with particular reference to the metalliferous mine wasteland in China: a review of research and practice. Sci Total Environ 357:38–53

Little D, Herbert B, Kunagalli S (2005) Ettringite formation in lime-treated soils: establishing thermodynamic foundations for engineering practice. Transp Res Record 1936:51–59

Lombi E, Gerzabek MH (1998) Determination of mobile heavy metal fraction in soil: results of a pot experiment with sewage sludge. Commun Soil Sci Plant Anal 29:2545–2556

Lombi E, Hamon RE, McGrath SP, McLaughlin MJ (2003) Lability of Cd, Cu, and Zn in polluted soils treated with lime, beringite, and red mud and identification of a non-labile colloidal fraction of metals using isotopic techniques. Environ Sci Technol 37:979–984

Matthes W, Madsen FT, Kahr G (1999) Sorption of heavy-metal cations by Al and Zr-hydroxy-intercalated and pillared bentonite. Clays Clay Miner 47:617–629

Mench M, Vangronsveld J, Beckx C, Ruttens A (2006) Progress in assisted natural remediation of an arsenic contaminated agricultural soil. Environ Pollut 144:51–61

Mench M, Schwitzguébel J-P, Schroeder P, Bert V, Gawronski S, Gupta S (2009) Assessment of successful experiments and limitations of phytotechnologies: contaminant uptake, detoxification and sequestration, and consequences for food safety. Environ Sci Poll Res 16:876–900

Neville AM, Brooks JJ (2010) Concrete technology. Prentice Hall, Harlow

Osborne-Lee IW, Conley TB, Huktt GA, Morris MI (1999) Demonstration results on the effects of mercury speciation on the stabilization of wastes (ORNL/TM-1999/120). United States Department of Energy, Oak Ridge

Palaia T (2007) Resurgence of in situ soil mixing for treating NAPL source areas. Paper presented at the Air force ESOH training symposium, Pittsburgh, March 18–23

Pantsar-Kallio M, Reinikainen S-P, Oksanen M (2001) Interactions of soil components and their effects on speciation of chromium in soils. Anal Chim Acta 439:9–17

Paria S, Yuet PK (2006) Solidification–stabilization of organic and inorganic contaminants using portland cement: a literature review. Environ Rev 14:217–255

Perera ASR, Al-Tabbaa A, Reid MJ, Stegemann JA, Shi C (2004) Testing and performance criteria for stabilised/solidified waste forms. In: Spence RD, Shi C (eds) Stabilization and solidification of hazardous, radioactive and mixed wastes. CRC Press, Boca Raton, pp 281–317

Perera ASR, Al-Tabbaa A, Reid JM, Johnson D (2005a) UK stabilisation/solidification treatment and remediation, Part V: long-term performance and environmental impact. In: Tabbaa AA, Stegemann JA (eds) Stabilisation/solidification treatment and remediation: advances in S/S for waste and contaminated land. Balkema, London, pp 437–458

Perera ASR, Al-Tabbaa A, Johnson D (2005b) UK stabilisation/solidification treatment and remediation, Part VI: quality assurance and quality control. In: Tabbaa AA, Stegemann JA (eds) Stabilisation/solidification treatment and remediation: advances in S/S for waste and contaminated land. Balkema, London, pp 459–468

Pierzynski GM, Sims JT, Vance GF (2005) Soils and environmental quality. Taylor & Francis, Boca Raton

Rahman IMM, Nazimuddin M, Hasan MT, Hossain MM (2008) Assimilation of arsenic into edible plants grown in soil irrigated with contaminated groundwater. In: Bundschuh J, Armienta MA, Birkle P, Bhattacharya P, Matschullat J, Mukherjee AB (eds) Natural arsenic in groundwaters of Latin America, vol 1, Arsenic in the environment. CRC Press/Balkema, Leiden, pp 351–358

Rahman IMM, Hossain MM, Begum ZA, Rahman MA, Hasegawa H (2011) Eco-environmental consequences associated with chelant-assisted phytoremediation of metal-contaminated soil. In: Golubev IA (ed) Handbook of phytoremediation. Nova, Hauppauge, pp 709–722

Rahman IMM, Begum ZA, Salehi-Lisar SY, Motafakkerazad R, Awual MR, Hasegawa H (2014) Distribution and abundance of arsenic in the soils and plants. In: Olson MJ (ed) Arsenic: detection, management strategies and health effects. Nova Science Publishers, Hauppauge, pp 117–129

Raicevic S, Kaludjerovic-Radoicic T, Zouboulis AI (2005) In situ stabilization of toxic metals in polluted soils using phosphates: theoretical prediction and experimental verification. J Hazard Mater 117:41–53

Roskam GD, Comans RNJ (2009) Availability and leaching of polycyclic aromatic hydrocarbons: controlling processes and comparison of testing methods. Waste Manag 29:136–142

Samsøe-Petersen L, Larsen EH, Larsen PB, Bruun P (2002) Uptake of trace elements and PAHs by fruit and vegetables from contaminated soils. Environ Sci Technol 36:3057–3063

Sánchez-Monedero MA, Mondini C, de Nobili M, Leita L, Roig A (2004) Land application of biosolids. Soil response to different stabilization degree of the treated organic matter. Waste Manag 24:325–332

Seaman JC, Arey JS, Bertsch PM (2001) Immobilization of nickel and other metals in contaminated sediments by hydroxyapatite addition. J Environ Qual 30:460–469

Sherwood PT (1993) Soil stabilization with cement and lime. HMSO, London

Sims JL, Sims RC, Matthews JE (1989) Bioremediaiton of contaminated surface soils (EPA 600/9-89/073). U.S. Environmental Protection Agency, Ada

Smith LA, Means JL, Chen A, Alleman B, Chapman CC, Tixier JS Jr, Brauning SE, Gavaskar AR, Royer MD (1995) Remedial options for metals-contaminated sites. CRC Press, Boca Raton

Spooner PA, Hunt GE, Hodge VE, Wagner PM (1984) Compatibility of grouts with hazardous wastes (EPA-600/2-84-015). United States Environmental Protection Agency, Cincinnati

Sposito G, Page AL (1984) Cycling of metal ions in the soil environment. In: Sigel H, Sigel A (eds) Metal Ions in biological systems, vol 18, Circulation of metals in the environment. Marcel Dekker, New York, pp 287–332

Stegemann JA, Cote PL (1990) Summary of an investigation of test methods for solidified waste evaluation. Waste Manag 10:41–52

Stegemann JA, Butcher EJ, Irabien A, Johnston P, de Miguel R, Ouki SK, Polettini A, Sassaroli G (2001) Neural network analysis for prediction of interactions in cement/waste systems. Commission of the European Community, Brussels

Su DC, Wong JWC (2004) Chemical speciation and phytoavailability of Zn, Cu, Ni and Cd in soil amended with fly ash-stabilized sewage sludge. Environ Int 29:895–900

Sukandar PT, Tanaka M, Aoyama I (2009) Chemical stabilization of medical waste fly ash using chelating agent and phosphates: heavy metals and ecotoxicity evaluation. Waste Manag 29:2065–2070

Taylor HFW (1992) Cement chemistry. Academic Press, London

Tordoff GM, Baker AJM, Willis AJ (2000) Current approaches to the revegetation and reclamation of metalliferous mine wastes. Chemosphere 41:219–228

Trussell S, Spence RD (1994) A review of solidification/stabilization interferences. Waste Manag 14:507–519

US EPA (1982) Guide to the disposal of chemically stabilized and solidified waste (SW-872). United States Environmental Protection Agency, Cincinnati

US EPA (1986a) Prohibition on the placement of bulk liquid hazardous waste in landfills-statutory interpretive guidance (OSWER policy directive no. 9487.00-2A; EPA 530-SW-86-016). United States Environmental Protection Agency, Washington, DC

US EPA (1986b) Handbook for stabilization/solidification of hazardous waste (EPA 540/2-861/001). United States Environmental Protection Agency, Cincinnati
US EPA (1989) Stabilization/solidification of CERCLA and RCRA waste (EPA 625/6-89/022). United States Environmental Protection Agency, Cincinnati
US EPA (1993) Clean water act (Sec. 503, Vol. 58, No. 32). United States Environmental Protection Agency, Washington, DC
US EPA (1997) Innovative site remediation design and application, vol 4, Stabilization/solidification (EPA 542-B-97-007). United States Environmental Protection Agency, Washington, DC
US EPA (1999) Solidification/stabilization resource guide (EPA 542-B-99-002). United States Environmental Protection Agency, Washington, DC
US EPA (2000) Solidification/stabilization use at superfund sites (EPA 542-R-00-010). United States Environmental Protection Agency, Washington, DC
US EPA (2004) Cleaning up the nation's waste sites: markets and technology trends, 4th edn, EPA 542-R-04-015. United States Environmental Protection Agency, Washington, DC
US EPA (2006) In situ treatment technologies for contaminated soil (EPA 542/F-06/013). United States Environmental Protection Agency, Washington, DC
US EPA (2007) Toxicity characteristic leaching procedure (Method 1311). Test methods for evaluating solid waste, physical/chemical methods (SW-846). United States Environmental Protection Agency, Washington, DC
US EPA (2012) The clean water act (Priority pollutants: Appendix A to 40 CFR Part 423). United States Environmental Protection Agency, Washington, DC
Vangronsveld J, Van Assche F, Clijsters H (1995) Reclamation of a bare industrial area contaminated by non-ferrous metals: in situ metal immobilization and revegetation. Environ Pollut 87:51–59
Vodyanitskii YN (2013) Contamination of soils with heavy metals and metalloids and its ecological hazard (analytic review). Eurasian Soil Sci 46:793–801
Wiles CC (1987) A review of solidification/stabilization technology. J Hazard Mater 14:5–21
Wiles CC, Barth E, de Percin P (1988) Status of solidification/stabilization in the United States and factors affecting its use. In: Wolf K, Van Den Brink WJ, Colon FJ (eds) Contaminated soil '88. Springer, Dordrecht, pp 947–956
Wong MH (2003) Ecological restoration of mine degraded soils, with emphasis on metal contaminated soils. Chemosphere 50:775–780
Zhao Q, Kaluarachchi JJ (2002) Risk assessment at hazardous waste-contaminated sites with variability of population characteristics. Environ Int 28:41–53

Chapter 7
Immobilization of Fluoride and Heavy-Metals in Polluted Soil

Masamoto Tafu and Atsushi Manaka

Abstract Estimation and immobilization of fluoride and heavy-metals in contaminated soil are important approaches for the remediation of fluoride and heavy-metal polluted soil. In this review, firstly, we described recent achievements about the on-site estimation of various pollutants by simple chemical reaction without special skills for operators. The in-situ immobilization of fluoride was also described. For immobilization of fluoride, dicalcium phosphate dihydrate (DCPD) was selected as functional material because it reacts with fluoride ion effectively and forms stable fluorapatite (FAp). Both laboratory and field tests showed that the DCPD is useful to immobilize fluoride in polluted soil.

Keywords Contaminated soil • Fluoride • On-site immobilization • Dicalcium phosphate dihydrate

7.1 Introduction

Immobilization of the heavy-metals is one of the effective methods used for the remediation of polluted soil, which is applied to avoid the dispersion of a pollutant to the surrounding soil and/or groundwater. In case of high concentration of heavy-metals in the soil, removal or washing treatment of the soil is required to be applied. When content of the heavy-metals is lower than the environmental standard, immobilization is the suitable and cost-effective approach to inhibit diffusion of the pollutants to the environment effectively.

We have investigated novel functional material to immobilize pollutants, especially fluoride and some heavy-metals. There has been much researches on immobilization of various heavy-metals in polluted soil (Guo et al. 2006); however, fluoride immobilization is limited to few applications, such as using magnesium salt (Booster et al. 2003; Maliyekkal et al. 2010), layered materials, and calcium phosphates (Suzuki et al. 1981). In Japanese law, the fluoride amount in the

M. Tafu (✉) • A. Manaka
Department of Applied Chemistry and Chemical Engineering, National Institute of Technology, Toyama College, 13 Hongo-machi, Toyama 939-8630, Japan
e-mail: tafu@toyama-nct.ac.jp

H. Hasegawa et al. (eds.), *Environmental Remediation Technologies for Metal-Contaminated Soils*, DOI 10.1007/978-4-431-55759-3_7

contaminated soil is regulated to 2000 mg kg^{-1}, and leachate amount in eluent form the contaminated soil is regulated to 0.8 mg L^{-1} (Japan MOE 2002). In this review, we focused on the use of calcium phosphate for the immobilization of fluoride in polluted soil. Our laboratory and field tests showed that calcium phosphate is promising for the immobilization of fluoride in polluted soil.

Estimation of heavy-metals eluted from polluted soil is also important to understand the environmental impact of the heavy-metals from the contaminated soil. We have introduced a novel and simple method for the estimation of pollutants eluted from soil that can be suitable for the people not having specialized skills.

In this review, first we summarized the recent achievements on simple estimation methods for heavy-metals in soil, and then, we introduced a newly developed functional material that can immobilize fluoride and heavy-metals in soil.

7.2 Assessment of Heavy-Metals in Polluted Soil

Dissolution test has been used to evaluate performance of immobilization technique for heavy-metals in soil. However, it is difficult to analyze target substance directly in eluted solution, because compared to co-existing substances the amount of target substance in the eluted solution is little. Therefore, complicated pre-treatment procedures such as solvent extraction and distillation are required. Based on the results of our laboratory and field tests, we introduced a simple and effective pre-treatment technique to analyze target metals directly in the eluted solution to evaluate performance of immobilization technique.

Pure water is used as an extraction solution in conventional dissolution technique, and a little amount of the heavy-metals was found to be dissolved in the pure water, and hence this technique lacks sensitivity (Hayakawa 1990). Washing with chelate solution is used for remediation of toxic metal contaminated soil (Peters 1999). Recently, as a lower environmental impact method, washing method with the biodegradable chelating agents has been reported (Begum et al. 2012). The cleaning method using the chelating agent, not only for cleaning of the soil, it is an effective technology to collect rare metals from the soil. Therefore, by utilizing the dissolution by chelating agent addition, effective separation for heavy-metal ion analysis in the soil could be possible compared with using pure water.

Another high sensitive approach is the analysis of the total amount of target metals in soil using hot plate, alkali fusion, and microwave-assisted acid-decomposition technique (US EPA 1996). Although this method can improve sensitivity and efficiency of heavy-metal extraction, there are concerns about the selectivity of the heavy-metals. In addition, as the decomposition is done by heating soil samples at the very high temperatures with strong acids, there are concerns about the loss of certain heavy-metals through evaporation and change of chemical forms of the heavy-metals during heating.

Solid extraction technique has been proposed as an alternative to acid-decomposition and conventional dissolution techniques to improve sensitivity and selectivity. The principle of solid extraction is that the target substance adsorbed by solid phase is eluted by passing sample solution through the solvent (Camel 2003). This method can separate and concentrate the target substances easily by passing through the solvent. Therefore, this method has been widely used as pre-treatment of several instrumental analyses. Various functional materials, such as the ion-exchange resin (Hasegawa et al. 2003), chelate resin (Lee et al. 2002), silica monolith made by sol-gel method (Seneviratne and Cox 2000), polytetrafluoroethylene (PTFE) fiber (Abe et al. 2006), a molecule recognition agent from highly selective resin composed of macrocyclic compound, such as crown ether (Rahman et al. 2013), have been developed. In the case of the environmental water analysis, solid phase extraction method has been adopted for some substance by Ministry of the Environment, Japan. Moreover, the effectiveness and performance of solid phase extraction for heavy-metal analysis in soil have been revealed in literatures (Furusho et al. 2008). For example, extraction efficiency solid phase technique was almost 100 % for Pb and Cd in a soil sample (Furusho et al. 2008). The X-ray fluorescence analysis is also a useful technique for the direct measurement of heavy-metals in soil samples (Stosnach 2006; Shibata et al. 2008). The amount of immobilized heavy-metals in the soil sample after an elution test can be measured directly by this method. However, this method is limited only to soil samples having high concentrations of heavy-metals due to the lack of sensitivity. Therefore, an effective solution to this limitation is to combine eluted solution analysis with solid extraction and direct soil analysis by X-ray fluorescence analysis.

7.3 Development of Functional Materials for Immobilization of Fluoride and Heavy-Metals in Polluted Soil

7.3.1 Immobilization of Heavy-Metals by Functional Materials

Various functional materials have been investigated for immobilization of fluoride and heavy-metals in polluted soil (Guo et al. 2006). Table 7.1 summarized the functional materials used for immobilization of fluoride and heavy-metals in soils. The functional materials are commercialized for immobilizing fluoride in waste water and/or polluted soil and are classified as magnesium compounds, cerium compound, and calcium phosphate. Cerium and other lanthanide compounds make complex with fluoride ion, and therefore, they are identified as promising to use them for immobilizing fluoride in the environments (Tokunaga et al. 1995; Wasay et al. 1996; Wu et al. 2007). Magnesium oxide and/or hydroxide make stable compounds with fluoride ion. However, the usage of magnesium salts results in increased soil alkalinity by changing soil pH, and therefore, magnesium compounds are unsuitable for soil remediation (Booster et al. 2003; Maliyekkal et al. 2010). In

Table 7.1 Functional materials for immobilizing heavy-metals

Material	Heavy-metals	References
Cerium compound	F, As, etc.	Tokunaga et al. (1995), Wasay et al. (1996), and Wu et al. (2007)
Hydroxyapatite	F, Cd, Pb, etc.	Spinelli et al. (1971), Rölla and Bowen (1978), Suzuki et al. (1981), and Reichert and Binner (1996)
Magnesium hydroxide	Various metals	Booster et al. (2003) and Maliyekkal et al. (2010)

Table 7.1, hydroxyapatite (HA, $Ca_{10}(PO_4)_6(OH)_2$) is used as adsorbent and/or seed crystal for precipitation method (Spinelli et al. 1971; Rölla and Bowen 1978; Reichert and Binner 1996).

7.3.2 *Immobilization of Heavy-Metals by Hydroxyapatite*

Hydroxyapatite is a stable calcium phosphate. Hydroxyapatite and its analogue fluorapatite (FAp, $Ca_{10}(PO_4)_6F_2$) are widely distributed in animal bone and teeth, natural phosphate rocks (Corbridge 2013). Calcium and hydroxyl ions in the apatite are easily exchangeable with other ions in their crystal lattice. For example, calcium ion in HA can be exchanged with various divalent ions, such as cadmium, lead, and copper (Stosnach 2006; Shibata et al. 2008). Hydroxide ion in HA crystal can be exchanged with fluoride ion (Rölla and Bowen 1978; Reichert and Binner 1996). By using these properties, HA and FAp have been widely investigated for precipitation and immobilization of fluoride and heavy-metals (Suzuki et al. 1981). However, as HA or FAp is stable on supersaturated condition in an aqueous solution, an excess amount of calcium and phosphate chemicals is required to precipitate HA or FAp (Eanes and Meyer 1978). At the supersaturated state, the HA or FAp precipitate particles are very small and difficult to separate and handle. To overcome this problem, we focused on the transformation reaction of the calcium phosphates. The transform reaction of the calcium phosphates was found in reaction of dental caries prevention (Chow and Brown 1973). Main inorganic phase in dental enamel is HA, which is transformed to dicalcium phosphate dihydrate (DCPD, $CaHPO_4 \cdot 2H_2O$) in acidic condition and to stable FAp by reacting with fluoride ion (Duff 1970). Therefore, DCPD is likely to be applicable for the immobilization of fluoride ion in aqueous solution. In the presence of another heavy-metal ion (such as lead, cadmium) in the solution, FAp is expected to contain fluoride and heavy-metal ions. To confirm this hypothesis, we investigated reaction of DCPD with fluoride and heavy-metal ions.

Reagent grade DCPD was mixed with fluoride solution. Fluoride concentration was adjusted to 20 mg/L. Figure 7.1 shows change of fluoride ion concentration in aqueous solution after adding DCPD. After 3–4 h of the addition of DCPD in fluoride solution, concentration of fluoride did not change. After a few hours of the

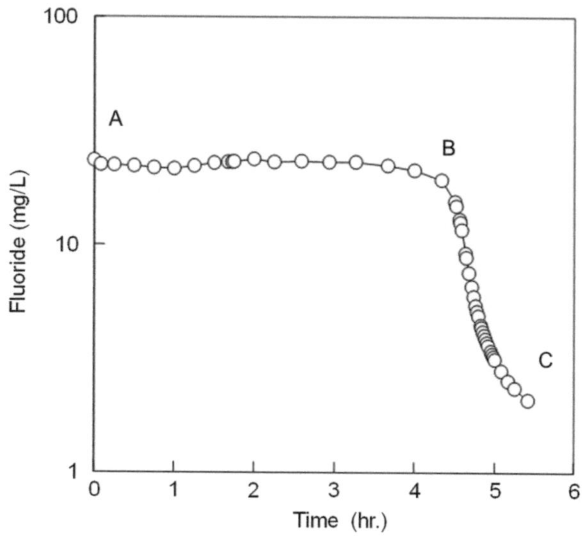

Fig. 7.1 Fluoride concentration after the addition of DCPD to the fluoride solution (Concentrations of fluoride ion and DCPD were 20 mg/L and 1 g/L, respectively)

"lag-time", fluorides were quickly removed by reaction with DCPD, and formed FAp. From the result, the lag-time is retention time required for reaction with DCPD (Tafu and Chohji 2006). In case of immobilizing fluoride in contaminated soil by using the DCPD, the lag-time is not a problem because mixture of the DCPD and the polluted soil is kept for long time. In previous studies, it was founded that fluorides in rain water (Tafu et al. 2001), ground water (Tafu and Chohji 2005a), and gypsum (Tafu and Chohji 2006) were immobilized by using DCPD. In this case, an excess amount of DCPD is not needed.

7.3.3 Improving the Reactivity of DCPD with Fluoride

To improve the reactivity of DCPD with fluoride, we investigated the reaction mechanism of DCPD with fluoride ion. Figure 7.2 shows scanning electron microscope (SEM) images of surface of DCPD particle. The figure demonstrates that nano-scale precursor particle is formed on the DCPD particle before it reacted with fluoride ion (Tafu and Chohji 2005b). The DCPD-induced nano-precursor quickly reacts with fluoride ion as shown in Fig. 7.3. The nano-scaled precursor particle could be induced by mixing the DCPD with water (Tafu et al. 2010; Masamoto and Tetsuji 2011). In a wide pH range, HA is more stable than DCPD. Therefore, the precursor particle was seemed to consist of HA or HA-like calcium phosphate phase. As shown in Fig. 7.4, fluoride ion was quickly immobilized by the mixture of HA and DCPD (Tafu et al. 2013). From this result, the existence of HA is likely to play an important role in initiating a reaction between DCPD and fluoride ion.

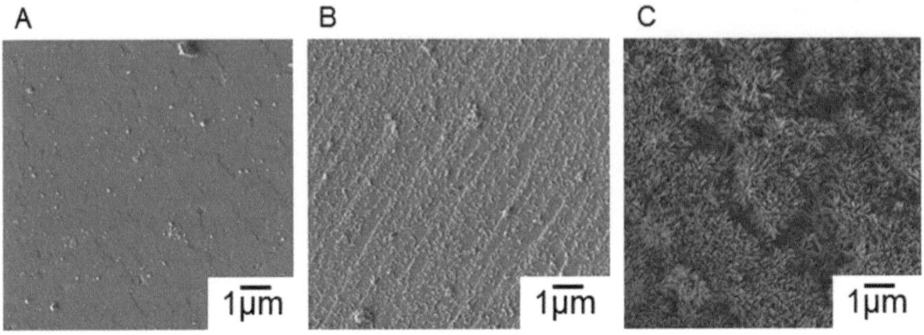

Fig. 7.2 Scanning electronic microscope (SEM) micrographs of surface of DCPD particle sampled at point A, B, and C indicated in Fig. 7.1

Fig. 7.3 Effect of induction of nano-precursor on surface of DCPD particle

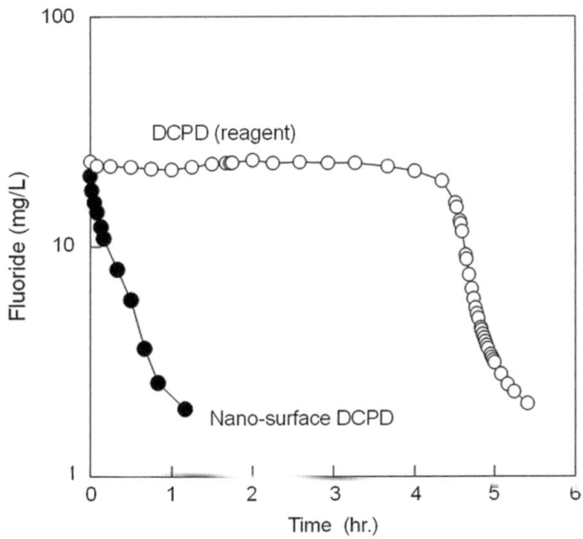

Reactivity of DCPD with other heavy-metal ions has also been studied. By mixing pure DCPD with an aqueous solution of fluoride and lead ions, fluoride and lead were immobilized into FAp. In the presence of cadmium ion in the system, formation of reactive layer on the surface of DCPD particle inhibited the reactivity of DCPD. Therefore, induction of precursor particle on DCPD particle is useful to avoid the formation reactive layer (Tafu et al. 2009). Table 7.2 shows immobilization of fluoride and heavy-metals by DCPD with and without nano-precursor particle. Induction of nano-precursor on DCPD was found to be useful to immobilize fluoride and cadmium ions in aqueous solution. The nano-surfaced DCPD seems to be convenient to immobilize fluoride and heavy-metals in the environments, including contaminated soil.

Table 7.3 shows efficiency of removal of fluoride by DCPD, and other compounds reported in previous studies (Spinelli et al. 1971; Rölla and Bowen 1978;

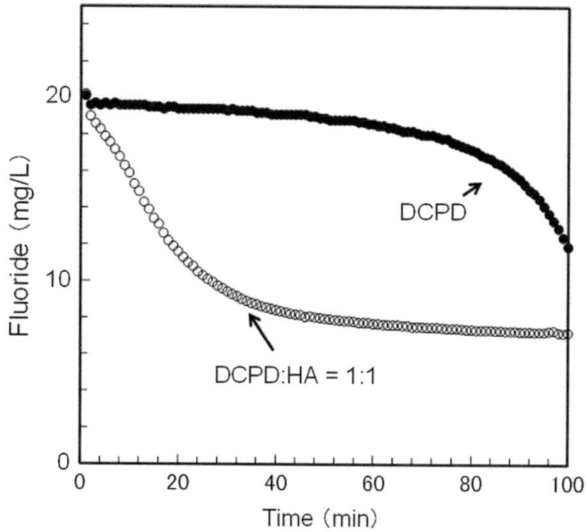

Fig. 7.4 Effect of mixing HA with DCPD (HA: DCPD = 1:1 in weight) on reactivity of fluoride ion

Table 7.2 Effect of induction of nano-precursor particle on DCPD for the immobilization of fluoride and heavy-metal ions[a]

	Fluoride and lead		Fluoride and cadmium	
DCPD	Fluoride	Lead	Fluoride	Cadmium
Without nano-precursor	100	94	0	29
With nano-precursor	100	94	100	98

[a]Initial concentration of fluoride ion was 20 mg/L and lead and cadmium ions were 10 mg/L

Table 7.3 Functional materials for immobilizing fluoride and heavy-metals

	Removal ability of fluoride ion	
Material	Fluoride in effluent (mg/L)	Adsorption capacity (mgF/g-material)
Cerium compound	<1	2
Hydroxyapatite (as seed crystal)	5	1–2
Hydrotalcite	<0.8	0.2
Magnesium hydroxide	<0.8	2
Dicalcium phosphate dihydrate	<0.8	22

Suzuki et al. 1981; Tokunaga et al. 1995; Reichert and Binner 1996; Wasay et al. 1996; Booster et al. 2003; Wu et al. 2007; Maliyekkal et al. 2010). The DCPD has the higher removal capacity of fluoride, and applicable to immobilize fluoride in the environments.

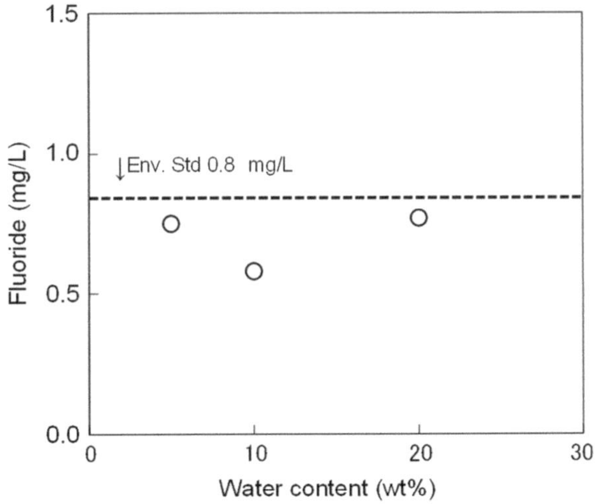

Fig. 7.5 Optimum water content in polluted soil required for the immobilization of fluoride by adding DCPD

7.3.4 *Application of DCPD to Immobilize Fluoride in Polluted Soil*

The DCPD is likely to be useful for the immobilization of fluoride and heavy-metals in contaminated soil. Firstly, we studied applicability of DCPD in the laboratory by mixing DCPD with contaminated soil that was collected from an industrial area in Japan. Leachate amount of fluoride from the polluted soil without treatment was around 7 mg/L. Figure 7.5 shows the optimum water content in polluted soil required for the immobilization of fluoride. We found that 10 wt% (wt %) of water in the soil are optimal for the immobilization of fluoride ion. When water content in polluted soil was zero, addition of DCPD was not able to immobilize fluoride. Figure 7.6 shows optimum amount of DCPD for the immobilization of fluoride in polluted soil. Fluoride was found to be immobilized to the environmental standard (0.8 mg/L) by the addition of over 5 wt% DCPD. Figure 7.7 shows the effect of aging time for the immobilization process. It was observed that the immobilization was completed within 10 days after the addition of DCPD to the aqueous solution. These results demonstrate that DCPD can be applicable to immobilize fluoride in polluted soil.

Based on the results of laboratory tests, we examined immobilization of fluoride using DCPD in the field (Tafu et al. 2014). Experimental site was a fluoride production plant in Japan. Fluoride leachate concentration in soil of the site was 24–58 mg/L. In Japanese law, a second regulation value of 24 mg/L is set for fluoride in the eluent from polluted soil (Japan MOE 2002). If fluoride concentration in soil of a site exceeded the second regulation value, treatment and handling of the polluted soil is quite limited. For the remediation, the polluted site by a conventional method, it is required to immobilize fluoride in the soil to satisfy the second regulation. In the field test, our objective was to immobilize fluoride in the

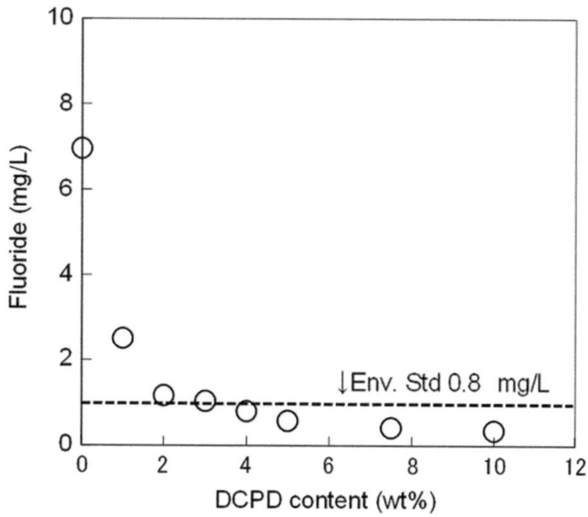

Fig. 7.6 Optimum content of DCPD to immobilize fluoride in polluted soil

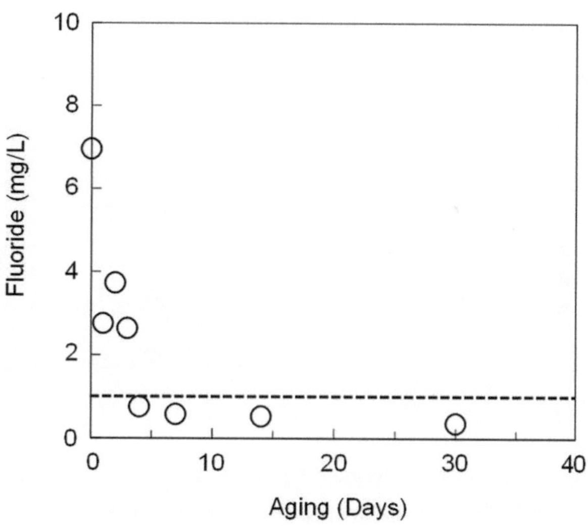

Fig. 7.7 Effect of aging time for the immobilization of fluoride in polluted soil by the addition of DCPD

soil below the second regulation value (24 mg/L). To achieve the objective of our study, DCPD was mixed with the polluted soil at 2.5 m depth as shown in Fig. 7.8. After 20 days, cementing material was mixed with the soil to increase hardness of the ground. Soil samples were collected from 2.5 m depth after 9, 20 and 200 days of DCPD application using a boring machine. Leachate amounts of fluoride from the soil samples were examined by an official analytical method (regulated by the Japanese Industrial Standard). Figure 7.9 shows results of immobilization of fluoride by DCPD at three experimental sites. Within 9 days of DCPD addition to the polluted soil, fluoride in all the three sites was found to be immobilized to a level

Fig. 7.8 On-site immobilization of fluoride-polluted soil using DCPD

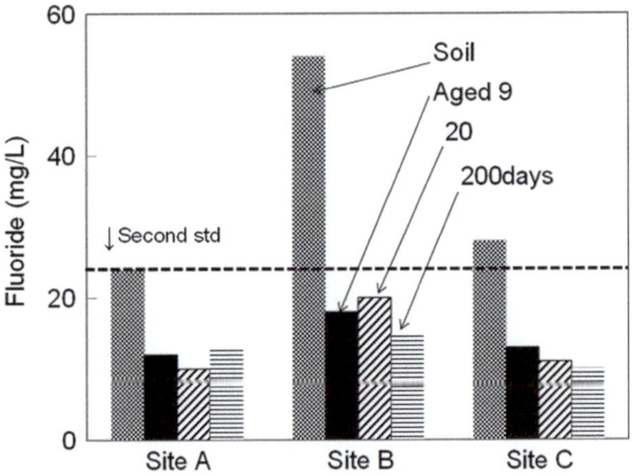

Fig. 7.9 Immobilization fluoride in polluted soil using DCPD at different experimental sites near a fluoride production plant in Japan

that satisfies the second regulation value of 24 mg/L. However, immobilization efficiency of fluoride was quite different at each site. This difference was likely due to the differences of chemical species of fluoride between the sites. In case of heavy-metals, such as lead, the difference in chemical species was estimated by elution with various chemicals (Tessier et al. 1979). However, this method is unsuitable for the estimation of fluoride species in soil. To improve the immobilization efficiency of DCPD for fluoride in polluted soil, it is necessary to develop an estimation method to determine the chemical species of fluoride.

7.4 Conclusion

In this work, we summarized an easy method for the estimation of heavy-metals, and fluoride pre-treatment by using calcium phosphate. The simple estimation method is one of the important solutions for analyzing heavy-metals that can be accessible to people not having specialized skill. The methods will help to understand and monitor soil pollution level. Immobilization of fluoride by calcium phosphate is very effective because of its high efficiency. This method is applicable not only to immobilize fluoride in polluted soil but also to avoid diffusion of fluoride from polluted soil and/or wastes to surrounding environment.

Acknowledgment This works was supported by JSPS KAKENHI (Grant-in-Aid for Scientific Research and Young Researcher, grant numbers were 14780451, 19310056, 23310058, 23710019, and 25740034), Industrial Technology Research Grant Program (2004), and Grant for Practical Application of University R&D Result under the Matching Fund Method (2007) from the New Industrial Technology Development Organization (NEDO), Japan. We would like to offer our special thanks to Daikin Industry Ltd., Fudo Tetra Corp., and Chiyoda Ute Co. Ltd. for enormous supports including on-site test.

References

Abe W, Isaka S, Koike Y, Nakano K, Fujita K, Nakamura T (2006) X-ray fluorescence analysis of trace metals in environmental water using preconcentration with an iminodiacetate extraction disk. X-Ray Spectrom 35:184–189

Begum ZA, Rahman IMM, Tate Y, Sawai H, Maki T, Hasegawa H (2012) Remediation of toxic metal contaminated soil by washing with biodegradable aminopolycarboxylate chelants. Chemosphere 87:1161–1170

Booster JL, Van Sandwijk A, Reuter MA (2003) Conversion of magnesium fluoride to magnesium hydroxide. Miner Eng 16:273–281

Camel V (2003) Solid phase extraction of trace elements. Spectrochim Acta Part B 58:1177–1233

Chow LC, Brown WE (1973) Reaction of dicalcium phosphate dihydrate with fluoride. J Dent Res 52:1220–1227

Corbridge DEC (2013) Phosphorus : chemistry, biochemistry and technology. Taylor & Francis, Boca Raton

Duff EJ (1970) Orthophosphates—I. Formation of apatites from calcium phosphates in potassium fluoride solutions. J Inorg Nucl Chem 32:3707–3708

Eanes ED, Meyer JL (1978) The influence of fluoride on apatite formation from unstable supersaturated solutions at pH 7.4. J Dent Res 57:617–624

Furusho Y, Ono M, Yamada M, Ohashi K, Kitade T, Kuriyama K, Ohta S, Inoue Y, Motomizu S (2008) Advanced solid phase extraction for inorganic analysis and its applications. Bunseki Kagaku 57:969–989

Guo G, Zhou Q, Ma LQ (2006) Availability and assessment of fixing additives for the in situ remediation of heavy metal contaminated soils: a review. Environ Monit Assess 116:513–528

Hasegawa S-I, Yamaguchi H, Itoh S, Ide K, Kobayashi T (2003) Determination of trace elements in high-purity aluminum by solid-phase extraction/ICP-MS. Bunseki Kagaku 52:533–537

Hayakawa R (1990) Extraction procedure for the evaluation of industrial solid wastes (In Japanese). Jpn J Water Pollut Res 13:155–159

Japan MOE (2002) Soil Contamination Countermeasures Act (Japanese Law: Act No. 53)

Lee K-H, Oshima M, Motomizu S (2002) Inductively coupled plasma mass spectrometric determination of heavy metals in sea-water samples after pre-treatment with a chelating resin disk by an on-line flow injection method. Analyst 127:769–774
Maliyekkal SM, Anshup AKR, Pradeep T (2010) High yield combustion synthesis of nanomagnesia and its application for fluoride removal. Sci Total Environ 408:2273–2282
Masamoto T, Tetsuji C (2011) Induction of nano-surface structure on the DCPD particles (Calcium hydrogenphosphate dihydrate). J Ecotech Res 16:65–69
Peters RW (1999) Chelant extraction of heavy metals from contaminated soils. J Hazard Mater 66:151–210
Rahman IMM, Begum ZA, Hasegawa H (2013) Selective separation of elements from complex solution matrix with molecular recognition plus macrocycles attached to a solid-phase: a review. Microchem J 110:485–493
Reichert J, Binner JGP (1996) An evaluation of hydroxyapatite-based filters for removal of heavy metal ions from aqueous solutions. J Mater Sci 31:1231–1241
Rölla G, Bowen WH (1978) Surface adsorption of fluoride and ionic exchange reactions on hydroxyapatite. Acta Odontol Scand 36:219–224
Seneviratne J, Cox JA (2000) Sol-gel materials for the solid phase extraction of metals from aqueous solution. Talanta 52:801–806
Shibata Y, Suyama J, Nakamura T, Hamamoto A, Yoshihara N, Tsuruta S, Nakano K (2008) Development of soil reference materials containing hazardous metals for X-ray fluorescence analysis. Bunseki Kagaku 57:477–483
Spinelli MA, Brudevold F, Moreno E (1971) Mechanism of fluoride uptake by hydroxyapatite. Arch Oral Biol 16:187–203
Stosnach H (2006) On-site analysis of heavy metal contaminated areas by means of total reflection X-ray fluorescence analysis (TXRF). Spectrochim Acta Part B 61:1141–1145
Suzuki T, Hatsushika T, Hayakawa Y (1981) Synthetic hydroxyapatites employed as inorganic cation-exchangers. J Chem Soc Faraday Trans 1 77:1059–1062
Tafu M, Chohji T (2005a) Simplified determination of trace amounts of fluoride in groundwater by transform reaction of calcium hydrogenphosphate dihydrate (DCPD). J Jpn Soc Water Environ 28:179–184
Tafu M, Chohji T (2005b) Reaction of calcium hydrogenphosphate dihydrate (DCPD) with a solution containing a small amount of fluoride. J Ceram Soc Jpn 113:363–367
Tafu M, Chohji T (2006) Reaction between calcium phosphate and fluoride in phosphogypsum. J Eur Ceram Soc 26:767–770
Tafu M, Kagaya S, Chohji T (2001) A separation method for simplified determination of fluoride by using calcium phosphate. Anal Sci 17:i753–i756
Tafu M, Nagamori T, Chujo Y (2009) Application of calcium hydrogen phosphate dihydrate (DCPD) to wastewater treatment. Arch Bioceram Res 9:191–194
Tafu M, Chohji T, Morioka I, Hiwasa M, Nakano H, Fujita T (2010) Stabilization of fluoride in waste gypsum by using surface-modified calcium phosphate particle. T MRS JAP 35:377–380
Tafu M, Masutani T, Takemura Y, Toshima T, Chohji T (2013) Effect of hydroxyapatite on reaction of dicalcium phosphate dihydrate (DCPD) and fluoride ion. Bioceram Develop Appl S1:015
Tafu M, Toshima T, Chohji T (2014) Eco-solutions for polluted soil by using calcium phosphate. J Soc Inorg Mater Jpn 21:421–425
Tessier A, Campbell PGC, Bisson M (1979) Sequential extraction procedure for the speciation of particulate trace metals. Anal Chem 51:844–851
Tokunaga S, Haron MJ, Wasay SA, Wong KF, Laosangthum K, Uchiumi A (1995) Removal of fluoride ions from aqueous solutions by multivalent metal compounds. Int J Environ Stud 48:17–28
US EPA (1996) Microwave assisted acid digestion of siliceous and organically based matrices (METHOD 3052; Revision 0). United States Environmental Protection Agency, Washington, DC

Wasay SA, Haran MJ, Tokunaga S (1996) Adsorption of fluoride, phosphate, and arsenate ions on lanthanum-impregnated silica gel. Water Environ Res 68:295–300

Wu X, Zhang Y, Dou X, Yang M (2007) Fluoride removal performance of a novel Fe–Al–Ce trimetal oxide adsorbent. Chemosphere 69:1758–1764

Chapter 8
Phytoremediation of Toxic Metals in Soils and Wetlands: Concepts and Applications

M. Azizur Rahman, Suzie M. Reichman, Luigi De Filippis, Seyedeh Belin Tavakoly Sany, and Hiroshi Hasegawa

Abstract Over centuries, industrial, mining and military activities, agriculture, farming, and waste practices have contaminated soils and wetlands in many countries with high concentrations of toxic metals. In addition to their negative effects on ecosystems and other natural resources, toxic metals pose a great danger to human health. Unlike organic compounds, metals cannot be degraded, and clean-up usually requires their removal. Most of the conventional remedial methods have lost economic favor and public acceptance because they are expensive and cause degradation of soil fertility that subsequently results in adverse impacts on the ecosystem. Conventional methods of environmental remediation do not solve the problem; rather they merely transfer it to future generation. Obviously, there is an urgent need for alternative, cheap, and efficient methods to clean-up sites contaminated with toxic metals.

Phytoremediation, a plant-based green technology, is cost effective, environmental friendly, aesthetically pleasing approach for the remediation of toxic metals. Due to its elegance and the extent of contaminated areas, phytoremediation approaches have already received significant scientific and commercial attention. Two approaches have been proposed for the phytoremediation of toxic metals

M.A. Rahman (✉)
Centre for Environmental Sustainability, School of the Environment, University of Technology Sydney, PO Box 123, Broadway, NSW 2007, Australia
e-mail: Mohammad.Rahman@uts.edu.au

S.M. Reichman
School of Civil, Environmental and Chemical Engineering, RMIT University, G.P.O. Box 2476, Melbourne 3001, Australia

L. De Filippis
School of the Environment, University of Technology Sydney, PO Box 123, Broadway, NSW 2007, Australia

S.B. Tavakoly Sany
Institute of Biological Sciences, University of Malaya, 50603 Kuala Lumpur, Malaysia

H. Hasegawa
Institute of Science and Engineering, Kanazawa University, Kakuma, Kanazawa 920-1192, Japan
e-mail: hhiroshi@se.kanazawa-u.ac.jp

H. Hasegawa et al. (eds.), *Environmental Remediation Technologies for Metal-Contaminated Soils*, DOI 10.1007/978-4-431-55759-3_8

from soils and wetlands: natural and induced phytoremediation. Natural phytoremediation refers to the use of hyper-accumulating plants and associated soil microbes, while the induced phytoremediation refers to the use chemicals, especially synthetic chelating ligands, for the increase of metal bioavailability and uptake in plants. Recently, genetically modified plants (GMPs) have been proposed to use in phytoremediation technology; however, this approach is being hindered by ideology-driven restrictive legislation over the use of GMPs. We will discuss the concepts and practical applications of phytoremediation technologies for the restoration of contaminated soils and wetlands.

Keywords Phytoremediation • Heavy metals • Soil • Wetland

8.1 Metal Contamination of Soils and Wetlands and Human Health Impact

Metals comprise about 75 % of the known elements and have been used from the beginning of ancient human civilization. Since the beginning of the industrial age, metals have been emitted to and deposited in the environment (Sparks 2005). In some cases, metals can be accumulated in terrestrial and aquatic environments in high concentrations and cause harm to living beings via ingestion of soil and/or dust, food, and water, inhalation of polluted air, and absorption via the skin from polluted soils, water, and air (Hillel 2005). Increasing use of metals with population and economic growth, especially in the developing countries, may contribute to soil and water contamination causing the deterioration of environmental quality and posing threats to human health (Sparks 2005).

Toxic metals can be derived from both natural and anthropogenic sources. Natural (geogenic) sources include rocks and minerals, and anthropogenic sources include agriculture (fertilizers, pesticides, herbicides, and animal manures), mining, smelting, and sewage sludge and scrap disposal (Adriano 2001). Anthropogenic deposition is a major mechanism for toxic metal input in the environment. Soil is the major recipient of trace elements in terrestrial environment, while sediments are the major sink in aquatic environment. Leaching of toxic metals or transport via mobile colloids can contaminate groundwater. On the other hand, runoff and drainage of toxic metals via sediments can contaminate freshwater environment (Adriano 2001; Hillel 2005).

For many years across the world, industrial, mining, military, farming, and waste management have contaminated large areas of soils and wetlands with high concentrations of toxic metals and organic pollutants (Peuke and Rennenberg 2005a; Li et al. 2001; Del Río et al. 2002). The problem of soil and water contamination of toxic metals is becoming more and more serious with increasing industrialization and disturbance of natural biogeochemical cycles by human activities and climate change (Ali et al. 2013). For example, in Europe, an estimated 52 million hectares of land – more than 16 % of the total land area of the continent – are affected by some level of soil degradation (Peuke and Rennenberg 2005a). The largest and probably most heavily contaminated areas are found near industrialized regions in north-western Europe, but many contaminated areas exist in the vicinity of major European cities (EEA 2003). In China, over 20 million acres of farmland (almost one fifth of the total arable farmland area) has been contaminated by toxic metals, such as Sn, Cr, Pb, and Zn, causing approximately 10 million tons of crop losses per year (Wu et al. 2010). A total area of 2.9×10^6 ha of degraded land has been produced in China as a result of mining and an additional mean area of 46,700 ha of destroyed land is produced annually. These degraded lands almost completely lack vegetation due to serious pollution and ultimately cause severe soil erosion and off-site pollution (Xia 2004). The Campine region in Belgium and the Netherlands with 700 km^2 is diffusely contaminated by atmospheric deposition of Cd, Zn and Pb (Meers et al. 2010).

Bioaccumulation of toxic elements in the food chain from contaminated soils and wetlands can be especially dangerous to human health. Toxic metals can enter the human body by either inhalation or ingestion (Islam et al. 2007). For the majority of people, the main route of exposure to toxic metals is diet except for workers with high levels of occupational exposure (Sharma et al. 2008). Exposure to toxic metals through the food chain has been reported in many countries, particularly in developing countries, and received significant attention from government and non-government agencies (Åkesson et al. 2008; Al Jassir et al. 2005; Demirezen and Aksoy 2006; Gulz et al. 2005). Concern over the accumulation of toxic metals in the food chain and environment has escalated in recent years. Once metals enter into biological systems they have the potential to disturb normal biochemical processes, and in extreme cases can be fatal (Pillay et al. 2003). Many countries have developed regulations for industries and other systems limiting discharges of pollutants into the environment in order to control the emission of trace elements and their subsequent health effects.

In fact, exposure to high content of toxic metals can cause significant adverse effects to humans, animals, microorganisms, and plants (Wagner 1993; Gaetke and Chow 2003; Hernández-Ochoa et al. 2005; Bodar et al. 2005). Regarding their toxicities, the most problematic toxic metals are Hg, Cd, Pb, As, Cu, Zn, Sn, and Cr (Ghosh 2010). Hg, Cd, Pb, and As are not essential for living organisms, while Cu and Zn are essential metals. Exposure of humans to toxic metals may results in neurobehavioral disorders, such as fatigue, insomnia, decreased concentration, depression, irritability, sensory, and motor symptoms. Exposure to toxic metals may also cause developmental retardation, various types of cancers, kidney damage, autoimmunity, and even death in some instances of exposure to very high concentrations (Glover-Kerkvliet 1995). For instance, at high concentration, Hg can damage vital organs, such as the lungs and kidneys, may cause fetal brain damage (Sharma 2003). Accumulation of Cd in human bodies (principally in the kidney and liver) can cause renal dysfunction and bone disease (e.g., Itai-Itai in Japan) (Nordberg 1996). Lead poisoning in children causes neurological damage leading to reduced intelligence, loss of short-term memory, learning disabilities, and coordination problems (Rai 2008a). The effects of As include cardiovascular problems, skin cancer and other skin effects, peripheral neuropathy, and kidney damage (Hughes 2002).

8.2 Remediation of Metal-Contaminated Soils and Wetlands

Low to medium range contamination of lands with toxic metals may induce their accumulation in the food chain. Therefore, strict environmental laws have been imposed in many countries to prevent any such occurrence of the toxic metals in foods by limiting the food production on contaminated lands. For example, a European Union Council Directive (EC 1986) limited the concentrations of toxic metals in agricultural soils to be 3 mg kg^{-1} for Cd, 140 mg kg^{-1} for Cu, 75 mg kg^{-1} for Ni, 300 mg kg^{-1} for Pb, 300 mg kg^{-1} for Zn, and 1.5 mg kg^{-1} for Hg (Grčman et al. 2001). According to the directive, several million hectares of agricultural lands are considered polluted by toxic metals in Europe (Flathman and Lanza 1998), and between 59 and 109 billion EUR is required to clean-up the contaminated sites using conventional techniques such as soil washing using particle size separation, chemical extraction with aqueous solutions of surfactants and mineral (EC 2004). Such high costs for land restoration using traditional remediation methods make the clean-up of many sites unaffordable even in the more developed nations. For instance, in Germany, only one-third of the total contaminated sites are cleaned up in soil remediation facilities while the remainder has been left untreated (Evangelou et al. 2007; Peuke and Rennenberg 2005a). This does not solve the problem, rather transferring it to future generation.

Aquatic ecosystems (including wetlands) are used directly or indirectly as recipients of potentially toxic liquids and solids from domestic, agricultural, and industrial wastes (Demirezen et al. 2007; Peng et al. 2008). Thus, wetland sediments are the major sink of toxic metals (Marchand et al. 2010). Clean-up of contaminated soils and wetlands is necessary to reduce the risk of metal toxicity to human and ecosystems. To date, different physical, chemical, and biological approaches have been employed for this purpose. The conventional remediation methods include *in situ* vitrification, soil incineration, excavation and landfill, soil washing, soil flushing, solidification, and stabilization of electro-kinetic systems (Sheoran et al. 2011; Wuana and Okieimen 2011). Generally, the physical and chemical methods suffer from limitations like high cost, intensive labor, irreversible changes in soil properties, and disturbance of native soil microflora. Chemical methods can also create secondary pollution problems. Therefore, there is an urgent need for alternative, cheap, and efficient methods to clean up heavily contaminated soils and wetlands. To improve how contaminated sites are remediated, there is a need to move beyond more traditional remediation practices and include some of the more novel remediation techniques like phytoremediation. Due to the potential of phytoremediation and the extent of contaminated sites, this technology has received significant scientific and commercial attention world-wide (McIntyre 2003; Gleba et al. 1999; Meagher 2000; Dietz and Schnoor 2001; Rahman et al. 2007; Salt et al. 1998).

8.3 Phytoremediation – A Green Technology for the Remediation of Contaminated Environment

8.3.1 Phytoremediation Strategies for the Restoration of Contaminated Soils and Wetlands

Phytoremediation (from ancient Greek "phyto" meaning "plant", and Latin "remedium" meaning "restoring balance") defines the remediation of contaminated environment (soils and wetlands) based on the idea of using natural bioaccumulation abilities of plants without excavating the contaminant material. Phytoremediating plants are ideally fast growing, tolerant of toxic metals, and efficient at transferring toxic metals from roots to above ground biomass. Sometimes, rhizospheric microorganisms or chemicals (e.g., chelating ligands) are used to increase of metal bioavailability and uptake. Based on the natural abilities of the phytoremediating plants, restoration of contaminated soils and wetlands can be achieved by employing the following phytoremediation strategies.

- Phytoextraction
- Phytostabilization
- Phytovolatilization
- Phytodegradation
- Phytofiltration
- Phytotransformation/detoxification

8.3.1.1 Phytoextraction

Phytoextraction is the uptake of contaminants from soils or waters by plant roots and their translocation to the harvestable biomass (Sekara et al. 2005; Yoon et al. 2006; Rafati et al. 2011). Translocation of toxic metals form roots to harvestable biomass (shoots) is necessary for an effective phytoextraction protocol because the harvest of root biomass is generally not feasible due to the location of roots within the soil (Zacchini et al. 2009; Tangahu et al. 2011). In general, phytoextraction has been tried more often for extracting toxic metals than for organics. Some of the examples of phytoextraction are sunflower (*Helianthus annuus*) (Marchiol et al. 2007) and Chinese brake fern (*Pteris vittata*) (Ma et al. 2001) for As; willow (*Salix viminalis*) Cd, Zn, Ni, Pb, and Cu (Greger and Landberg 1999; Boriŝev et al. 2009); Indian mustard (*Brassica juncea*) for Pb (Blaylock et al. 1997). The main advantage of phytoextraction is environmental friendliness. Traditional methods that are used commonly used for cleaning up metal-contaminated soil disrupt soil structure and reduce soil productivity, whereas phytoextraction can clean up soil without causing harm to soil quality. Another benefit of phytoextraction is that it is less expensive than the traditional clean-up process.

Phytoextraction can be used for phytoremediation of toxic and hazardous metals as well as for phytomining of precious metals (such as Au, Pd, Tl, and Pt) (Ali et al. 2013; Brooks et al. 1998). Phytomining has the potential to be an economically feasible solution to the disposal of used phytoremediation biomass. Plant biomass containing metals can be combusted to obtain energy and the remaining ash is considered as "bio-ore", which can be processed for the recovery or extraction of precious metals (Rascio and Navari-Izzo 2011). An advantage of phytomining is that this is a cost-effective and ecofriendly option as compared to the conventional metal extraction methods (Ali et al. 2013; Rascio and Navari-Izzo 2011). The commercial feasibility of phytomining, however, depends on many factors like phytoextraction efficiency of precious metals and current market value of the processed metals. In addition, this technique can only be used to extract metals from the rhizosphere. Phytomining has been commercially used for Ni and it was found to be less expensive than the conventional extraction methods for Ni (Robinson et al. 1997; Nicks and Chambers 1998). Commercial phytomining has also been used for Au and Tl (Anderson et al. 1999).

The chemical composition and sorption properties of soil influence metal mobility and bioavailability (Kłos et al. 2012). The bioavailability of toxic metals in soils is a critical factor affecting the efficiency of phytoextraction of target metals. Low bioavailability is a major limiting factor for phytoextraction of contaminants such as Pb (Ali et al. 2013). Strong binding of toxic metals to soil particles or precipitation causes a significant fraction of the toxic metals non-bioavailable, and therefore remain unavailable for uptake by phytoremediating plants (Sheoran et al. 2011). Based on bioavailability, toxic metals in soils can be categorised into three groups: readily bioavailable (Cd, Ni, Zn, As, Se, and Cu); moderately bioavailable (Co, Mn, and Fe) and least bioavailable (Pb, Cr, and U) (Prasad 2003). However, many plants have developed mechanisms for solubilizing heavy metals in soil such as the secretion of metal-mobilizing "phytosiderophores" into the rhizosphere by members of the Poaceae (Lone et al. 2008; Reichman and Parker 2005), for solubilizing toxic metals in soil.

Phytoextraction is classified into natural and induced based on the bioaccumulation process of the plants species involved. Natural phytoextraction is based on the idea of the use of natural hyperaccumulators that have exceptionally high metal-accumulation ability and tolerance to toxic metals (Baker et al. 2000). In induced phytoextraction, a conditioning fluid containing a chelator or another agent is added to soil to increase metal solubility or mobilization so that the plants can absorb higher concentrations of metals.

In the past decade, chelant-enhanced phytoextraction has received much attention from the scientific community. Chelants, when added to soil, are capable of forming soluble complexes with both "labile" and "non-labile" metal in the soil solution via desorption of sorbed species and dissolution of precipitated compounds (Norvell 1984). Re-precipitation and re-sorption of metals are prevented by the chelant-metal-complex formation, and the metals become bioavailable (Salt et al. 1995a). The drawback of metal phytoextraction due to limited bioavailability of metals is minimized after the unearthing that the translocation of metals from soil to plants can be increased to maximum with the addition of certain chelants (Blaylock and Haung 1999). However, the soil properties and the nature of the applied chelant determine the amounts of bioavailable metals in soil matrix (Kos and Lestan 2004; Tandy et al. 2004; Luo et al. 2005). Stability constants, K_s, of chelant-metal-complexes are the decisive feature to select a chelant or rank different chelants for the extraction of metals from the metal-contaminated soil. The chemical characteristics of the chelant itself and the metal speciation in the soil matrix also influence the effectiveness of different chelant in the separation process (Elliott and Brown 1989; Luo et al. 2005; Huang et al. 1997).

A wide range of synthetic chelants have been tested for chelant-induced phytoextraction with aminopolycarboxylate chelants (APCs) among the most used chelant type. Typically used APCs for metal phytoextraction include Ethylenediaminetetraacetic acid (EDTA), diethylenetriaminepentaacetic acid (DTPA), N-(hydroxyethyl)ethylenediaminetriacetic acid (HEDTA), nitrilotriacetic acid (NTA), S,S-ethylenediaminedisuccinic acid (S,S-EDDS), methylglycine-diacetic acid (MGDA), and glutamic acid diacetic acid (GLDA) . In particular, EDTA has most often been utilized among the APCs, since EDTA forms strong water-soluble chelant complexes with most toxic metals (Egli 2001; Nowack and VanBriesen 2005; Leštan et al. 2008; Salt et al. 1998). Despite the success of the use of EDTA in phytoextraction of toxic metals, the enhanced mobility of the metals in soil by EDTA and their potential risk of leaching are important concerns (Cooper et al. 1999; Wu et al. 2004; Sun et al. 2001). Several studies showed that EDTA enhanced the leaching of heavy metals during the phytoextraction process (Sun et al. 2001; Wu et al. 2004; Grčman et al. 2001; Kedziorek et al. 1998). For example, Wu et al. (2004) found that the mobility and leaching of Cu, Zn, and Cd increased significantly during the EDTA-enhanced phytoextraction process. Increased mobility and leaching of Zn, Cu, Cd, and Pb during the EDTA-enhanced phytoextraction process has also been reported by Sun et al. (2001).

In addition to enhanced mobility and leaching of heavy metals by EDTA, the persistence of metal-EDTA complexes in contaminated soils and their effects on soil microbial community are important drawbacks of this technology. Several studies have indicated that EDTA-metal complexes are resistant to microbial degradation (Nörtemann 1999; Oviedo and Rodríguez 2003). Palumbo et al. (1994) found that the bacterial ability to degrade EDTA is rare, since they could not obtain degrading consortia from places polluted with the chelate. Other studies also found EDTA to be slowly biodegraded to CO_2 in soil, with only 6.7 % degraded after 4 weeks and a lower rate of degradation in the subsoil than in surface soil (Tiedje 1975). Means et al. (1980) reported that the EDTA degradation rate was not rapid enough, even under optimal laboratory conditions, to stop disquiet about its release into the environment. Therefore, biodegradable chelating ligands can be alternatives to the EDTA for the phytoextraction of toxic metals.

The method of chelant application significantly affects the efficiency of the phytoextraction process. The chelant can be applied to the soil matrix either in a single dose after the optimum growth of the accumulator crop, or in small multiple doses gradually during the growth cycle. Studies showed that application of chelant in multiple doses is more effective than that of single dose (Wenzel et al. 2003). Phytoextraction efficiency can also be improved by the combined application of different chelants to the metal-contaminated soil (Leštan et al. 2008). For example, Blaylock et al. (1997) showed that the application of EDTA and acetic acid results in a twofold accumulation of Pb in Indian mustard shoots compared with the application of EDTA alone. Luo et al. (2006) also found that the combined application of EDTA and EDDS results in a higher level of efficiency in the phytoextraction of Cu, Pb, Zn, and Cd than could be obtained by the application of either chelant alone.

Although phytoextraction of toxic metals can be achieved by using hyperaccumulators, the technology may not be applicable for remediating sites with multiple contaminants. Wu et al. (2006) proposed a solution to the phytoremediation of soils with multiple contaminants using a combination of microbe-plant symbiosis within the plant rhizosphere. They showed that inoculation of sunflower roots with the engineered rhizobacterium resulted in a marked decrease in Cd phytotoxicity and a 40 % increase in Cd accumulation in the plant root. Owing to the significantly improved growth characteristics of both the rhizobacterium and plant, they proposed the use of a metal-binding peptide (EC20) in a rhizobacterium (*Pseudomonas putida*) with organic-degrading capabilities as a promising strategy to remediate mixed organic-metal-contaminated sites.

8.3.1.2 Phytostabilization

Phytostabilization is the immobilization of a contaminant in soil through adsorption onto roots, absorption and accumulation by roots, or precipitation within the root zone of plants (Brunner et al. 2008). Unlike phytoextraction, phytostabilization focuses mainly on sequestering pollutants in soil near the roots. This technique is used to reduce the mobility and bioavailability of pollutants in the environment, thus preventing their leaching in groundwater and entry into the food chain (Erakhrumen and Agbontalor 2007).

Phytostabilization occurs through contaminant accumulation in plant tissue and in the soil around the roots because of changes in the chemistry of the contaminants, which become insoluble and/or immobilized on soil components. Plants used for phytostabilization will need to be tolerant of the metals present in the particular site, but the accumulation of metals in their aerial parts may be positively disadvantageous. If the objective of phytostabilization is purely to prevent erosion and improve the visual amenity of a derelict site, then the accumulation of metals in the plants may be irrelevant (Macnair et al. 2000).

Metal immobilizing chemicals (soil amendments) are used for phytostabilization technology to improve soil conditions for plant growth and to reduce the chemical mobility in soil and toxicity of the metals to biota (Vangronsveld et al. 2009). Different soil amendments are used in phytostabilization technology for different toxic metals. For example, the most promising amendments for phytostabilization of Pb are phosphate materials, materials containing hydrous iron oxides, steel shot, inorganic clay minerals, and organic material (Cunningham and Berti 2000). By excreting special proteins and/or redox enzymes, certain plant species can convert metals to relatively less bioavailable forms and decrease possible metal bioavailability and toxicity to biota. For example, Cr(III) is less mobile and toxic than Cr(VI), and the reduction of Cr(VI) to Cr(III) can be a strategy of phytostabilization (Wu et al. 2010).

Phytostabilization technology influences the mobility of toxic metals in soils in different ways:

- The amendments directly alter the soil conditions (acidic and/or alkaline conditions, organic matter, oxygen levels) in the rhizosphere that influence metal mobility.
- Proteins and/or enzymes are released by the roots into the rhizosphere soil, leading to precipitation and immobilization of the toxic metals either in the soil or on the root surface.
- The toxic metals are taken up by the plants and sequestered in the root system.
- The surface of the soil is vegetated, and the vegetation acts as a barrier to erosion and exposure of the contaminated soil to wind, water, and direct contact with humans or animals.

An example of the application phytostabilization technology is the use of vegetation cap to stabilize and contain mine tailings (Mendez and Maier 2008; Conesa et al. 2007).

Phytostabilization technology for the remediation of metal-contaminated soils has both the advantage and disadvantage. Advantages of this technology include:

- This technology reduces the mobility, and therefore the risk, of toxic metals without removing them from their location.
- This technology does not generate secondary contamination that needs treatment.
- Usually this technology enhances the soil fertility. It may combine treatment with ecosystem restoration.

Disadvantages of phytostabilization technology may include:

- The contaminants are left in place, so the site must be monitored perpetually to make sure the stabilizing conditions continue.
- If the contaminant concentrations are very high, toxic effects may prevent the growth of plants until extensive amendments application reduce their bioavailability to plants.

- If soil additives are used, they may need to be periodically reapplied to maintain the effectiveness of the immobilization.

8.3.1.3 Phytovolatilization

Phytovolatilization is the uptake of pollutants from soil by plants, their conversion to volatile form and subsequent release into the atmosphere (Ali et al. 2013). This method can be used for organic pollutants and some toxic metals like Hg, Se, and As that have gaseous forms.

There is some evidence that certain plant species have the ability to accumulate Hg both from the atmospheric and soil sources; however, no plant species with Hg hyperaccumulating properties has been identified (Raskin and Ensley 2000). Therefore, transgenic plants such as Arabidopsis (*Arabidopsis thaliana* L.) and tobacco (*Nicotiana tobacum*) containing bacterial mercuric ion reductase (*merA*) and organomercurial lyase (*merB*) genes, responsible for detoxifying methyl-mercury, have been investigated for their ability of Hg phytovolatilization (Heaton et al. 1998; Bizily et al. 1999). The advantage of this technology is that the plant may transform toxic methyl-mercury to a less toxic volatile elemental Hg. However, the important limitation of Hg phytovolatilization is that the released elemental Hg into the atmosphere is likely to be recycled by precipitation and then redeposit back into ecosystem (Henry 2000).

Bacteria containing the As (III) S-adenosylmethionine methyltransferase (*arsM*) gene were able to sequentially methylate toxic inorganic As to less toxic pentavalent methylated arsenicals such as methylarsenate (MAs(V)), dimethylarsenate (DMAs(V)), and trimethylarsine oxide (TMAs(III)) (Qin et al. 2006, 2009). The phytovolatilization of the final product, gaseous TMAs(III), could remove arsenic from polluted water and soil using engineered hyperaccumulator such as Chinese fern *Pteris vittata* (Sakakibara et al. 2010; Zhu and Rosen 2009).

The major disadvantage of phytovolatilization is that it does not remove the pollutants completely from the environment; rather it transfers the pollutants from soils/waters to atmosphere from where it can be re-deposited through atmospheric precipitation. Therefore, the use of phytovolatilization for the remediation of environmental contaminants remains controversial (Padmavathiamma and Li 2007).

8.3.1.4 Phytodegradation

Phytodegradation refers to the microbial breakdown of pollutants, particularly organic pollutants, in the rhizosphere (Mukhopadhyay and Maiti 2010; Newman and Reynolds 2004). The main reason for the enhanced degradation of organic pollutants in the rhizosphere compared to the bulk soil? Is likely the increase in the numbers and metabolic activities of the microbes in the rhizosphere. Plants can stimulate microbial activity by 10–100 times higher in the rhizosphere compared to the bulk soil by the secretion of exudates containing substances such as carbohydrates, amino acids, and flavonoids (Ali et al. 2013). The release of nutrient-containing exudates by plant roots provides carbon and nitrogen sources to the soil microbes and creates a nutrient-rich environment in which microbial activity is stimulated. In addition to secreting organic substrates that facilitate the growth and activity of rhizospheric microbes, plants also release enzymes that are directly capable of degrading organic contaminants in rhizosphere (Kuiper et al. 2004; Yadav et al. 2010).

8.3.1.5 Phytofiltration

Phytofiltration (also known as rhizofiltration), which is related to phytoextraction, is the removal of pollutants from contaminated wetlands by aquatic plants (Mukhopadhyay and Maiti 2010; Dushenkov et al. 1995). Gardea-Torresdey et al. (2004) reviewed phytofiltration technology for the removal of toxic metals form contaminants from aqueous effluents. Phytofiltration may be rhizofiltration (use of plant roots) or blastofiltration (use of seedlings) or caulofiltration (use of excised plant shoots) (Mesjasz-przybyłowicz et al. 2004).

The phytofiltration of toxic metals from contaminated waters using aquatic plants has been extensively studied (Selvapathy and Sreedhar 1991; Sen and Bhattacharyya 1993; Low et al. 1994; Alam et al. 1995; Ingole and Ting 2002; Sen and Mondal 1990; Dushenkov et al. 1995). This cleanup process involves biosorption and accumulation of pollutants. Many aquatic plants (floating and sub-merged) have been investigated for the remediation of wastewater contaminated with Cu(II), Cd(II), and Hg(II) (Sen and Mondal 1987; Selvapathy and Sreedhar 1991; Alam et al. 1995). Water fern (*Salvinia natans* L.) is a free-floating freshwater macrophyte that has been tested for remediation of Hg(II) (Sen and Mondal 1987), and Cu(II) (Sen and Mondal 1990) and As(V) (Rahman et al. 2008c). Other examples of phytofiltration of toxic metals by aquatic plants are – *Medicago sativa* (Alfalfa) for Cd, Cr Pb, and Zn (Gardea-Torresdey et al. 1998); ferns (*Pteris vittata* and *Pteris cretica*) for As (Huang et al. 2004); Yellow burrhead (*Limnocharis flava*) for Cd (Abhilash et al. 2009); water hyacinth (*Eichchornia crassipes*) for Cd and Zn (Hasan et al. 2007); and duckweed (*Spirodela polyrhiza*) for As (Rahman et al. 2007) .

8.3.1.6 Phytotransformation/Detoxification

Phytotransformation of toxic metals is not a direct remediation technique, rather it reduce/detoxify the toxicity of toxic metals to the organisms. There are some microbes (e.g., bacteria, phytoplankton, fungi, etc.) that in the soil and aquatic environment that have the ability/mechanisms to transform more toxic forms of the toxic metals and metalloids to their less toxic form (Summers and Silver 1978; Raab and Feldmann 2003; Bender et al. 1995). This detoxification process of microbes is considered as a promising method for bioremediation of heavy metals and metalloids.

Higher plants also have detoxification mechanisms of toxic metals (Zenk 1996). A set of toxic-metal-complexing peptides has been isolated from plants and plant suspension cultures. The structure of these peptides was established as (γ-glutamic acid-cysteine)$_n$-glycine and are called phytochelatins (PC) (Zenk 1996; Cobbett and Goldsbrough 2002). The biosynthesis of PCs proceeds by metal activation of a constitutive enzyme that uses glutathione (−GSH) as a substrate. In a recent review, Rahman and Hassler (2014) discussed the rules of PCs and GSH in As resistance and detoxification by photosynthetic organisms. Other studies also reported the phytotransformation/detoxification of toxic metals by plants and photosynthetic organisms (Cobbett and Goldsbrough 2002; Lytle et al. 1998; Ow et al. 1998). However, more research and knowledge on the natural detoxification mechanisms of toxic metals by plants are required to improve plant's performance in removing these toxicants from the environment.

8.4 Tolerance and Detoxification of Toxic Metals in Plants

8.4.1 Mechanisms of Heavy Metal Uptake in Plants

Plants uptake heavy metals from soil solution and waters into their roots, and then a fraction of the heavy metal ions are stored in the roots while the rest are translocated to the aboveground parts primarily through xylem vessels (Prasad 2004; Jabeen et al. 2009). The uptake of heavy metal ions from soil solution by plants' roots and subsequent translocation to the shoots and vacuoles is controlled and regulated by a variety of molecules. Some molecules are involved in the cross-membrane transport of the heavy metal ions and others are involved in their complexation with chelating compounds and subsequent sequestration in the vacuoles (Ali et al. 2013; Tong et al. 2004). Uptake of heavy metal ions into plant's roots is mediated by several classes of specialized transporter proteins (channel proteins) in the plasma membrane (Seth 2012). These include the CPx-type heavy metal ATPases, the natural resistance-associated macrophage (Nramp) family of proteins, action diffusion facilitator (CDF) family proteins, and zinc-iron permease (ZIP) family proteins (Williams et al. 2000). The ZIP family proteins contribute to the uptake of Zn^{2+} and Fe^{2+} (Clemens 2001), while CPx-type heavy metal ATPases have been involved in the transport of essential as well as potentially toxic metals like Cu, Zn, Cd, and Pb across the cell membranes (Williams et al. 2000). The Nramp family proteins play an important role in transport of divalent metal ions into the plant's roots (Seth 2012).

8.4.2 Mechanisms of Metal Tolerance and Detoxification in Plants

Plants have a range of potential cellular mechanisms in order to tolerate and detoxify heavy metal stress. These include metal binding to cell walls, exudation of metal chelating compounds, and a network of processes that take up metals, chelate them, and transport these complexes to above-ground tissues where they are sequestered into vacuoles (Peuke and Rennenberg 2005a). Based on these mechanisms, plants can be classified into two groups: (i) non-accumulators that achieve metal tolerance by preventing toxic metals uptake into roots cells passively through binding the metal ions onto the cell walls; (ii) accumulators that evolved specific mechanisms for high levels of metal accumulation and detoxification in cells. The second group of plants uptake heavy metals in shoots and sequester them in cellular vacuoles to remove excess metal ions from the cytosol to reduce their interactions with cellular metabolic processes (Assunção et al. 2003).

The hyperaccumulating or metal-tolerant plant species such as *Silene vulgaris*, *Thlaspi caerulescens*, *Alyssum lesbiacum*, *Arabidopsis halleri*, and *Brassica* spp. have been investigated by several researchers (Clemens et al. 2002; Kraemer 2003). The ability of these plants to accumulate high concentrations of metals was observed for both essential nutrients, such as Cu, Fe, Zn, and Se, as well as non-essential metals, such as Cd, Hg, Pb, Al, and As (Salt et al. 1998; Meagher 2000; Clemens et al. 2002; McGrath and Zhao 2003). Metal concentrations in the shoots of accumulating plants can be 100–1000-fold higher than in non-accumulating plants (Peuke and Rennenberg 2005a). Frequent generation of free radicals are taking place during heavy metal stress in accumulating plants and it leads to oxidative stress. Plants have been developed a good tolerance mechanisms against these heavy metal induced oxidative stress via significant synthesis of antioxidants and chelating compounds (Mishra et al. 2006; Seth et al. 2007, 2008). The transformation of toxic forms to less harmful forms is also an approach to detoxifying heavy metals, particularly As, Hg, Fe, Se, and Cr, which exist in a variety of cationic and oxyanionic species and thio- and organo-metallic forms (Meagher 2000; Guerinot and Salt 2001).

In antioxidant system, plants have shown significant synthesis of various types of antioxidants subjected to metal stress, indicating a possible role in defense mechanisms (Cobbett 2000; Mishra et al. 2006; Seth et al. 2007). The functional significance of a compound as antioxidant achieved through different mechanisms, such as metal chelation, activated oxygen species scavenging, recycling of other antioxidant, inhibition of lipid per oxidation, and repair of damaged DNA molecules caused by oxidative stress (Allen 1995; Seth et al. 2008). Among these, scavenging and/or removal of free radicals are the most likely mechanisms for antioxidants mediated tolerance strategy in plants (Seth 2012). Several enzymatic antioxidants such as super oxide dismutase (SOD), ascorbate peroxidase (APX), and catalase (CAT) are major types of reactive oxygen species (ROS)-scavenger that play significant roles in these mechanisms (Mishra et al. 2006). Other than enzymatic antioxidant, some non-enzymatic antioxidants such as cysteine, non-protein thiols (NP-SH), ascorbic acid and GSH are also playing a very important role in plant resistance against oxidative stress (Seth et al. 2007; Hammond-Kosack and Jones 1996).

Chelating peptides, most notably metallothioneins (MTs) and PCs, have a significant role in the detoxification of metals, and their synthesis in the plant is induced by exposure of root cells to heavy metals (Rauser 1999; Cobbett and Goldsbrough 2002; Cobbett 2000; Hall 2002). These cysteine-rich polypeptides exploit the property of heavy metals to bind to the thiol-groups of proteins – one of the toxic effects of heavy metals – for detoxification. Metallothioneins are S-rich proteins of 60–80 amino acids that contain 9–16 cysteine residues and are found in plants, animals, and some prokaryotes (Cobbett and Goldsbrough 2002; Cobbett 2000; Rauser 1999). Phytochelatins are a family of γ-glutamylcysteine oligopeptides with glycine or other amino acids at the carboxy-terminal end, in which γ-Glu-Cys units are repeated 2–11 times. They are synthesized from GSH and its derivates by phytochelatine synthase in the presence of heavy metal ions (Cobbett 2000). It is reported that cells and tissues exposed to a range of heavy metal ions, such as Cd, Ni, Cu, Zn, Ag, Hg, and Pb, are rapidly synthesized PCs, with the fact that Cd was the strongest inducer for PCs synthesis (Rauser 1995; Yang and Yang 2001; Pinto et al. 2003). The functional significance of PCs can be attributed due to the presence of thiol groups (–SH) which make co-ordination bond with toxic metal ions (Seth 2012), which are then sequestered into the cellular vacuoles.

Sequestration of heavy metal ions in cellular vacuoles is an important detoxification/tolerance mechanism in metal hyperaccumulators (Tong et al. 2004). Complexation to low molecular weight organic chelators such as organic acids (malate, citrate), amino acids (O-acetylserine, histidine), and nicotinamine have also shown to play significant roles in metal detoxification/tolerance in plants (Salt et al. 1995b; Clemens 2001; Cobbett and Goldsbrough 2002; Hall 2002; Kraemer 2003).

8.5 Phytoremediation of Toxic Metals Using Transgenic Plants

The efficiency of a remediation strategies depends on many issues including biotic (e.g., the plant species concerned) and abiotic (e.g., bioavailability of the heavy metals, speciation of the heavy metals, physico-chemical conditions of the contaminated sites, level of contamination) factors. The abiotic factors can be changed to conditions favorable for a phytoremediation strategy. For example, bioavailability of heavy metals has been reported to be increased by chelating ligands (Blaylock et al. 1997; Evangelou et al. 2007). The main challenge of an effective phytoremediation strategy for the removal of heavy metals from contaminated sites is the choice of a potential plant species that has desirable characteristics such as fast growth rate, high above-ground biomass, widely distributed root systems, tolerance to the toxic effects of the target heavy metals, adaptive to the target sites, and easy to harvest (Adesodun et al. 2010; Sakakibara et al. 2011; Ali et al. 2013). It is difficult to find a plant species with all of these characteristics. However, some researchers proposed that the phytoextraction potential of a plant species should be determined mainly by two key factors: (i) shoot metal concentration and (ii) shoot biomass (Li et al. 2010). Other researchers proposed hyperaccumulation and hypertolerance as more important characteristics than shoot biomass for a phytoremediation strategy (Chaney et al. 1997).

Most scientific and commercial interest in phytoremediation now focuses on phytoextraction and phytodegradation, which use selected plant species grown on contaminated sites. In phytoextraction, the plant species are harvested to remove the plants together with the pollutants that have accumulated in their tissues. Two different approaches have been tested for phytoextraction of heavy metals (Robinson et al. 1998; Tlustoš et al. 2006).

1. The use of hyperaccumulating plant species. In this technique, the idea is to use hyperaccumulators that produce comparatively less aboveground biomass but accumulate high amount of the target heavy metals.
2. The use of high biomass producing plant species. This technique aims to use plant species which is not a hyperaccumulator but produce more aboveground biomass than the hyperaccumulators so that overall metal accumulation in the aboveground biomass is comparable to that of hyperaccumulators.

However, in selecting a plant species for phytoremediation based on different desired characteristics discussed above, it should be carefully considered that the use of hyperaccumulators will yield a metal-rich, low-volume biomass, which is economical and easy to handle in case of both metal recovery and safe disposal. On the other hand, use of non-accumulators will yield a metal-poor, large-volume biomass, which will be uneconomical to process for recovery of metals and also costly to safely dispose. However, high biomass yielding plants are usually not hyperaccumulators. Use of genetically modified plants (GMPs) has been proposed to be a solution to overcome the limitations of fast growing non-hyperaccumulators (Rugh et al. 1998; Pilon-Smits and Pilon 2002; Cunningham and Ow 1996; Bennett et al. 2003).

Unlike plant growth, which depends on numerous genetic and non-genetic factors, the accumulation of heavy metals is controlled by only a few gene loci and is more easily accessible for genetic manipulation (Clemens et al. 2002). Therefore, phytoremediation strategies that have been put into consideration are the genetic manipulation of GSH and PC production in plant tissues (Song et al. 2003; Noctor et al. 1998; Cobbett 2000; Yadav 2010).

Initial experiments with transgenic plants have shown that they are indeed efficient in drawing metals from heavily contaminated soils (Rugh et al. 1998; Cherian and Oliveira 2005; Tong et al. 2004). Trees are probably the best-suited plants for transgenic approaches to improve the heavy-metal accumulation. Tree biotechnology is thus becoming an increasingly important tool for the remediation of contaminated environments (Peuke and Rennenberg 2005b). Fast-growing trees, such as *Populus* spp., are good candidates for phytoremediation due to their extensive root systems, high rates of water uptake and transpiration that is helpful in efficient transport of contaminants from roots to shoot, rapid growth and large biomass production (Rugh et al. 1998; Taghavi et al. 2005). Poplars can be grown in a wide range of climatic conditions and are used with increasing frequency in 'short-rotation forestry' systems for pulp and paper production. This raises the possibility of using plantations of transgenic poplars across several multiyear cycles to remove heavy metals from contaminated soils (Robinson et al. 2000; Peuke and Rennenberg 2005a). In addition, a dense tree cover would also prevent erosion and the spread of contaminated soil by wind. After the first planting, the costs for field management are relatively low and the products (biomass/wood) can be used for the production of electricity and heat by burning in wood power stations. Another important point is that it is very unlikely that poplars will enter the human food chain or end up as feedstock for animals.

The transformation of gray poplar trees (*Populus tremula* x *P. alba*) to overexpress γ-ECS from *Escherichia coli* resulted in higher levels of GSH and its precursor γ-L-glutamyl-L-cysteine compared with wild type (Noctor et al. 1998), and an elevated capacity for PC production. These new transgenic trees have been shown a high potential for the uptake and detoxification of heavy metals (Peuke and Rennenberg 2005b). Results from preliminary trials showed that the transgenic poplars are genetically stable and there are no indications so far of any impact on the environment. The transgenic trees have a higher capacity than wild-type trees for accumulating heavy metals on the heavily contaminated sites (Peuke and Rennenberg 2005a).

Despite this and other advantages, the progress and application of GMPs in phytoremediation technology to tackle widespread environmental contamination problems is being hampered by ideology-driven, restrictive legislation over the use and release of GMPs in Europe, and many other countries (Peuke and Rennenberg 2005a). However, if genetic engineering is eventually successful in producing plants that are able to restore/remediate contaminated sites without any potential impact on the environment in general and in particular on agriculture and human health, then we may also see a better public acceptance of GMPs in the future.

8.6 Phytoremediation of Toxic Metals by Aquatic Plants

Phytoremediation of the heavy metals can be achieved by aquatic plants since the process involves biosorption and bioaccumulation of the soluble and bioavailable metals from water (Brooks and Robinson 1998). The aquatic plants can be floating, emergent, and submerged. The floating aquatic plants accumulate metals by their roots from water, while the submerged plants accumulate metals from the sediments by their roots and from the water by their shoots (Rahman and Hasegawa 2011; Rahman et al. 2011).

Provably, Hutchinson (1975) reviewed, for the first time, the ability of aquatic macrophytes to concentrate elements from the aquatic environment and described that the levels of toxic elements in these plants were at least an order of magnitude higher than that in the supporting aqueous medium. Later on, Outridge and Noller (1991) reviewed the accumulation of toxic trace elements by aquatic vascular plants and discussed the pathways and rates of elemental uptake and excretion, environmental factors that control uptake of elements, and the significance of trace elements uptake for the field of wastewater treatment and biomonitoring of pollutants. To date, numerous papers have been published in leading international journals on different aspects of biogeochemistry, mechanisms and uptake of toxic metals by different aquatic macrophytes. The aim of these studies was to develop an efficient and cost-effective phytoremediation technology. A list of aquatic plants that have been studied for the phytoremediation of toxic metals is listed in Table 8.1.

Microspora and *Lemna minor* were studied for Pb and Ni phytoremediation (Axtell et al. 2003). Five common aquatic plant species (*Typha latifolia, Myriophyllum exalbescens, Potamogeton epihydrus, Sparganium angustifolium*, and *Sparganium multipedunculatum*) were tested for Al phytoremediation (Gallon et al. 2004). Parrot feather (*Myriophyllum aquaticum*), creeping primrose (*Ludwigina palustris*), and water mint (*Mentha aquatic*) have shown to remove Fe, Zn, Cu, and Hg from contaminated water effectively (Kamal et al. 2004). *L. minor* was reported to accumulate Cu and Cd from contaminated wastewater (Kara 2004; Hou et al. 2007). The submerged aquatic plant *Myriophyllum spicatum* L. was found to be efficient for metal-contaminated industrial wastewater treatment (Lesage et al. 2007). The aquatic plants *Rorippa nasturtium-aquaticum* (L.) and *Mentha* spp. have been reported to accumulate significant amount of As from contaminated freshwater (Robinson et al. 2006). Based on the outcomes of many studies, aquatic plants have been used for the remediation of contaminated constructed wetlands.

8.6.1 Constructed Wetlands for Phytoremediation of Toxic Metals

Wetlands are often considered sinks for contaminants, and there are many cases in which wetland plants are utilized for removal of pollutants, including metals. Constructed wetlands offer a cost-effective and technically feasible method and have proven effective and successful in remediation of heavy metal pollution (Weis and Weis 2004; Williams 2002). Aquatic macrophytes have been shown to play important roles in wetland biogeochemistry through their active and passive circulation of elements including heavy metals (Weis and Weis 2004). Active uptake into the wetland plant tissues may promote phytofiltration and immobilization of heavy metals in plant tissues, as seen in constructed wetlands for wastewater

Table 8.1 Aquatic plants studied for the phytoremediation of toxic elements from wetlands

Common name	Scientific name	Trace elements	References
Duckweed	*Lemna gibba* L.	As, U, Zn	Mkandawire and Dudel (2005), Mkandawire et al. (2004a, b), and Fritioff and Greger (2003)
Lesser duckweed	*Lemna minor* L.	As, Zn, Cu, Hg	Alvarado et al. (2008), Fritioff and Greger (2003), Kara (2004), Miretzky et al. (2004), Mishra et al. (2008), and Robinson et al. (2005)
Star duckweed	*Lemna trisulca* L.	Zn	Huebert and Shay (1992)
Water hyacinth	*Eichhornia crassipes*	As, Fe, Cu, Zn, Pb, Cd, Cr, Ni, Hg	Alvarado et al. (2008), Vesk et al. (1999), Wolverton and McDonald (1978), Chandra and Kulshreshtha (2004), Cordes et al. (2000), Delgado et al. (1993), Dixit and Dhote (2010), Espinoza-Quiñones et al. (2008), Junior et al. (2008), Mishra et al. (2008), Odjegba and Fasidi (2007), and Muramoto and Oki (1983)
Water-starwort	*Callitriche cophocarpa*	Cr(V)	Augustynowicz et al. (2010)
Petries starwort	*Callitriche petriei*	As	Robinson et al. (2005)
Common reed	*Phragmites australis*	Cr, Cu, Ni, Pb, S, V, Zn, Cd	Baldantoni et al. (2009), Deng et al. (2004), and Ghassemzadeh et al. (2008)
Butterfly fern	*Salvinia rotundifolia*	Pb(II)	Banerjee and Sarker (1997) and Dhir (2009)
	Salvinia natans	As, Ni, Cu, Hg(II)	Rahman et al. (2008c), Sen and Bhattacharyya (1993), and Sen and Mondal (1987, 1990)
	Salvinia minima	As, Pb, Cd, Cr	Sanchez-Galvan et al. (2008), Hoffmann et al. (2004), and Olguin et al. (2003)
	Salvinia herzogii	Cd, Cr	Maine et al. (2004) and Suñe et al. (2007)
Eared watermoss	*Salvinia auriculata*	Zn, Hg, Cr	Wolff et al. (2009), Molisani et al. (2006), and Espinoza-Quiñones et al. (2008)
Greater duckweed	*Spirodela intermedia*	Cu, Zn, Mn,Cr, Pb	Miretzky et al. (2004)
	Spirodela polyrhiza L.	As, Hg	Rahman et al. (2007, 2008b), and Mishra et al. (2008)
Indian/Sacred lotus	*Nelymbium speciosum*	Cr, Cu, Ba, Ti, Co, Pb	Vardanyan and Ingole (2006)
	Ludwigia perennis L.		Vardanyan and Ingole (2006)

(continued)

Table 8.1 (continued)

Common name	Scientific name	Trace elements	References
Arrowhead	*Sagittaria sagittiflia* L.		Vardanyan and Ingole (2006)
	Nymphoides ceristatum		Vardanyan and Ingole (2006)
Shoreline seapurslane	*Sasuvium portulacastrum* L.		Vardanyan and Ingole (2006)
-	*Nymphae stellata*		Vardanyan and Ingole (2006)
Water spinach	*Ipomoea aquatica*	As, Cd, Pb, Hg, Cu, Zn	Wang et al. (2008), Göthberg et al. (2002, 2004), Hu et al. (2008), and Lee et al. (1991)
Eelgrass/ Eelweed	*Vallisneria spiralis L.*	Cu, Cd, Hg	Wang et al. (2010) and Rai and Tripathi (2009)
Esthwaite waterweed	*Hydrilla verticillata*	As, Pb, Zn, Cr	Dixit and Dhote (2010) and Lee et al. (1991)
Mosquito fern	*Azolla caroliniana*	As	Zhang et al. (2008)
Water fern	*Azolla filiculoides* *Azolla pinnata*	As, Hg, Cd	Zhang et al. (2008), Rahman et al. (2008a), Rai and Tripathi (2009), and Rai (2008b)
Elephant's ear	*Colocasia esculenta*	Cd, Cu, Pb, Zn	Cardwell et al. (2002)
Umbrella sedge	*Cyperus eragrostis*		Cardwell et al. (2002)
Spike rush	*Eleocharis equisitina*		Cardwell et al. (2002)
Parrot's feather	*Myriophyllum aquaticum*		Cardwell et al. (2002)
Miriophyllum	*Myriophyllum propinquum*	As	Robinson et al. (2005)
Water lily	*Nymphaea violacea*	Cd, Cu, Pb, Zn	Cardwell et al. (2002)
	Nymphaea aurora	Cd	Schor-Fumbarov et al. (2003)
Marshwort	*Nymphoides germinata*	Cd, Cu, Pb, Zn	Cardwell et al. (2002)
Knotweeds	*Persicaria attenuatum*		Cardwell et al. (2002)
-	*Persicaria orientalis*		Cardwell et al. (2002)
-	*Persicaria subsessilis*		Cardwell et al. (2002)
-	*Potamogeton orchreatus*	As	Robinson et al. (2005)

(continued)

Table 8.1 (continued)

Common name	Scientific name	Trace elements	References
Willow smartweed	*Persicaria lapathifolium*	Cd, Cu, Pb, Zn	Cardwell et al. (2002)
-	*Potamogeton javanicus*		Cardwell et al. (2002)
Fennel pondweed	*Potamogeton pectinatus*	Cd, Pb, Cr, Ni, Zn, Cu	Demirezen and Aksoy (2004)
Curled dock	*Rumex crispus*	Cd, Cu, Pb, Zn	Cardwell et al. (2002)
River clubrush	*Schoenoplectus validus*		Cardwell et al. (2002)
Cumbungi	*Typha domingensis*		Cardwell et al. (2002)
Cumbung	*Typha orientalis*		Cardwell et al. (2002)
Lesser Bulrush	*Typha angustifolia*	Cd, Pb, Cr, Ni, Zn, Cu	Chandra and Kulshreshtha (2004) and Demirezen and Aksoy (2004)
Bulrush	*Typha latifolia*	Cr, As, Zn, Pb, Cd, Cu. Ni	Chandra and Kulshreshtha (2004), Ye et al. (1997, 1998), Blute et al. (2004), Deng et al. (2004), Hozhina et al. (2001), Pratas et al. (2007), and Sasmaz et al. (2008)
Waterweed/ Pondweed	*Elodea canadensis*	As, Pb, Cr, Zn, Cu, Cd	Chandra and Kulshreshtha (2004), Dogan et al. (2009), Fritioff and Greger (2003), Mal et al. (2002), Mayes et al. (1977), and Robinson et al. (2005)
Brazilian Waterweed	*Veronica aquatica*	As	Robinson et al. (2005)
Water-milfoil	*Myriophyllum spicatum*	Co, Cr, Cu, Pb, Zn, Ni	Chandra and Kulshreshtha (2004), Keskinkan et al. (2003), and Lesage et al. (2007)
Fragrant water lily	*Nymphaea odorata*	Cr	Chandra and Kulshreshtha (2004)
Pickerelweed	*Pontederia cordata*		Chandra and Kulshreshtha (2004)
Tape grass/ Eel grass	*Vallisneria spiralis*	Hg	Gupta and Chandra (1998)
Wild celery	*Vallisneria americana*	Cr	Chandra and Kulshreshtha (2004)
-	*Nymphaea spontanea*	Cr(VI)	Choo et al. (2006)
Shichito matgrass	*Cyperus malaccensis* Lam.	Pb, Zn, Cu, Cd	Deng et al. (2004)
Swamp rice grass	*Leersia hexandra* Swartz.		Deng et al. (2004)
Burma reed	*Neyraudia reynaudiana*		Deng et al. (2004)

(continued)

Table 8.1 (continued)

Common name	Scientific name	Trace elements	References
Flagroot	*Acorus calamus* L.		Deng et al. (2004)
-	*Eleocharis valleculosa*		Deng et al. (2004)
Water pepper	*Polygonum hydropiper*	As	Robinson et al. (2005)
Reed canary grass	*Phalaris arundinacea* L.	Pb, Zn, Cu, Cd	Deng et al. (2004)
-	*Equisetum ramosisti* Desf.		Deng et al. (2004)
Soft rush	*Juncus effusus* L.		Deng et al. (2004)
-	*Polypogon fugax* Steud.		Deng et al. (2004)
-	*Egeria densa*	As	Robinson et al. (2005)
Alligatorweed	*Althernanthera philoxeroides*	As, Pb	Elayan (1999)
Water lettuce	*Pistia stratiotes*	As, Cr, Pb, Ag, Cd, Cu, Hg, Ni, Zn	Espinoza-Quiñones et al. (2008, 2009), Lee et al. (1991), Maine et al. (2004), and Miretzky et al. (2004)
Floating pondweed	*Potamogeton natans*	Zn, Cu, Cd, Pb	Fritioff and Greger (2003, 2006)
Willow moss	*Fontinalis antipyretica*	Cu, Zn	Goncalves and Boaventura (1998) and Martins and Boaventura (2002)
Needle spikerush	*Eleocharis acicularis*	As, In, Ag, Pb, Cu, Cd, Zn, Sb, Ni, Mg	Ha et al. (2009a, b, 2011)
Rigid hornwort	*Ceratophyllum demersum*	As, Pb, Zn, Cu	Keskinkan et al. (2004) and Robinson et al. (2005)
New Zealand watercress	*Lepidium sativum* L.	As	Robinson et al. (2003)
-	*Najas indica*	Pb	Singh et al. (2010)
Watercresses	*Nasturtium officinale*	Cu, Zn, Ni	Kara (2005)
Curly waterweed	*Lagarosiphon major*	As	Robinson et al. (2005)

This table was previously published in Rahman and Hasegawa (2011), and reused here with the permission of the publisher

treatment and in the use of wetland plants in phytoremediation. Aquatic macrophytes are more suitable for restoration of water quality of wetlands due to their faster growth rate and relatively higher ability of pollutant uptake than terrestrial plants (Ali et al. 2013; Sood et al. 2012).

Water hyacinth (*Eichhornia crassipes*) has been one of the widely studied wetland plants for the phytoremediation of heavy metals in constructed wetlands (Jayaweera et al. 2007, 2008; Zhu et al. 1999; Liao and Chang 2004). It is a fast growing and easily adaptable to various aquatic conditions floating plant with a well-developed fibrous root system and large biomass that can accumulate significant amount of heavy metals from water (Liao and Chang 2004).

Cheng et al. (2002) investigated a twin-shaped constructed wetland comprising a vertical flow (inflow) chamber with *Cyperus alternifolius* followed by a reverse-vertical flow (outflow) chamber with *Villarsia exaltata* for phytoremediation of artificial wastewater polluted by heavy metals. Results showed that the system was very effective in removing toxic heavy metals from wastewater. From a field study with 12 emergent-rooted wetland plant species including different populations of *Leersia hexandra*, *Juncus effusus* and *Equisetum ramosisti*, Deng et al. (2004) proposed that these plants can be used in constructed wetlands for effective removal of toxic metals like Pb, Zn, Cu, and Cd. Removal of Cu, Ni, and Zn by *Phragmites australis* using a horizontal subsurface flow constructed wetland model for domestic wastewater treatment was studied by Galletti et al. (2010). Several other studies showed that engineered/constructed wetlands can be an effective model for phytoremediation of toxic metals (Liu et al. 2007, 2010; Rai 2008a; Sobolewski 1999; Yang and Ye 2009; Zhang et al. 2010).

8.7 Conclusion

Over the past two decades, phytoremediation technology has become increasingly popular and has been employed to restore sites including soils and wetlands contaminated with toxic metals. While this technology has the advantages that environmental concerns may be treated without harming the ecosystems; one major disadvantage of phytoremediation is that it requires relatively longer time compared to traditional physical or chemicals methods as the process is dependent on a plant's ability to grow and thrive in an environment that is not ideal for normal plant growth. There are other limitations (listed below) of phytoremediation approaches for the restoration of contaminated soils and wetlands, which need to be considered for commercial application of this technology:

- Long operational time required for clean-up.
- Phytoremediation efficiency of most metal hyperaccumulators is usually limited by their slow growth rate and biomass production.
- Difficulty in mobilization (bioavailability) of tightly bound fraction of metal ions from soil.
- Phytoremediation is limited to the surface area and depth occupied by the roots of the hyperaccumulators.

- With plant-based systems of remediation, especially in the case of chemically-induced phytoremediation, it is not possible to completely prevent the leaching of contaminants into the groundwater.
- It is ideal for sites with low to moderate levels of metal contamination. In heavily contaminated sites, high concentrations of toxic metals can hamper the normal growth of the hyperaccumulators.
- The survival of the hyperaccumulating plants is affected by the toxicity of the contaminants and the general condition of the soil. Therefore, this approach may not be applicable an environment where the conditions are completely unfavorable for plant growth.
- There is a risk of food chain contamination in case of mismanagement and lack of proper care. Bioaccumulation of contaminants, especially the toxic metals, into primary producers (e.g., phytoplankton) can be passed on to the higher trophic levels of the food chain. Therefore, safe disposal of the used plants or organisms is required.

References

Abhilash PC, Pandey VC, Srivastava P, Rakesh PS, Chandran S, Singh N, Thomas AP (2009) Phytofiltration of cadmium from water by *Limnocharis flava* (L.). Buchenau grown in free-floating culture system. J Hazard Mater 170:791–797

Adesodun JK, Atayese MO, Agbaje TA, Osadiaye BA, Mafe O, Soretire AA (2010) Phytoremediation potentials of sunflowers (*Tithonia diversifolia* and *Helianthus annuus*) for metals in soils contaminated with zinc and lead nitrates. Water Air Soil Pollut 207:195–201

Adriano DC (ed) (2001) Trace elements in the terestrial environment. Springer, New york

Åkesson A, Julin B, Wolk A (2008) Long-term dietary cadmium intake and postmenopausal endometrial cancer incidence: a population-based prospective cohort study. Cancer Res 68:6435–6441

Al Jassir MS, Shaker A, Khaliq MA (2005) Deposition of heavy metals on green leafy vegerables sold on roadsides of Riyadh City. Saudi Arab Bull Environ Contam Toxicol 75:1020–1027

Alam B, Chatterjee AK, Duttagupta S (1995) Bioaccumulation of Cd(II) by water lettuce. Pollut Res 14:59–64

Ali H, Khan E, Sajad MA (2013) Phytoremediation of heavy metals-concepts and applications. Chemosphere 91:869–881

Allen RD (1995) Dissection of oxidative stress tolerance using transgenic plants. Plant Physiol 107:1049

Alvarado S, Guédez M, Lué-Merú MP, Nelson G, Alvaro A, Jesús AC, Gyula Z (2008) Arsenic removal from waters by bioremediation with the aquatic plants water hyacinth (*Eichhornia crassipes*) and lesser duckweed (*Lemna minor*). Bioresour Technol 99:8436–8440

Anderson CWN, Brooks RR, Chiarucci A, Lacoste CJ, Leblanc M, Robinson BH, Simcock R, Stewart RB (1999) Phytomining for nickel, thallium and gold. J Geochem Explor 67:407–415

Assunção AGL, Schat H, Aarts MGM (2003) Thlaspi caerulescens, an attractive model species to study heavy metal hyperaccumulation in plants. New Phytol 159:351–360

Augustynowicz J, Grosicki M, Hanus-Fajerska E, Lekka M, Waloszek A, Koloczek H (2010) Chromium(VI) bioremediation by aquatic macrophyte *Callitriche cophocarpa* Sendtn. Chemosphere 79:1077–1083

Axtell NR, Sternberg SPK, Claussen K (2003) Lead and nickel removal using *Microspora* and *Lemna minor*. Bioresour Technol 89:41–48

Baker AJM, Mcgrath SP, Reeves RD, Smith JAC (2000) Metal hyperaccumulator plants: a review of the ecology and physiology of a biochemical resource for phytoremediation of metal-polluted soils. In: Terry N, Banuelos G (eds) Phytoremediation of contaminated soil and water. Lewis Publication, Boca Raton

Baldantoni D, Ligrone R, Alfani A (2009) Macro- and trace-element concentrations in leaves and roots of *Phragmites australis* in a volcanic lake in Southern Italy. J Geochem Explor 101:166–174

Banerjee G, Sarker S (1997) The role of *Salvinia rotundifolia* in scavenging aquatic Pb (II) pollution: a case study. Bioprocess Eng 17:295–300

Bender J, Lee RF, Phillips P (1995) Uptake and transformation of metals and metalloids by microbial mats and their use in bioremediation. J Ind Microbiol 14:113–118

Bennett LE, Burkhead JL, Hale KL, Terry N, Pilon M, Pilon-Smits EAH (2003) Analysis of transgenic Indian mustard plants for phytoremediation of metal-contaminated mine tailings. J Environ Qual 32:432–440

Bizily SP, Rugh CL, Summers AO, Meagher RB (1999) Phytoremediation of methylmercury pollution: merB expression in Arabidopsis thaliana confers resistance to organomercurials. Proc Natl Acad Sci 96:6808–6813

Blaylock MJ, Haung JW (1999) Phytoextraction of metals. In: Raskin I, Ensley BD (eds) Phytoremediation of toxic metals: using plants to clean up the environment. Wiley, New York

Blaylock MJ, Salt DE, Dushenkov S, Zakharova O, Gussman C, Kapulnik Y, Ensley BD, Raskin I (1997) Enhanced accumulation of Pb in Indian mustard by soil-applied chelating agents. Environ Sci Technol 31:860–865

Blute NK, Brabander DJ, Hemond HF, Sutton SR, Newville MG, Rivers ML (2004) Arsenic sequestration by ferric iron plaque on cattail roots. Environ Sci Technol 38:6074–6077

Bodar CWM, Pronk MEJ, Sijm DTHM (2005) The european union risk assessment on zinc and zinc compounds: the process and the facts. Integr Environ Assess Manag 1:301–319

Borišev M, Pajević S, Nikolić N, Pilipović A, Krstić B, Orlović S (2009) Phytoextraction of Cd, Ni, and Pb using four willow clones (*Salix* spp.). Pol J Environ Stud 18:553

Brooks RR, Robinson BH (1998) Aquatic phytoremediation by accumulator plants. In: Brooks RR (ed) Plants that hyperaccumulate heavy metals: their role in archaeology, microbiology, mineral exploration, phytomining and phytoremediation. CAB International, Wallingford

Brooks RR, Chambers MF, Nicks LJ, Robinson BH (1998) Phytomining. Trends Plant Sci 3:359–362

Brunner I, Luster J, Günthardt-Goerg MS, Frey B (2008) Heavy metal accumulation and phytostabilisation potential of tree fine roots in a contaminated soil. Environ Pollut 152:559–568

Cardwell AJ, Hawker DW, Greenway M (2002) Metal accumulation in aquatic macrophytes from southeast Queensland, Australia. Chemosphere 48:653–663

Chandra P, Kulshreshtha K (2004) Chromium accumulation and toxicity in aquatic vascular plants. Bot Rev 70:313–327

Chaney RL, Malik M, Li YM, Brown SL, Brewer EP, Angle JS, Baker AJM (1997) Phytoremediation of soil metals. Curr Opin Biotechnol 8:279–284

Cheng S, Grosse W, Karrenbrock F, Thoennessen M (2002) Efficiency of constructed wetlands in decontamination of water polluted by heavy metals. Ecol Eng 18:317–325

Cherian S, Oliveira MM (2005) Transgenic plants in phytoremediation: recent advances and new possibilities. Environ Sci Technol 39:9377–9390

Choo TP, Lee CK, Low KS, Hishamuddin O (2006) Accumulation of chromium(VI) from aqueous solutions using water lilies (*Nymphaea spontanea*). Chemosphere 62:961–967

Clemens S (2001) Developing tools for phytoremediation: towards a molecular understanding of plant metal tolerance and accumulation. Int J Occup Med Environ Health 14:235–239

Clemens S, Palmgren MG, Krämer U (2002) A long way ahead: understanding and engineering plant metal accumulation. Trends Plant Sci 7:309–315
Cobbett CS (2000) Phytochelatins and their roles in heavy metal detoxification. Plant Physiol 123:825–832
Cobbett C, Goldsbrough P (2002) Phytochelatins and metallothioneins: roles in heavy metal detoxification and homeostasis. Annu Rev Plant Biol 53:159–182
Conesa HM, Faz Á, Arnaldos R (2007) Initial studies for the phytostabilization of a mine tailing from the Cartagena-La Union Mining District (SE Spain). Chemosphere 66:38–44
Cooper EM, Sims JT, Cunningham SD, Huang JW, Berti WR (1999) Chelate-assisted phytoextraction of lead from contaminated soils. J Environ Qual 28:1709–1719
Cordes KB, Mehra A, Farago ME, Banerjee DK (2000) Uptake of Cd, Cu, Ni and Zn by the water hyacinth, *Eichhornia crassipes* (Mart.) solms from pulverised fuel ash (PFA) leachates and slurries. Environ Geochem Health 22:297–316
Cunningham SD, Berti WR (2000) Phytoextraction and phytostabilization: technical, cconomic, and regulatory considerations of the soil–lead issue. In: Terry N, Bañuelos G (eds) Phytoremediation of contaminated soil and water. Taylor & Francis, Boca Raton
Cunningham SD, Ow DW (1996) Promises and prospects of phytoremediation. Plant Physiol 110:715–719
Del Río M, Font R, Almela C, Vélez D, Montoro R, De Haro Bailón A (2002) Heavy metals and arsenic uptake by wild vegetation in the Guadiamar river area after the toxic spill of the Aznalcóllar mine. J Biotechnol 98:125–137
Delgado M, Bigeriego M, Guardiola E (1993) Uptake of Zn, Cr and Cd by water hyacinths. Water Res 27:269–272
Demirezen D, Aksoy A (2004) Accumulation of heavy metals in *Typha angustifolia* (L.) and *Potamogeton pectinatus* (L.) living in Sultan Marsh (Kayseri, Turkey). Chemosphere 56:685–696
Demirezen D, Aksoy A (2006) Heavy metal levels in vegetables in Turkey are within safe limits for Cu, Zn, Ni and exceeded for Cd and Pb. J Food Qual 29:252–265
Demirezen D, Aksoy A, Uruç K (2007) Effect of population density on growth, biomass and nickel accumulation capacity of *Lemna gibba* (Lemnaceae). Chemosphere 66:553–557
Deng H, Ye ZH, Wong MH (2004) Accumulation of lead, zinc, copper and cadmium by 12 wetland plant species thriving in metal-contaminated sites in China. Environ Pollut 132:29–40
Dhir B (2009) *Salvinia*: an aquatic fern with potential use in phytoremediation. Environ We Int J Sci Technol 4:23–27
Dietz AC, Schnoor JL (2001) Advances in phytoremediation. Environ Health Perspect 109:163–168
Dixit S, Dhote S (2010) Evaluation of uptake rate of heavy metals by *Eichhornia crassipes* and *Hydrilla verticillata*. Environ Monit Assess 169:367–374
Dogan M, Saygideger S, Colak U (2009) Effect of lead toxicity on aquatic macrophyte *Elodea canadensis* Michx. Bull Environ Contam Toxicol 83:249–254
Dushenkov V, Kumar PBAN, Motto H, Raskin I (1995) Rhizofiltration: the use of plants to remove heavy metals from aqueous streams. Environ Sci Technol 29:1239–1245
EC (1986) On the protection of the environment, and in particular of the soil, when sewage sludge is used in agriculture. In: EC (ed) Council directive 86/278/EEC, In: EC Official Journal L181. European Community, Brussels
EC (2004) Communication from the Commission to the Council, the European Parliament, the European Economic and Social Committee and the Committee of the Regions: towards a thematic strategy on the urban environment. In: EC (ed) COM(2004)60 final. Commission of the European Communities, Brussels
EEA (2003) Europe's environment: the third assessment. European Environment Agency, Copenhagen
Egli T (2001) Biodegradation of metal-complexing aminopolycarboxylic acids. J Biosci Bioeng 92:89–97

Elayan NMS (1999) Phytoremediation of arsenic and lead from contaminated waters by the emergent aquatic macrophyte Althernanthera philoxeroides (alligatorweed). Southern University, Louisiana, USA

Elliott HA, Brown GA (1989) Comparative evaluation of NTA and EDTA for extractive decontamination of Pb-polluted soils. Water Air Soil Pollut 45:361–369

Erakhrumen A, Agbontalor A (2007) Phytoremediation: an environmentally sound technology for pollution prevention, control and remediation in developing countries. Educ Res Rev 7:151–156

Espinoza-Quiñones F, Da Silva E, De Almeida Rizzutto M, Palácio S, Módenes A, Szymanski N, Martin N, Kroumov A (2008) Chromium ions phytoaccumulation by three floating aquatic macrophytes from a nutrient medium. World J Microbiol Biotechnol 24:3063–3070

Espinoza-Quiñones F, Módenes A, Costa I, Palácio S, Szymanski N, Trigueros D, Kroumov A, Silva E (2009) Kinetics of lead bioaccumulation from a hydroponic medium by aquatic macrophytes *Pistia stratiotes*. Water Air Soil Pollut 203:29–37

Evangelou MWH, Ebel M, Schaeffer A (2007) Chelate assisted phytoextraction of heavy metals from soil. Effect, mechanism, toxicity, and fate of chelating agents. Chemosphere 68:989–1003

Flathman PE, Lanza GR (1998) Phytoremediation: current views on an emerging green technology. J Soil Contam 7:415–432

Fritioff A, Greger M (2003) Aquatic and terrestrial plant species with potential to remove heavy metals from stormwater. Int J Phytorem 5:211–224

Fritioff A, Greger M (2006) Uptake and distribution of Zn, Cu, Cd, and Pb in an aquatic plant *Potamogeton natans*. Chemosphere 63:220–227

Gaetke LM, Chow CK (2003) Copper toxicity, oxidative stress, and antioxidant nutrients. Toxicology 189:147–163

Galletti A, Verlicchi P, Ranieri E (2010) Removal and accumulation of Cu, Ni and Zn in horizontal subsurface flow constructed wetlands: contribution of vegetation and filling medium. Sci Total Environ 408:5097–5105

Gallon C, Munger C, Premont S, Campbell PGC (2004) Hydroponic study of aluminum accumulation by aquatic plants: effects of fluoride and pH. Water Air Soil Pollut 153:135–155

Gardea-Torresdey JL, Gonzalez JH, Tiemann KJ, Rodriguez O, Gamez G (1998) Phytofiltration of hazardous cadmium, chromium, lead and zinc ions by biomass of *Medicago sativa* (Alfalfa). J Hazard Mater 57:29–39

Gardea Torresdey JL, De La Rosa G, Peralta-Videa JR (2004) Use of phytofiltration technologies in the removal of heavy metals: a review. Pure Appl Chem 76:801–813

Ghassemzadeh F, Yousefzadeh H, Arbab-Zavar MH (2008) Arsenic phytoremediation by *Phragmites australis*: green technology. Int J Environ Stud 65:587–594

Ghosh S (2010) Wetland macrophytes as toxic metal accumulators. Int J Environ Sci 1:523–528

Gleba D, Borisjuk NV, Borisjuk LG, Kneer R, Poulev A, Skarzhinskaya M, Dushenkov S, Logendra S, Gleba YY, Raskin I (1999) Use of plant roots for phytoremediation and molecular farming. Proc Natl Acad Sci 96:5973–5977

Glover-Kerkvliet J (1995) Environmental assault on immunity. Environ Health Perspect 103:236–239

Goncalves EP, Boaventura RAR (1998) Uptake and release kinetics of copper by the aquatic moss *Fontinalis antipyretica*. Water Res 32:1305–1313

Göthberg A, Greger M, Bengtsson BE (2002) Accumulation of heavy metals in water spinach (*Ipomoea aquatica*) cultivated in the Bangkok region, Thailand. Environ Toxicol Chem 21:1934–1939

Göthberg A, Greger M, Holm K, Bengtsson BE (2004) Influence of nutrient levels on uptake and effects of mercury, cadmium, and lead in water spinach. J Environ Qual 33:1247–1255

Grčman H, Velikonja-Bolta Š, Vodnik D, Kos B, Leštan D (2001) EDTA enhanced heavy metal phytoextraction: metal accumulation, leaching and toxicity. Plant Soil 235:105–114

Greger M, Landberg T (1999) Use of willow in phytoextraction. Int J Phytorem 1:115–123

Guerinot ML, Salt DE (2001) Fortified foods and phytoremediation. Two sides of the same coin. Plant Physiol 125:164–167

Gulz PA, Gupta S-K, Schulin R (2005) Arsenic accumulation of common plants from contaminated soils. Plant Soil 272:337–347

Gupta M, Chandra P (1998) Bioaccumulation and toxicity of mercury in rooted-submerged macrophyte *Vallisneria spiralis*. Environ Pollut 103:327–332

Ha NTH, Sakakibara M, Sano S (2009a) Phytoremediation of Sb, As, Cu, and Zn from contaminated water by the aquatic macrophyte *Eleocharis acicularis*. Clean: Soil, Air, Water 37:720–725

Ha NTH, Sakakibara M, Sano S, Hori RS, Sera K (2009b) The potential of *Eleocharis acicularis* for phytoremediation: case study at an abandoned mine site. Clean: Soil, Air, Water 37:203–208

Ha NTH, Sakakibara M, Sano S (2011) Accumulation of Indium and other heavy metals by *Eleocharis acicularis*: an option for phytoremediation and phytomining. Bioresour Technol 102:2228–2234

Hall JL (2002) Cellular mechanisms for heavy metal detoxification and tolerance. J Exp Bot 53:1–11

Hammond-Kosack KE, Jones J (1996) Resistance gene-dependent plant defense responses. Plant Cell 8:1773

Hasan SH, Talat M, Rai S (2007) Sorption of cadmium and zinc from aqueous solutions by water hyacinth (*Eichchornia crassipes*). Bioresour Technol 98:918–928

Heaton AC, Rugh CL, Wang N-J, Meagher RB (1998) Phytoremediation of mercury-and methylmercury-polluted soils using genetically engineered plants. J Soil Contam 7:497–509

Henry JR (2000) An overview of phytoremediation of lead and mercury. U.S. Environmental Protection Agency, Washington, DC

Hernández-Ochoa I, García-Vargas G, López-Carrillo L, Rubio-Andrade M, Morán-Martínez J, Cebrián ME, Quintanilla-Vega B (2005) Low lead environmental exposure alters semen quality and sperm chromatin condensation in northern Mexico. Reprod Toxicol 20:221–228

Hillel D (2005) Heavy metals. In: Adriano DC, Bolan NS, Vangronsveld J, Wenzel WW (eds) Encyclopedia of soils in the environment. Elsevier, Amsterdam

Hoffmann T, Kutter C, Santamaria J (2004) Capacity of *Salvinia minima* Baker to tolerate and accumulate As and Pb. Eng Life Sci 4:61–65

Hou WH, Chen X, Song GL, Wang QH, Chang CC (2007) Effects of copper and cadmium on heavy metal polluted waterbody restoration by duckweed (*Lemna minor*). Plant Physiol Biochem 45:62–69

Hozhina EI, Khramov AA, Gerasimov PA, Kumarkov AA (2001) Uptake of heavy metals, arsenic, and antimony by aquatic plants in the vicinity of ore mining and processing industries. J Geochem Explor 74:153–162

Hu MH, Ao YS, Yang XE, Li TQ (2008) Treating eutrophic water for nutrient reduction using an aquatic macrophyte (*Ipomoea aquatica* Forsskal) in a deep flow technique system. Agric Water Manag 95:607–615

Huang JW, Chen J, Berti WB, Cunningham SD (1997) Phytoremediation of lead-contaminated soils: role of synthetic chelates in lead phytoextraction. Environ Sci Technol 31:800–805

Huang JW, Poynton CY, Kochian LV, Elless MP (2004) Phytofiltration of arsenic from drinking water using arsenic-hyperaccumulating ferns. Environ Sci Technol 38:3412–3417

Huebert DB, Shay JM (1992) Zinc toxicity and its interaction with cadmium in the submerged aquatic macrophyte *Lemna trisulca* L. Environ Toxicol Chem 11:715–720

Hughes MF (2002) Arsenic toxicity and potential mechanisms of action. Toxicol Lett 133:1–16

Hutchinson GE (1975) A treatise on limnology. Wiley, London

Ingole NW, Ting JP (2002) Study on nutrient removal potential of selected aquatic macrophytes. J Inst Eng (India) Environ Eng Div 83:1–6

Islam E, Yang X-E, He Z-L, Mahmood Q (2007) Assessing potential dietary toxicity of heavy metals in selected vegetables and food crops. J Zhejiang Univ (Sci B) 8:1–13

Jabeen R, Ahmad A, Iqbal M (2009) Phytoremediation of heavy metals: physiological and molecular mechanisms. Bot Rev 75:339–364

Jayaweera MW, Kasturiarachchi JC, Kularatne RKA, Wijeyekoon SLJ (2007) Removal of aluminium by constructed wetlands with water hyacinth (Eichhornia crassipes (Mart.) Solms) grown under different nutritional conditions. J Environ Sci Health A Tox Hazard Subst Environ Eng 42:185–193

Jayaweera MW, Kasturiarachchi JC, Kularatne RKA, Wijeyekoon SLJ (2008) Contribution of water hyacinth (Eichhornia crassipes (Mart.) Solms) grown under different nutrient conditions to Fe-removal mechanisms in constructed wetlands. J Environ Manag 87:450–460

Junior ACG, Lindino CA, Da Rosa MF, Bariccatti R, Gomes GD (2008) Removal of toxic heavy metals cadmium, lead and chromium from swine biofertilizer, using an aquatic macrophyte (*Eichornia crassipes*) as a bioindicator. Acta Sci Technol 30:9–14

Kamal M, Ghaly AE, Mahmoud N, Cote R (2004) Phytoaccumulation of heavy metals by aquatic plants. Environ Int 29:1029–1039

Kara Y (2004) Bioaccumulation of copper from contaminated wastewater by using *Lemna minor*. Bull Environ Contam Toxicol 72:467–471

Kara Y (2005) Bioaccumulation of Cu, Zn and Ni from the wastewater by treated *Nasturtium officinale*. Int J Environ Sci Technol 2:63–67

Kedziorek MAM, Dupuy A, Bourg ACM, Compère F (1998) Leaching of Cd and Pb from a polluted soil during the percolation of EDTA: laboratory column experiments modeled with a non-equilibrium solubilization step. Environ Sci Technol 32:1609–1614

Keskinkan O, Goksu MZL, Yuceer A, Basibuyuk M, Forster CF (2003) Heavy metal adsorption characteristics of a submerged aquatic plant (*Myriophyllum spicatum*). Process Biochem 39:179–183

Keskinkan O, Goksu MZL, Basibuyuk M, Forster CF (2004) Heavy metal adsorption properties of a submerged aquatic plant (*Ceratophyllum demersum*). Bioresour Technol 92:197–200

Kłos A, Czora M, Rajfur M, Wacławek M (2012) Mechanisms for translocation of heavy metals from soil to epigeal mosses. Water Air Soil Pollut 223:1829–1836

Kos B, Lestan D (2004) Chelator induced phytoextraction and in situ soil washing of Cu. Environ Pollut 132:333–339

Kraemer U (2003) Phytoremediation to phytochelatin–plant trace metal homeostasis. New Phytol 158:4–6

Kuiper I, Lagendijk EL, Bloemberg GV, Lugtenberg BJJ (2004) Rhizoremediation: a beneficial plant microbe interaction. Mol Plant-Microbe Interact 17:6–15

Lee CK, Low KS, Hew NS (1991) Accumulation of arsenic by aquatic plants. Sci Total Environ 103:215–227

Lesage E, Mundia C, Rousseau DPL, Van De Moortel AMK, Du Laing G, Meers E, Tack FMG, De Pauw N, Verloo MG (2007) Sorption of Co, Cu, Ni and Zn from industrial effluents by the submerged aquatic macrophyte *Myriophyllum spicatum* L. Ecol Eng 30:320–325

Leštan D, Luo C-L, Li X-D (2008) The use of chelating agents in the remediation of metal-contaminated soils: a review. Environ Pollut 153:3–13

Li X, Poon C-S, Liu PS (2001) Heavy metal contamination of urban soils and street dusts in Hong Kong. Appl Geochem 16:1361–1368

Li JT, Liao B, Lan CY, Ye ZH, Baker AJM, Shu WS (2010) Cadmium tolerance and accumulation in cultivars of a high-biomass tropical tree (*Averrhoa carambola*) and its potential for phytoextraction. J Environ Qual 39:1262–1268

Liao S-W, Chang W-L (2004) Heavy metal phytoremediation by water hyacinth at constructed wetlands in Taiwan. J Aquat Plant Manag 42:60–68

Liu J, Dong Y, Xu H, Wang D, Xu J (2007) Accumulation of Cd, Pb and Zn by 19 wetland plant species in constructed wetland. J Hazard Mater 147:947–953

Liu JG, Li GH, Shao WC, Xu JK, Wang DK (2010) Variations in uptake and translocation of copper, chromium and nickel among nineteen wetland plant species. Pedosphere 20:96–103

Lone MI, He Z-L, Stoffella PJ, Yang X-E (2008) Phytoremediation of heavy metal polluted soils and water: progresses and perspectives. J Zhejiang Univ Sci B 9:210–220

Low KS, Lee CK, Tai CH (1994) Biosorption of copper by water hyacinth roots. J Environ Sci Health, Part A: Environ Sci Eng 29:171–188

Luo C, Shen Z, Li X (2005) Enhanced phytoextraction of Cu, Pb, Zn and Cd with EDTA and EDDS. Chemosphere 59:1–11

Luo C, Shen Z, Li X, Baker AJM (2006) Enhanced phytoextraction of Pb and other metals from artificially contaminated soils through the combined application of EDTA and EDDS. Chemosphere 63:1773–1784

Lytle CM, Lytle PW, Yang N, Qian JH, Hansen D, Zayed A, Terry N (1998) Reduction of Cr (VI) to Cr(III) by wetland plants: potential for in situ heavy metal detoxification. Environ Sci Technol 32:3087–3093

Ma LQ, Komar KM, Tu C, Zhang W, Cai Y (2001) A fern that hyperaccumulates arsenic. Nature (-London) 409:579

Macnair MR, Tilstone GH, Smith SE (2000) The genetics of metal tolerance and accumulation in higher plants. In: Terry N, Bañuelos G (eds) Phytoremediation of contaminated soil and water. Taylor & Francis, Boca Raton

Maine MA, Suñé NL, Lagger SC (2004) Chromium bioaccumulation: comparison of the capacity of two floating aquatic macrophytes. Water Res 38:1494–1501

Mal TK, Adorjan P, Corbett AL (2002) Effect of copper on growth of an aquatic macrophyte, *Elodea canadensis*. Environ Pollut 120:307–311

Marchand L, Mench M, Jacob DL, Otte ML (2010) Metal and metalloid removal in constructed wetlands, with emphasis on the importance of plants and standardized measurements: a review. Environ Pollut 158:3447–3461

Marchiol L, Fellet G, Perosa D, Zerbi G (2007) Removal of trace metals by *Sorghum bicolor* and *Helianthus annuus* in a site polluted by industrial wastes: a field experience. Plant Physiol Biochem 45:379–387

Martins RJE, Boaventura RAR (2002) Uptake and release of zinc by aquatic bryophytes (*Fontinalis antipyretica* L. ex. Hedw.). Water Res 36:5005–5012

Mayes RA, Macintosh AW, Anderson VL (1977) Uptake of cadmium and lead by a rooted aquatic macrophyte (*Elodea canadensis*). Ecology 58:1176–1180

Mcgrath SP, Zhao F-J (2003) Phytoextraction of metals and metalloids from contaminated soils. Curr Opin Biotechnol 14:277–282

Mcintyre T (2003) Phytoremediation of heavy metals from soils. *Phytoremediation*. Springer, Berlin/Heidelberg

Meagher RB (2000) Phytoremediation of toxic elemental and organic pollutants. Curr Opin Plant Biol 3:153–162

Means JL, Kucak T, Crerar DA (1980) Relative degradation rates of NTA, EDTA and DTPA and environmental implications. Environ Pollut Ser B Chem Phys 1:45–60

Meers E, Slycken SV, Adriaensen K, Ruttens A, Vangronsveld J, Laing GD, Witters N, Thewys T, Tack FMG (2010) The use of bio-energy crops (Zea mays) for 'phytoremediation' of heavy metals on moderately contaminated soils: a field experiment. Chemosphere 78:35–41

Mendez MO, Maier RM (2008) Phytostabilization of mine tailings in arid and semiarid environments-an emerging remediation technology. Environ Health Perspect 116:278–283

Mesjasz-Przybyłowicz J, Nakonieczny M, Migula P, Augustyniak M, Tarnawska M, Reimold W, Koeberl C, Przybyłowicz W, Głowacka E (2004) Uptake of cadmium, lead nickel and zinc from soil and water solutions by the nickel hyperaccumulator *Berkheya coddii*. Acta Biol Cracov Ser Bot 46:75–85

Miretzky P, Saralegui A, Cirelli AF (2004) Aquatic macrophytes potential for the simultaneous removal of heavy metals (Buenos Aires, Argentina). Chemosphere 57:997–1005

Mishra S, Srivastava S, Tripathi RD, Kumar R, Seth CS, Gupta DK (2006) Lead detoxification by coontail (*Ceratophyllum demersum* L.) involves induction of phytochelatins and antioxidant system in response to its accumulation. Chemosphere 65:1027–1039

Mishra V, Upadhyay A, Pathak V, Tripathi B (2008) Phytoremediation of mercury and arsenic from tropical opencast coalmine effluent through naturally occurring aquatic macrophytes. Water Air Soil Pollut 192:303–314

Mkandawire M, Dudel EG (2005) Accumulation of arsenic in *Lemna gibba* L. (ducweed) in tailing waters of two abandoned uranium mining sites in Saxony, Germany. Sci Total Environ 336:81–89

Mkandawire M, Lyubun YV, Kosterin PV, Dudel EG (2004a) Toxicity of arsenic species to Lemna gibba L. and the influence of phosphate on arsenic bioavailability. Environ Toxicol 19:26–35

Mkandawire M, Taubert B, Dudel EG (2004b) Capacity of *Lemna gibba* L. (Duckweed) for uranium and arsenic phytoremediation in mine tailing waters. Int J Phytorem 6:347–362

Molisani MM, Rocha R, Machado W, Barreto RC, Lacerda LD (2006) Mercury contents in aquatic macrophytes from two reservoirs in the Paraíba do Sul: Guandú river system, SE Brazil. Braz J Biol 66:101–107

Mukhopadhyay S, Maiti SK (2010) Phytoremediation of metal enriched mine waste: a review. Glob J Environ Res 4:135–150

Muramoto S, Oki Y (1983) Removal of some heavy metals from polluted water by water hyacinth (*Eichhornia crassipes*). Bull Environ Contam Toxicol 30:170–177

Newman LA, Reynolds CM (2004) Phytodegradation of organic compounds. Curr Opin Biotechnol 15:225–230

Nicks LJ, Chambers MF (1998) A pioneering study of the potential of phytomining for nickel. In: Brooks RR (ed) Plants that hyperaccumulate heavy metals. CAB International, Oxford

Noctor G, Arisi A-CM, Jouanin L, Kunert KJ, Rennenberg H, Foyer CH (1998) Glutathione: biosynthesis, metabolism and relationship to stress tolerance explored in transformed plants. J Exp Bot 49:623–647

Nordberg GF (1996) Current issues in low-dose cadmium toxicology: nephrotoxicity and carcinogenicity. Environ Sci 4:133–147

Nörtemann B (1999) Biodegradation of EDTA. Appl Microbiol Biotechnol 51:751–759

Norvell WA (1984) Comparison of chelating agents as extractants for metals in diverse soil materials. Soil Sci Soc Am J 48:1285–1292

Nowack B, Vanbriesen JM (2005) Chelating agents in the environment. In: Nowack B, Vanbriesen JM (eds) Biogeochemistry of chelating agents. American Chemical Society, Washington, DC

Odjegba VJ, Fasidi IO (2007) Phytoremediation of heavy metals by *Eichhornia crassipes*. Environmentalist 27:349–355

Olguin E, Rodriguez D, Sanchez G, Hernandez E, Ramirez M (2003) Productivity, protein content and nutrient removal from anaerobic effluents of coffee wastewater in *Salvinia minima* ponds, under subtropical conditions. Acta Biotechnol 23:259–270

Outridge PM, Noller BN (1991) Accumulation of toxic trace elements by freshwater vascular plants. Rev Environ Contam Toxicol 121:1–63

Oviedo C, Rodríguez J (2003) EDTA: the chelating agent under environmental scrutiny. Quim Nova 26:901–905

Ow D, Shewry P, Napier J, Davis P (1998) Prospects of engineering heavy metal detoxification genes in plants. Symposium of the Industrial Biochemistry and Biotechnology Group of the Biochemical Society, Bristol

Padmavathiamma PK, Li LY (2007) Phytoremediation technology: hyper-accumulation metals in plants. Water Air Soil Pollut 184:105–126

Palumbo AV, Lee SY, Boerman P (1994) The effect of media composition on EDTA degradation by *Agrobacterium* sp. Appl Biochem Biotechnol 45–46:811–822

Peng K, Luo C, Lou L, Li X, Shen Z (2008) Bioaccumulation of heavy metals by the aquatic plants *Potamogeton pectinatus* L. and *Potamogeton malaianus* Miq. and their potential use for contamination indicators and in wastewater treatment. Sci Total Environ 392:22–29

Peuke AD, Rennenberg H (2005a) Phytoremediation. EMBO Rep 6:497–501

Peuke AD, Rennenberg H (2005b) Phytoremediation with transgenic trees. Z Naturforsch C 60c:199–207

Pillay AE, Williams JR, El Mardi MO, Al-Lawati SMH, Al-Hadabbi MH, Al-Hamdi A (2003) Risk assessment of chromium and arsenic in date palm leaves used as livestock feed. Environ Int 29:541–545

Pilon-Smits E, Pilon M (2002) Phytoremediation of metals using transgenic plants. Crit Rev Plant Sci 21:439–456

Pinto E, Sigaud-Kutner T, Leitao MA, Okamoto OK, Morse D, Colepicolo P (2003) Heavy metal-induced oxidative stress in algae. J Phycol 39:1008–1018

Prasad MNV (2003) Phytoremediation of metal-polluted ecosystems: hype for commercialization. Russ J Plant Physiol 50:686–701

Prasad MNV (2004) Phytoremediation of metals in the environment for sustainable development. Proc Indian Natl Sci Acad Part B 70:71–98

Pratas J, Favas P, Rodrigues N, Prasad M (2007) Arsenic accumulation in aquatic plants (Central Portugal). In: The 6th WSEAS international conference, 2007 Arcachon, World Science and Engineering Academy and Society (WSEAS), France, pp 73–76

Qin J, Rosen BP, Zhang Y, Wang G, Franke S, Rensing C (2006) Arsenic detoxification and evolution of trimethylarsine gas by a microbial arsenite S-adenosylmethionine methyltransferase. Proc Natl Acad Sci U S A 103:2075–2080

Qin J, Lehr CR, Yuan C, Le XC, Mcdermott TR, Rosen BP (2009) Biotransformation of arsenic by a Yellowstone thermoacidophilic eukaryotic alga. Proc Natl Acad Sci 106:5213–5217

Raab A, Feldmann J (2003) Microbial transformation of metals and metalloids. Sci Prog 86:179–202

Rafati M, Khorasani N, Moattar F, Shirvany A, Moraghebi F, Hosseinzadeh S (2011) Phytoremediation potential of *Populus alba* and *Morus alba* for cadmium, chromuim and nickel absorption from polluted soil. Int J Environ Res 5:961–970

Rahman MA, Hasegawa H (2011) Aquatic arsenic: phytoremediation using floating macrophytes. Chemosphere 83:633–646

Rahman MA, Hassler C (2014) Is arsenic biotransformation a detoxification mechanism for microorganisms? Aquat Toxicol 146:212–219

Rahman MA, Hasegawa H, Ueda K, Maki T, Okumura C, Rahman MM (2007) Arsenic accumulation in duckweed (*Spirodela polyrhiza* L.): a good option for phytoremediation. Chemosphere 69:493–499

Rahman MA, Hasegawa H, Kitahara K, Maki T, Ueda K, Rahman MM (2008a) The effects of phosphorous on the accumulation of arsenic in water fern (*Azolla pinnata* L.). J Ecotechnol Res 14:21–24

Rahman MA, Hasegawa H, Ueda K, Maki T, Rahman MM (2008b) Arsenic uptake by aquatic macrophyte *Spirodela polyrhiza* L.: interactions with phosphate and iron. J Hazard Mater 160:356–361

Rahman MA, Hasegawa H, Ueda K, Maki T, Rahman MM (2008c) Influence of phosphate and iron ions in selective uptake of arsenic species by water fern (*Salvinia natans* L.). Chem Eng J 145:179–184

Rahman MA, Ismail MMR, Rahman MM, Hasegawa H (2011) Arsenic in the environment: phytoremediation using aquatic macrophytes. In: Golubev IA (ed) Handbook of phytoremediation. Nova, Hauppauge

Rai PK (2008a) Heavy metal pollution in aquatic ecosystems and its phytoremediation using wetland plants: an ecosustainable approach. Int J Phytorem 10:133–160

Rai PK (2008b) Phytoremediation of Hg and Cd from industrial effluents using an aquatic free floating macrophyte *Azolla pinnata*. Int J Phytorem 10:430–439

Rai PK, Tripathi BD (2009) Comparative assessment of *Azolla pinnata* and *Vallisneria spiralis* in Hg removal from G.B. Pant Sagar of Singrauli industrial region, India. Environ Monit Assess 148:75–84

Rascio N, Navari-Izzo F (2011) Heavy metal hyperaccumulating plants: how and why do they do it? And what makes them so interesting? Plant Sci 180:169–181

Raskin I, Ensley BD (2000) Phytoremediation of toxic metals: using plants to clean up the environment. John Wiley & Sons, New York

Rauser WE (1995) Phytochelatins and related peptides. Structure, biosynthesis, and function. Plant Physiol 109:1141–1149

Rauser WE (1999) Structure and function of metal chelators produced by plants. Cell Biochem Biophys 31:19–48

Reichman SR, Parker DR (2005) Metal complexation by phytosiderophores in the rhizosphere. In: Huang PM, Gobran GR (eds) Biogeochemistry of trace elements in the rhizophere. Elsevier, Toronto

Robinson BH, Brooks RR, Howes AW, Kirkman JH, Gregg PEH (1997) The potential of the high-biomass nickel hyperaccumulator *Berkheya coddii* for phytoremediation and phytomining. J Geochem Explor 60:115–126

Robinson BH, Leblanc M, Petit D, Brooks RR, Kirkman JH, Gregg PEH (1998) The potential of *Thlaspi caerulescens* for phytoremediation of contaminated soils. Plant Soil 203:47–56

Robinson BH, Mills TM, Petit D, Fung LE, Green SR, Clothier BE (2000) Natural and induced cadmium-accumulation in poplar and willow: implications for phytoremediation. Plant Soil 227:301–306

Robinson B, Duwing C, Bolan N, Kannathasan M, Saravanan A (2003) Uptake of arsenic by New Zealand watercress (*Lepidium sativum* L.). Sci Total Environ 301:67–73

Robinson B, Marchetti M, Moni C, Schroeter L, Van Den Dijssel C, Milne G, Bolan N, Mahimairaja S (2005) Arsenic accumulation by aquatic and terrestrial plants. In: Naidu R, Smith E, Owens G, Bhattacharya P, Nadebaum P (eds) Managing arsenic in the environment: from soil to human health. CSIRO, Collingwood

Robinson B, Kim N, Marchetti M, Moni C, Schroeter L, Van Den Dijssel C, Milne G, Clothier B (2006) Arsenic hyperaccumulation by aquatic macrophytes in the Taupo Volcanic Zone, New Zealand. Environ Exp Bot 58:206–215

Rugh CL, Senecoff JF, Meagher RB, Merkle SA (1998) Development of transgenic yellow poplar for mercury phytoremediation. Nat Biotechnol 16:925–928

Sakakibara M, Watanabe A, Inoue M, Sano S, Kaise T (2010) Phytoextraction and phytovolatilization of arsenic from as-contaminated soils by Pteris vittata. In: Proceedings of the annual international conference on soils, sediments, water and energy, vol 12, Article 26. Available at: http://scholarworks.umass.edu/soilsproceedings/vol12/iss1/26

Sakakibara M, Ohmori Y, Ha NTH, Sano S, Sera K (2011) Phytoremediation of heavy metal-contaminated water and sediment by *Eleocharis acicularis*. Clean: Soil, Air, Water 39:735–741

Salt DE, Blaylock M, Kumar NPBA, Dushenkov V, Ensley BD, Chet I, Raskin I (1995a) Phytoremediation: a novel strategy for the removal of toxic metals from the environment using plants. Nat Biotechnol 13:468–474

Salt DE, Prince RC, Pickering IJ, Raskin I (1995b) Mechanisms of cadmium mobility and accumulation in Indian mustard. Plant Physiol 109:1427–1433

Salt DE, Smith RD, Raskin I (1998) Phytoremediation. Annu Rev Plant Physiol Plant Mol Biol 49:643–668

Sanchez-Galvan G, Monroy O, Gomez J, Olguin EJ (2008) Assessment of the hyperaccumulating lead capacity of *Salvinia minima* using bioadsorption and intracellular accumulation factors. Water Air Soil Pollut 194:77–90

Sasmaz A, Obek E, Hasar H (2008) The accumulation of heavy metals in *Typha latifolia* L. grown in a stream carrying secondary effluent. Ecol Eng 33:278–284

Schor-Fumbarov T, Keilin Z, Tel-Or E (2003) Characterization of cadmium uptake by the water lily *Nymphaea aurora*. Int J Phytorem 5:169–179

Sekara A, Poniedzialeek M, Ciura J, Jedrszczyk E (2005) Cadmium and lead accumulation and distribution in the organs of nine crops: implications for phytoremediation. Pol J Environ Stud 14:509–516

Selvapathy P, Sreedhar P (1991) Heavy metals removal by water hyacinth. J Indian Publ Health Eng 3:11–17

Sen AK, Bhattacharyya M (1993) Studies on uptake and toxic effects of lead on *Salvinia natans*. Indian J Environ Health 35:308–320

Sen AK, Mondal NG (1987) *Salvinia natans* as the scavenger of Hg (II). Water Air Soil Pollut 34:439–446

Sen AK, Mondal NG (1990) Removal and uptake of copper (II) by *Salvinia natans* from wastewater. Water Air Soil Pollut 49:1–6

Seth CS (2012) A review on mechanisms of plant tolerance and role of transgenic plants in environmental clean-up. Bot Rev 78:32–62

Seth CS, Chaturvedi PK, Misra V (2007) Toxic effect of arsenate and cadmium alone and in combination on giant duckweed (*Spirodela polyrrhiza* L.) in response to its accumulation. Environ Toxicol 22:539–549

Seth CS, Kumar Chaturvedi P, Misra V (2008) The role of phytochelatins and antioxidants in tolerance to Cd accumulation in *Brassica juncea* L. Ecotoxicol Environ Saf 71:76–85

Sharma DC (2003) Concern over mercury pollution in India. Lancet 362:1050

Sharma RK, Agrawal M, Marshall FM (2008) Heavy metal (Cu, Zn, Cd and Pb) contamination of vegetables in urban India: a case study in Varanasi. Environ Pollut 154:254–263

Sheoran V, Sheoran A, Poonia P (2011) Role of hyperaccumulators in phytoextraction of metals from contaminated mining sites: a review. Crit Rev Environ Sci Technol 41:168–214

Singh R, Tripathi RD, Dwivedi S, Kumar A, Trivedi PK, Chakrabarty D (2010) Lead bioaccumulation potential of an aquatic macrophyte *Najas indica* are related to antioxidant system. Bioresour Technol 101:3025–3032

Sobolewski A (1999) A review of processes responsible for metal removal in wetlands treating contaminated mine drainage. Int J Phytorem 1:19–51

Song W-Y, Sohn EJ, Martinoia E, Lee YJ, Yang Y-Y, Jasinski M, Forestier C, Hwang I, Lee Y (2003) Engineering tolerance and accumulation of lead and cadmium in transgenic plants. Nat Biotechnol 21:914–919

Sood A, Uniyal PL, Prasanna R, Ahluwalia AS (2012) Phytoremediation potential of aquatic macrophyte, *Azolla*. Ambio 41:122–137

Sparks DL (2005) Toxic metals in the environment: the role of surfaces. Elements 1:193–197

Summers AO, Silver S (1978) Microbial transformations of metals. Ann Rev Microbiol 32:637–672

Sun B, Zhao FJ, Lombi E, Mcgrath SP (2001) Leaching of heavy metals from contaminated soils using EDTA. Environ Pollut 113:111–120

Suñe N, Sánchez G, Caffaratti S, Maine MA (2007) Cadmium and chromium removal kinetics from solution by two aquatic macrophytes. Environ Pollut 145:467–473

Taghavi S, Barac T, Greenberg B, Borremans B, Vangronsveld J, Van Der Lelie D (2005) Horizontal gene transfer to endogenous endophytic bacteria from poplar improves phytoremediation of toluene. Appl Environ Microbiol 71:8500–8505

Tandy S, Bossart K, Mueller R, Ritschel J, Hauser L, Schulin R, Nowack B (2004) Extraction of heavy metals from soils using biodegradable chelating agents. Environ Sci Technol 38:937–944

Tangahu BV, Sheikh Abdullah SR, Basri H, Idris M, Anuar N, Mukhlisin M (2011) A review on heavy metals (As, Pb, and Hg) uptake by plants through phytoremediation. Int J Chem Eng 2011:31 pp. Article ID 939161. doi: 10.1155/2011/939161

Tiedje JM (1975) Microbial degradation of ethylenediaminetetraacetate in soils and sediments. Appl Microbiol 30:327–329

Tlustoš P, Száková J, Hrubý J, Hartman I, Najmanová J, Nedělník J, Pavlíková D, Batysta M (2006) Removal of As, Cd, Pb, and Zn from contaminated soil by high biomass producing plants. Plant Soil Environ 52:413–423

Tong Y-P, Kneer R, Zhu Y-G (2004) Vacuolar compartmentalization: a second-generation approach to engineering plants for phytoremediation. Trends Plant Sci 9:7–9

Vangronsveld J, Herzig R, Weyens N, Boulet J, Adriaensen K, Ruttens A, Thewys T, Vassilev A, Meers E, Nehnevajova E (2009) Phytoremediation of contaminated soils and groundwater: lessons from the field. Environ Sci Pollut Res 16:765–794

Vardanyan LG, Ingole BS (2006) Studies on heavy metal accumulation in aquatic macrophytes from Sevan (Armenia) and Carambolim (India) lake systems. Environ Int 32:208–218

Vesk PA, Nockolds CE, Allaway WG (1999) Metal localization in water hyacinth roots from an urban wetland. Plant Cell Environ 22:149–158

Wagner GJ (1993) Accumulation of cadmium in crop plants and its consequences to human health. In: Donald LS (ed) Advances in agronomy. Academic, New York

Wang KS, Huang LC, Lee HS, Chen PY, Chang SH (2008) Phytoextraction of cadmium by *Ipomoea aquatica* (water spinach) in hydroponic solution: effects of cadmium speciation. Chemosphere 72:666–672

Wang QA, Li Z, Cheng SP, Wu ZB (2010) Influence of humic acids on the accumulation of copper and cadmium in *Vallisneria spiralis* L. from sediment. Environ Earth Sci 61:1207–1213

Weis JS, Weis P (2004) Metal uptake, transport and release by wetland plants: implications for phytoremediation and restoration. Environ Int 30:685–700

Wenzel WW, Unterbrunner R, Sommer P, Sacco P (2003) Chelate-assisted phytoextraction using canola (*Brassica napus* L.) in outdoors pot and lysimeter experiments. Plant Soil 249:83–96

Williams JB (2002) Phytoremediation in wetland ecosystems: progress, problems, and potential. Crit Rev Plant Sci 21:607–635

Williams LE, Pittman JK, Hall J (2000) Emerging mechanisms for heavy metal transport in plants. Biochim Biophys Acta Biomembr 1465:104–126

Wolff G, Assis LR, Pereira GC, Carvalho JG, Castro EM (2009) Effects of zinc toxicity on leaves of *Salvinia auriculata* cultivated in nutrient solution. Planta Daninha 27:133–137

Wolverton BC, Mcdonald RC (1978) Bioaccumulation and detection of trace levels of cadmium in aquatic systems by *Eichhornia crassipes*. Environ Health Perspect 27:161–164

Wu LH, Luo YM, Xing XR, Christie P (2004) EDTA-enhanced phytoremediation of heavy metal contaminated soil with Indian mustard and associated potential leaching risk. Agric Ecosyst Environ 102:307–318

Wu CH, Wood TK, Mulchandani A, Chen W (2006) Engineering plant-microbe symbiosis for rhizoremediation of heavy metals. Appl Environ Microbiol 72:1129–1134

Wu G, Kang H, Zhang X, Shao H, Chu L, Ruan C (2010) A critical review on the bio-removal of hazardous heavy metals from contaminated soils: issues, progress, eco-environmental concerns and opportunities. J Hazard Mater 174:1–8

Wuana RA, Okieimen FE (2011) Heavy metals in contaminated soils: a review of sources, chemistry, risks and best available strategies for remediation. ISRN Ecol 4:1–20. Article ID 402647. doi:10.5402/2011/402647

Xia HP (2004) Ecological rehabilitation and phytoremediation with four grasses in oil shale mined land. Chemosphere 54:345–353

Yadav SK (2010) Heavy metals toxicity in plants: an overview on the role of glutathione and phytochelatins in heavy metal stress tolerance of plants. S Afr J Bot 76:167–179

Yadav R, Arora P, Kumar S, Chaudhury A (2010) Perspectives for genetic engineering of poplars for enhanced phytoremediation abilities. Ecotoxicology 19:1574–1588

Yang XE, Yang MJ (2001) Some mechanisms of zinc and cadmium detoxification in a zinc and cadmium hyperaccumulating plant species (*Thlaspi*). In: Orst W (ed) Plant nutrition-food security and sustainability of agro-ecosystems. Kluwer Academic Publishers, Dordrecht

Yang J, Ye Z (2009) Metal accumulation and tolerance in wetland plants. Front Biol China 4:282–288

Ye ZH, Baker AJM, Wong MH, Willis AJ (1997) Copper and nickel uptake, accumulation and tolerance in *Typha latifolia* with and without iron plaque on the root surface. New Phytol 136:481–488

Ye Z, Baker AJM, Wong M-H, Willis AJ (1998) Zinc, lead and cadmium accumulation and tolerance in *Typha latifolia* as affected by iron plaque on the root surface. Aquat Bot 61:55–67

Yoon J, Cao X, Zhou Q, Ma LQ (2006) Accumulation of Pb, Cu, and Zn in native plants growing on a contaminated Florida site. Sci Total Environ 368:456–464

Zacchini M, Pietrini F, Mugnozza GS, Iori V, Pietrosanti L, Massacci A (2009) Metal tolerance, accumulation and translocation in poplar and willow clones treated with cadmium in hydroponics. Water Air Soil Pollut 197:23–34

Zenk MH (1996) Heavy metal detoxification in higher plants-a review. Gene 179:21–30

Zhang X, Lin A-J, Zhao F-J, Xu G-Z, Duan G-L, Zhu Y-G (2008) Arsenic accumulation by the aquatic fern *Azolla*: comparison of arsenate uptake, speciation and efflux by *Azolla caroliniana* and *Azolla filiculoides*. Environ Pollut 156:1149–1155

Zhang BY, Zheng JS, Sharp RG (2010) Phytoremediation in engineered wetlands: mechanisms and applications. Procedia Environ Sci 2:1315–1325

Zhu Y-G, Rosen BP (2009) Perspectives for genetic engineering for the phytoremediation of arsenic-contaminated environments: from imagination to reality? Curr Opin Biotechnol 20:220–224

Zhu YL, Zayed AM, Qian JH, De Souza M, Terry N (1999) Phytoaccumulation of trace elements by wetland plants: II. Water hyacinth. J Environ Qual 28:339–344

Chapter 9
Chemical-Induced Washing Remediation of Metal-Contaminated Soils

Zinnat A. Begum, Ismail M.M. Rahman, Hikaru Sawai, and Hiroshi Hasegawa

Abstract The immobilization or removal of toxic components using aqueous extractants, with or without additives, is one of the commonly practiced techniques for the treatment of metal-contaminated soils. However, rather than the use of water alone, the solution with chemical-additives is preferred due to the less time requirement and better separation effectiveness. There is a long-favored list of additives that have been used for the chemical-induced washing remediation of soils, which include acids, bases, chelants, surfactants, and so forth. The objective of this chapter is to provide a brief overview of the chemical-assisted soil washing approaches.

Keywords Soil • Metal contamination • Chemical-assisted washing • Acid • Chelant • Surfactant • Remediation

Z.A. Begum (✉)
Department of Civil Engineering, Southern University, 739/A Mehedibag Road, Chittagong 4000, Bangladesh
e-mail: Zinnat.Ara@gmail.com

I.M.M. Rahman (✉)
Department of Applied and Environmental Chemistry, Faculty of Science, University of Chittagong, Chittagong 4331, Bangladesh

Graduate School of Natural Science and Technology, Kanazawa University, Kakuma, Kanazawa 920-1192, Japan
e-mail: I.M.M.Rahman@gmail.com

H. Sawai
Graduate School of Natural Science and Technology, Kanazawa University, Kakuma, Kanazawa 920-1192, Japan
e-mail: hikaru.sawai@gmail.com

H. Hasegawa
Institute of Science and Engineering, Kanazawa University, Kakuma, Kanazawa 920-1192, Japan
e-mail: hhiroshi@se.kanazawa-u.ac.jp

H. Hasegawa et al. (eds.), *Environmental Remediation Technologies for Metal-Contaminated Soils*, DOI 10.1007/978-4-431-55759-3_9

9.1 Introduction

Metal contamination of soils, which has been known as threatening for human health and the environment, has been caused from manufacturing discharges, effluents from service industries or wood preserving operations (Sposito and Page 1984; Basta et al. 2005; Khan et al. 2008). The sites of metal-contaminated soils can either be declared abandoned restricting for future exploitation, or the soils can be excavated and transported to secured disposal (Abumaizar and Smith 1999). However, the leaching possibility of the toxic metals from the contaminated soil cannot be avoided even it confined and has been considered rather as a provisional and an economically less-viable option (Leštan et al. 2008). Instead, the depollution of metal-contaminated soils has been preferred as it not only minimizes any future contamination risk but also offers an option to re-exploit the restricted locations (Abumaizar and Smith 1999; Dermont et al. 2008; Pavel and Gavrilescu 2008).

The 'soil washing' technique used for the treatment of metal-contaminated soils is a physico-chemical approach based on mining and mineral processing principles (Mann 1999). The target contaminant usually remains in specific particle fractions of the metal-contaminated soils, which can be concentrated into a much smaller volume of contaminated residue via washing treatment of soil (ITRC 1997). However, the effectiveness of the washing treatment is closely related to the ability of the extracting solution to separate out the metals in soil (Peters 1999). The solubility of metals in water is too limited for removing a high amount of cations in the leachates and, hence, the washing solution includes various chemical agents (e.g., acids, bases, surfactants, chelating, or sequestering agents) to enhance the separation of contaminants from soils (Davis and Singh 1995; ITRC 1997; Davis and Hotha 1998).

Soil decontamination by washing treatment can be accomplished either on the excavated (i.e., physically removed) soil (*ex situ*) or on-site (*in situ*). However, the effectiveness of *in situ* washing treatment is limited due to the restricted mobility of the extractants while the soil is in the intact state. Furthermore, it is necessary to maintain site-specific control measures to prevent subsequent leaching occurrences (Abumaizar and Smith 1999). Therefore, the soil washing technique is generally performed as an *ex situ* method (Peters 1999; Pavel and Gavrilescu 2008), which have been discussed in detail in this chapter.

9.2 General Outline of the *Ex Situ* Soil Washing Process

In the metal-contaminated soils, the toxic components tend to be attached to the fine fractions (silt and clay) either chemically or physically, which are further bind to the coarse fraction consisting of sand and gravel (US EPA 1996). The cumulative target of the soil washing process is to treat the entire volume of a contaminated soil site, including the separation of the fine soil parts from the coarse ones. Hence, the

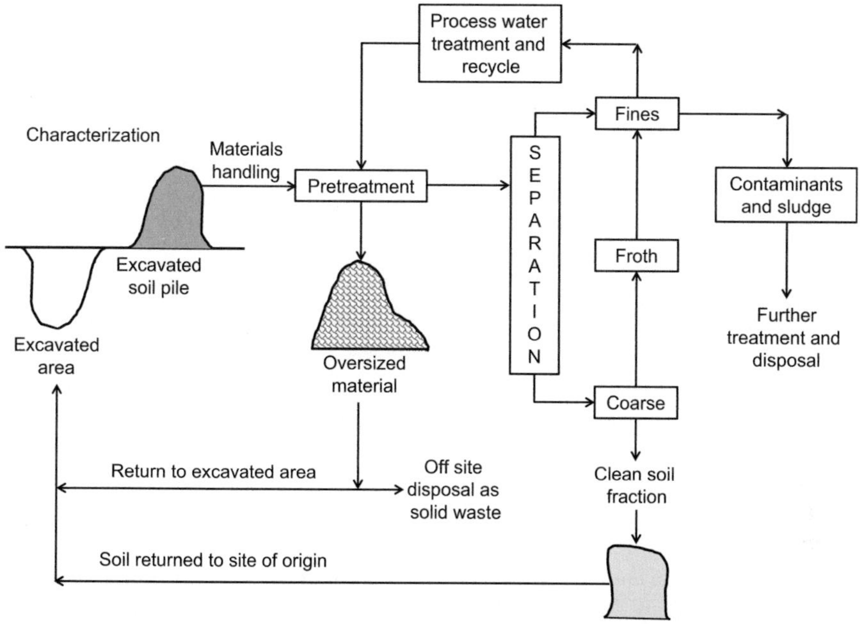

Fig. 9.1 A schematic diagram of a basic soil washing process (Adapted from ITRC 1997)

total process of soil washing can be said to be a combination of the following basic steps: (a) separation of the contaminated zone by excavation; (b) segmentation of the unearthed soil to fine, sand and gravel fractions; (c) treatment of the sand fraction using suitable extractant; (d) rinsing with water to remove residual contaminants and extracting agents; (e) re-deposition of the cleaned sand fraction along with the gravel parts to the site; (f) further treatment of the fine fractions or disposed of according to the regulatory guidelines (Griffiths 1995; US EPA 1996; ITRC 1997; Abumaizar and Smith 1999; Mann 1999; Ramamurthy et al. 2008).

The volume reduced during the washing treatment of soils is a typical performance indication of process application in a particular metal-contaminated site, which is calculated from the metal content reduction in the coarse and sand fraction in accordance with the regulatory standards using the following equation (Mann 1999).

$$\text{Volume reduction}\,(\%) = 1 - \left(\frac{\text{Feed soil (tons)} - \text{Clean products (tons)}}{\text{Feed soils (tons)}}\right)$$

A typical *ex situ* soil washing process (ITRC 1997) is illustrated in Fig. 9.1.

9.3 Factors Limiting the Effectiveness of Soil Washing Technology

There are several factors, which limit the effectiveness of the soil washing technology during the treatment of metal-contaminated soil. The factors include the percent distribution of soil particle sizes, organic content in the soil, the ratio of hydrophobic contaminants, the percentage of co-contaminants other than the metals, and the treatment of spent washing fluid.

The application of soil washing will not be cost-effective if the percentage of the fine-fractions of soils (silt/clay, < 63–74 microns) is in excess of 30–50 %. The high organic content, such as humic substances, in soils make the separation of metal-contaminants difficult because it provides additional binding sites for metals. An increased ratio of hydrophobic contaminants in soil requires extra additives and, in addition, a supplementary problem is created during the removal or recycling of the additives from the residual washing liquid. The volume requirements of washing liquid and the operating parameters for soil washing are determined depending on the comparative nature and concentration of metals and co-contaminants in soil, and a huge variation among those can alter the washing effectiveness of solvents to a considerable extent. Moreover, the after-use concentration of washing solvents (e.g., acids, chelating agents, surfactants, or other additives) in the treated soils evokes concerns regarding application of soil washing technology due to the environmental issues related to the disposal of residuals (US EPA 1995; ITRC 1997).

9.4 *Ex Situ* Soil Washing: Pros and Cons

The *ex situ* soil washing process has several advantages, such as quantitative removal of the contaminants, rapid cleanup of a contaminated site, reduce or elimination of long-term liability, the possibility of producing recyclable material or energy (Evanko and Dzombak 1997; Hester and Harrison 1997). Furthermore, it is one of the few permanent treatment alternatives for soils contaminated with metals and radionuclides. In addition to the metals, organic contaminants can be treated in the same system using soil washing technology. Besides, the clean coarse fractions of soils can be returned to the site at a very low cost after the soil washing treatment depending upon soil matrix characteristics (ITRC 1997).

The disadvantages of *ex situ* soil washing process include a must requirement of further treatment or disposal of the spent washing liquid, the risk of spreading contaminated soil and dust particles during removal and transportation of excavated soils. The soil excavation can also be expensive when a large amount of soil is required to be removed, or disposal as hazardous or toxic waste is required. There are possibilities of a complication during the treatment process due to the high soil-humic contents, elevated percentage of soil fines, complex mixtures of contaminants, or excessively variable influent contaminant concentrations. Moreover, the space requirement for the installation of the treatment system is also an issue of concern (Evanko and Dzombak 1997; Hester and Harrison 1997; ITRC 1997; Peters 1999; Dermont et al. 2008).

9.5 Extractants for Soil Washing

Acids, bases, chelants, surfactants, alcohols, reducing agents, or other solvents are used as the extracting agent in the soil washing processes either individually or as an additive to the aqueous mixtures. The solubilization, exchange, and/or extraction of metals by washing solutions differ considerably with the soil characteristics as well as the types and concentration of co-contaminants other than the metals (Wood et al. 1990; Yu and Klarup 1994; Griffiths 1995; Chu and Chan 2003; Gao et al. 2003; Maturi and Reddy 2008). Hence, the selection of extractants is decided on a case-by-case basis depending on the various factors as mentioned Sect. 9.3.

The metal immobilization in soils occurs either by forming insoluble precipitates or incorporating into the soil-crystalline structures, if the metal sorption ability of soils exceeds the limit due to the high input (Davis and Singh 1995; Pichtel and Pichtel 1997). To treat such a soils, the acids and chelants have been often studied at laboratory scale and suggested for the commercial-scale remediation practices (Dermont et al. 2008). The selection of acids or chelants as the washing liquid is attributable to their better-responsive ability towards the metal-mobilization factors, e.g., acidity, ionic strength, redox potential and complex formation (Pickering 1986; Rampley and Ogden 1998). The acid-induced leaching of metals in soil takes place through ion exchange and/or soil matrix dissolution (Bricka et al. 1993; Peters 1999). The ability of the chelants to form stable water-soluble complexes with the metal ions is exploited during the chelant-assisted soil washing of metal-contaminated soils (Davis and Singh 1995; Pichtel and Pichtel 1997; Davis and Hotha 1998; Tejowulan and Hendershot 1998; Abumaizar and Smith 1999; Peters 1999). The application of surfactants becomes an attractive option for the extraction of contaminants from soil (Wang and Mulligan 2004; Conte et al. 2005), due to less acute toxicity relative to that of the organic solvents and considerable rate of environmental degradability to produce non-toxic substrates (Mulligan et al. 2001c; Roundhill 2001; Ehsan et al. 2006a, b). The capacity of surfactants to increase the aqueous solubility of hydrophobic organic compounds at concentrations above the critical micelle concentration (CMC) is the key factor in the surfactant-enhanced soil washing (Deshpande et al. 1999; Wen and Marshall 2011). In this work, we have concentrated our discussion on the remediation of metal-contaminated soils using acids, chelants and surfactants considering the increasing and continued research focus on the use of those extractants.

9.6 Acid-Induced Washing Remediation of Metal-Contaminated Soils

The acid-leaching treatment of metal-contaminated soils, sediments, and sludges is an established remediation approach, which exploit the pH of the washing fluids. The mechanisms involved, by far, can be either desorption of metal cations via ion exchange or the dissolution of metal compounds and/or metal contaminant containing soil mineral components (Tampouris et al. 2001; Kuo et al. 2006). The protons in solution, at low pH, reacts with the layer silicate minerals and/or surface functional groups (e.g., Al–OH, Fe–OH and –COOH) of soils, and the desorption rate of metal ions increased (Isoyama and Wada 2007). The dissolution of Fe- and Al-oxides and phyllosilicates occurs when strong acidic fluid is added to the soils, and it replaces the ion-exchange process during metal extraction at $pH < 2$ (Kuo et al. 2006).

The acid-leaching treatment usually employs strong mineral acids, such as hydrochloric (HCl), sulfuric (H_2SO_4), nitric (HNO_3), phosphoric (H_3PO_4), and so forth. Although the use of weak organic acid, such as acetic acid (CH_3COOH) is attempted (ESTCP 1997), the efficiency was proved limited because of relative low strength and foul-smelling odors (Dermont et al. 2008). The leaching of toxic metals (As, Cu, Pb, and Zn) from soils contaminated with metallurgical materials (Moutsatsou et al. 2006) can effectively be achieved with HCl compared to the H_2SO_4 and HNO_3. Furthermore, a significant Pb-leaching (65 – 100 %) from artificially or naturally contaminated soils is possible with HCl (Cline and Reed 1995; Reed et al. 1996; Abumaizar and Smith 1999). However, similar rates of Zn and Ni-extraction have been observed with HCl, H_2SO_4, and H_3PO_4, while a higher As-extraction rate was achieved with H_2SO_4 and H_3PO_4 than that of HCl (Ko et al. 2005; 2006). In brief, it can be concluded that the metal-leaching efficiency of the acid-variants strongly depends on the metal-types, the geochemistry of soils, as well as the reagent concentrations.

The metal-leaching treatment of contaminated soils using acids alters soil structure and induces co-dissolution of soil components causing approximately 50 % losses of soil minerals (Tampouris et al. 2001) and organic matters (Ko et al. 2005). The co-dissolution of the soil matrix is an issue of concern in terms of both environmental and economic point of view, because it not only increases the consumption of acid reagent and the complexity of the effluent management but also the acidity of treated soil is increased (Tampouris et al. 2001; Ko et al. 2005).

To minimize the destructive impact from the leaching treatment using high-concentrated acid, the diluted acidic solutions containing chloride salts (e.g., $CaCl_2$, NaCl) have been proposed as the effective alternatives. The chloride salt solutions have been applied either in a mixed solution of the mineral acids of lower concentration (Kuo et al. 2006), or individually at a very high concentrations (>1 M) at a pH-controlled condition (Nedwed and Clifford 2000; Lin et al. 2001). A subsequent application of chloride salt solutions after the acid-leaching has also been evaluated, which in addition help to prevent the re-adsorption of acid-extracted metals to soils (Nedwed and Clifford 2000; Wasay et al. 2002; Isoyama and Wada 2007). However, the monitoring of Eh and pH parameters should be conducted to achieve and maintain the optimum thermodynamic conditions as well as to prevent the formation of insoluble compounds (Lin et al. 2001). The processes involved in the removal of metal ions (e.g., Pb^{2+}, Cd^{2+}) with chloride salt solutions (e.g., $CaCl_2$, NaCl) can either be ion exchange of Ca^{2+}/Na^{+} with Pb^{2+}/Cd^{2+} on the reactive surface sites of the soil matrix, or the formation of stable and soluble metal chloro-complexes with Cl^{-} ions (e.g., $Cd^{2+} + yCl^{-} \Leftrightarrow CdCl_y^{2-y}$) (Nedwed and Clifford 2000; Tampouris et al. 2001; Kuo et al. 2006). It has been observed that such a saline leaching treatment of metal-contaminated soil, with or without acid, can minimize the co-dissolution of soil matrix, and maintain the physico-chemistry and microbiology of soils close to that of source soil (Tampouris et al. 2001; Kuo et al. 2006; Makino et al. 2007).

The instances of the application of acid-leaching for the washing remediation of contaminated soils, both at laboratory and full-scale field-tests, are available from VanBenschoten et al. (1997), Steele and Pichtel (1998), Lin et al. (2001), Ko et al. (2005), Kuo et al. (2006), Moutsatsou et al. (2006), Isoyama and Wada (2007), and Dermont et al. (2008), and are recommended for further reading.

9.7 Chelant-Assisted Washing Remediation of Metal-Contaminated Soils

A multi-protic chelant (H_nL), which typically contains multiple coordination sites available for complexation with a metal center, undergoes acid–base equilibrium reactions in the aqueous phase, e.g.,

$$H_nL = H^{+} + H_{n-1}L^{-} \tag{9.1}$$

There will be subsequent reaction steps followed by the Eq. (9.1). The total solubility of metal ion (M_{Tot}) in the presence of chelant in solution can be computed using the following relation:

$$M_{Tot} = M_{aq} + \sum M_pH_qL_r = M_{aq} + ML_{Tot} \tag{9.2}$$

In Eq. (9.2), p, q, and r are used to denote the coefficients for metal ions, protons, and chelants, respectively, and indicate that each conjugate acid or base of the chelants may form a strong complex with the metals in the contaminated soil when added to the washing solution. The complexation ability and comparative interaction quotient of the chelants towards the metals in soils can be evaluated assuming the equilibrium computation procedures formulated in Eq (9.2). If the chelant is strong in interacting with the metals in soils, the ML_{Tot} will be much higher than that of M_{aq}. In addition, performance of a chelant can be evaluated based on their interaction with and partition potential to soil surfaces according to soil texture, particle size distribution, clay content, humic matter contents, metal and waste characteristics, mineralogy, and solution pH (Peters 1999).

A suitable chelant for the treatment of contaminated soil may be required to possess several of the following criteria (Peters 1999; Hong and Jiang 2005; Leštan et al. 2008):

(a) The chelant should have higher metal complexing abilities, as indicated by the equilibrium complexation constants, towards the heavy and transition metals compared to the hard sphere cations (e.g., Ca^{II} or Mg^{II}).
(b) The chelant is better to possess extraction selectivity towards the target metals. The donor atoms in the chelant decide its comparative selectivity behavior. For example, chelants having sulfur and nitrogen as donor atoms show higher selectivity toward the transition metals (e.g., Cu^{II}, Ni^{II}) and soft sphere cations (e.g., Zn^{II}, Cd^{II}, Pb^{II}, Hg^{II}), while chelants containing oxygen as the donor atoms are more selective to the hard sphere cations.
(c) Chelants having multiple coordinating sites (i.e., multidentate) are capable of forming more stable metal-chelant complexes, therefore, preferable.
(d) The adsorption affinity of metal-chelant complexes towards solid surfaces of soils should be low.
(e) The reusability of chelant, including low toxicity in the environment, is desirable to design a cost-effective separation scheme.

Aminopolycarboxylate chelants (APCs), such as ethylenediaminetetraacetic acid (EDTA) and its homologs, are commonly utilized in the *ex situ* soil washing processes due to their ability to interact with the majority of toxic metals (Leštan et al. 2008; Hasegawa et al. 2010, 2011). However, the free-form of classical APCs (e.g., EDTA) exhibit poor photo-, chemo- and biodegradability in the environment (Means et al. 1980; Bolton Jr. et al. 1993; Kari and Giger 1995; Kari et al. 1995; Egli 2001; Nowack 2002; Nörtemann 2005) and, in most cases, metal complexation raises the threshold values for toxic effects of metals (Sillanpää and Oikari 1996; Sorvari and Sillanpää 1996; Sillanpää 2005). The requirement of using an excess amount of chelant to ensure the adequate desorption of metal-contaminants from soil, as well as to minimize the competition effect due to the coexisting elements in the soil (e.g., $Ca^{II,}$ Mg^{II}, Fe^{III}, Al^{III}) further enhance the problem (Leštan et al. 2008). The consequence raise concern regarding eco-safety issues, and increasingly stringent legislative regulations regarding the disposal of soil washing fluid containing APCs have been proposed or imposed (Grundler et al. 2005; Begum et al. 2013a).

The search for eco-friendly biodegradable variants to replace the classical APCs, thus, became a topic of interest for the treatment of heavy metal-contaminated soils (Tandy et al. 2004; Begum et al. 2012a, c; Pinto et al. 2014). Nitrilotriacetic acid (NTA) and [S,S]-ethylenediaminedisuccinic acid (EDDS) have been evaluated as the biodegradable and environmental-friendly replacement for EDTA in soil washing in the beginning phase of such works (Elliott and Brown 1989; Pichtel and Pichtel 1997; Vandevivere et al. 2001; Tandy et al. 2004; Polettini et al. 2006). The work of Vandevivere et al. (2001) confirms that a comparable rate of efficiency for Pb, Zn, Cu, and Cd extraction with EDDS to that of EDTA is possible if the contact time is sufficient and solution pH is maintained above 7. The result is, however, contradicts in the work of Yang et al. (2012), who proposes the use of pH 5.5 for Pb or Cd extraction. The performance of NTA, EDDS and EDTA for the extraction of Cd, Cu, Pb, and Zn from soils by Polettini et al. (2006) and, among the biodegradable options, EDDS performed superior than the NTA. The effectiveness of NTA, EDTA, IDSA (iminodisuccinic acid) and MGDA (methylglycine diacetic acid) as potential alternatives of EDTA was investigated by Tandy et al. (2004) for the extraction of Cu, Zn, and Pb from contaminated soils, which indicate EDDS as the best option among all. The removal of Cu, Pb, and Zn by the action of the EDDS and MGDA has been reported by Arwidsson et al. (2010). The DL-2-(2-carboxymethyl) nitrilotriacetic acid (GLDA) and 3-hydroxy-2,2′-iminodisuccinic acid (HIDS) have been introduced as the biodegradable alternatives to EDTA, along with EDDS, IDSA, and MGDA by Begum et al. (2012a). The performance of GLDA is found better than other options, in some cases even better than EDTA, at pH 4 and 7 for the extraction of Cd, Cu, Ni, Pb, and Zn from contaminated soils.

The solution pH seems to be a prime deciding factor during the chelant-assisted washing remediation, and the optimal pH condition for the treatment of metal-contaminated soils is frequently varied with the change in soil characteristics, the incorporation of metals within the soil phases and the chelant employed (Begum et al. 2012b, 2013b). In addition, the relative stability of metal-chelant complexes in the solution is often altered due to the variation in the formation efficiency of the soluble dominant species in solution, re-sorption of the metal-chelant complexes in the active surface site of the soil solids and so forth (Nowack 2002; Begum et al. 2012b, 2013b).

The basic information about the chelants (EDTA, EDDS, IDSA, MGDA, GLDA, and HIDS) by far explored for the washing remediation of metal-contaminated soils is given in Table 9.1. The protonation and complexation characteristics of those chelants with Cd, Cu, Ni, Pb and Zn are listed in Table 9.2, while the changes in the conditional stability constants of the corresponding metal–chelant complex as a function of pH are shown graphically in Fig. 9.2. Some instances of chelant-assisted washing remediation of metal-contaminated soils are summarized by Peters (1999), Tandy et al. (2004), Dermont et al. (2008), and Begum et al. (2012b), and are recommended for further reading.

9.8 Surfactant-Enhanced Washing Remediation of Metal-Contaminated Soils

Surfactants are heterogeneous and long-chain molecules containing both hydrophilic (head) and hydrophobic (tail) moieties (Mao et al. 2015), and these are can be classified as anionic, cationic, zwitter-ionic, and non-ionic depending on the nature

Table 9.1 Basic information about the chelants (EDTA, EDDS, IDSA, NTA, MGDA, GLDA, and HIDS) that have been explored for washing remediation of metal-contaminated soils, such as chemical structure, acid dissociation constants (pK_a) and stability constants ($\log K_{ML}$) of metal-chelant (ML) complexes with selected toxic metals (Cd, Cu, Ni, Pb and Zn)[a]

APCs	Structure	pK_a				Metal	$\log K_{ML}$
		pK_{a1}	pK_{a2}	pK_{a3}	pK_{a4}		
EDTA[b]	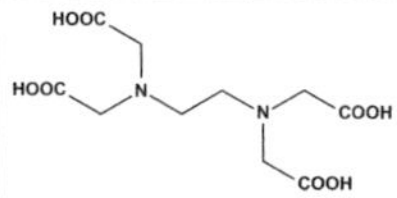	2.00	2.69	6.13	10.37	–	–
						Cd	16.5
						Cu	18.78
						Ni	18.4
						Pb	18
						Zn	16.5
EDDS[b]	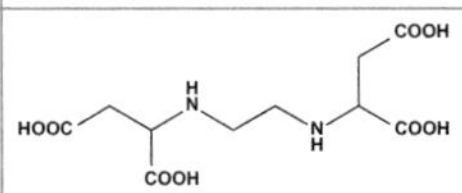	2.95	3.86	6.84	10.01	–	–
						Cd	10.9[c]
						Cu	18.36
						Ni	16.7
						Pb	12.7[c]
						Zn	13.4[c]
IDSA[b]	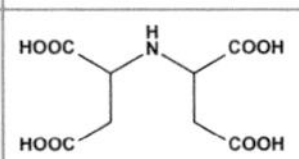	1.97	3.24	4.24	10.00	–	–
						Cd	8.33
						Cu	12.69
						Ni	11.68
						Pb	9.75
						Zn	9.88
NTA[c]	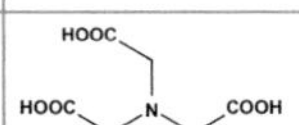	1.89	2.49	9.73		–	–
						Cd	9.78
						Cu	12.94
						Ni	11.50
						Pb	11.34
						Zn	10.66
MGDA[d]	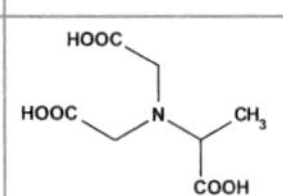	1.5	2.45	10.43		–	–
						Cd	10.61
						Cu	13.88
						Ni	11.99
						Pb	12.07
						Zn	10.98
GLDA[e]	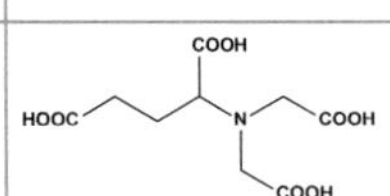	2.56	3.49	5.01	9.39	–	–
						Cd	10.31
						Cu	13.03
						Ni	12.74
						Pb	11.6
						Zn	11.52

(continued)

Table 9.1 (continued)

APCs	Structure	pK_a				Metal	$\log K_{ML}$
		pK_{a1}	pK_{a2}	pK_{a3}	pK_{a4}		
HIDS[e]	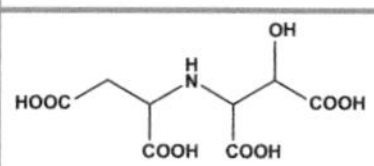	2.14	3.08	4.07	9.61	–	–
						Cd	7.58
						Cu	12.58
						Ni	11.3
						Pb	10.21
						Zn	9.76

'–' stands for 'no metal added'
[a]A partial adaptation from Begum et al. (2012b)
[b]At 25 °C ($\mu = 0.1$ M) (Martell et al. 2004)
[c]At 25 °C ($\mu = 0.1$ M) (Martell and Smith 1974)
[d]At 20 °C ($\mu = 0.1$ M) (Martell et al. 2004)
[e]At 25 °C ($\mu = 0.1$ M) (Begum et al. 2012a)

of the hydrophilic group (Rosen and Kunjappu 2012). In an aqueous medium, the monomer molecules of surfactant create aggregates of a large number of molecules called 'micelles' when the surfactant concentration exceeds the critical micelle concentration (CMC) (Fig. 9.3). Accordingly, the lowering of surface and interfacial tensions between the contaminants occurs followed by the displacement of contaminants (Mulligan et al. 2001c; Paria 2008). The application of surfactant-enhanced remediation is more suitable for the treatment of organic contaminants in soils. Hence, the washing by surfactants can be more effective when the metals are closely associated with organic contaminants (US EPA 1997; Dermont et al. 2008). The removal of metal contaminants from soils occurs either due to the surfactant-associated complexation (Ochoa-Loza et al. 2001) and/or ionic exchange (Swarnkar et al. 2011). A list of surfactants by far employed for the washing remediation of contaminated soils is provided in Table 9.3.

Several comparative studies have been conducted to explore and select the best surfactant types for the enhanced remediation of heavy metal-contaminated soils. For example, cationic surfactant DPC, nonionic surfactant Ammonyx KP, and anionic surfactant JBR-425 have been used for the treatment of metal-contaminated soils, and the JBR-425 demonstrated the best elution effect towards Zn, Cu, Pb, and Cd among the surfactant variants (Slizovskiy et al. 2011). There was a study to evaluate the utility of 11 different kinds of surfactants, which includes 4 non-ionic, 4 anionic, one zwitter-ionic, and two charge-unknown surfactants, for the remediation of As, Cd, Cu, Ni, Pb, and Zn-contaminated soil. The maximum remediation effectiveness has been achieved with Texapon N-40 anionic surfactant for most of the metals (Torres et al. 2012).

In comparison with the synthetic surfactants, the bio-surfactants are often preferred due to their larger molecular structure with more ligand groups, which facilitate usually high surface activity for the decontamination of both hydrophobic organics and heavy metals (Sachdev and Cameotra 2013). The potency of bio-surfactants in enhancing metal removal either as an individual solvent or as

Table 9.2 The protonation and complexation characteristics of the chelants (EDTA, EDDS, IDSA, NTA, MGDA, GLDA, and HIDS) with selected toxic metals (Cd, Cu, Ni, Pb, and Zn) in the aqueous medium[a, b]

Equilibria	EDTA	EDDS	IDSA	NTA	MGDA[d, f]	GLDA[g]	HIDS[g]
[HL]/[H][L]	9.52–10.37	10.01	10	9.46–9.84	10.43	9.36	9.61
$[H_2L]/[HL][H]$	6.13	6.84	4.24	2.52	2.45	5.01	4.07
$[H_3L]/[H_2L][H]$	2.69	3.86	3.24	(1.81)	1.5	3.49	3.08
$[H_4L]/[H_3L][H]$	2	2.95	1.97	(1.0)	–	2.56	2.14
$[H_5L]/[H_4L][H]$	(1.5)	–	–	–	–	–	1.6
$[H_6L]/[H_5L][H]$	(0.0)	–	–	–	–	–	–
Cd^{2+}							
[ML]/[MOHL][H]	(13.2)[c]	–	–	11.25	–	10.25	10.2
[ML]/[M][L]	16.5	10.9[e]	8.33	9.76	10.6	10.31	7.58
[MHL]/[ML][H]	2.9	4.5	4.68	–	–	4.72	5.11
$[MH_2L]/[MHL][H]$	(1.6)[c]	–	3.28	–	–	3.46	3.77
$[ML_2]/[M][L]^2$	–	–	–	14.47	–	–	–
$[M_2L]/[ML][M]$	–	–	–	–	–	–	2.64
Cu^{2+}							
[ML]/[MOHL][H]	(11.4)	10.38	–	9.2	–	9.91	8.9
[ML]/[M][L]	18.78	18.4	12.69	13	13.9	13.03	12.58
[MHL]/[ML][H]	3.1	3.48	4.01	1.6	–	4.13	3.65
$[MH_2L]/[MHL][H]$	2	1.95	2.65	–	–	–	2.57
$[ML_2]/[M][L]^2$	–	–	–	17.4	–	–	–
Ni^{2+}							
[ML]/[MOHL][H]	(11.9)	–	–	10.86	–	–	9.5
[ML]/[M][L]	18.4	16.7	11.68	11.51	12.0	12.74	11.3
[MHL]/[ML][H]	3.1	3.22	4.14	–	–	4.38	3.52
$[MH_2L]/[MHL][H]$	(0.9)[c]	–	–	–	–	2.19	2.24
$[ML_2]/[M][L]^2$	–	–	–	16.32	–	–	–

Pb^{2+}							
[ML]/[MOHL][H]	–	–	–	–	–	10.65	9.34
[ML]/[M][L]	18	12.7[d]	9.75	11.48	12.1	11.6	10.21
[MHL]/[ML][H]	2.8	5.9	–	2.3[e]	–	4.69	4.13
$[MH_2L]/[MHL][H]$	(1.7)[c]	–	–	–	–	2.11	2.41
$[MH_3L]/[MH_2L][H]$	(1.2)[c]	–	–	–	–	–	–
$[ML_2]/[M][L]^2$	–	–	16.27	12.8[d]	–	–	–
Zn^{2+}							
[ML]/[MOHL][H]	(11.6)	–	–	10.06	–	10.64	8.96
[ML]/[M][L]	16.5	13.4[d]	9.88	10.65	10.9	11.52	9.76
[MHL]/[ML][H]	3	6.68	4.29	–	–	4.6	3.92
$[MH_2L]/[MHL][H]$	(1.2)[c]	2.48	–	–	–	2.58	–
$[ML_2]/[M][L]^2$	–	–	–	14.27	–	–	–

[a]Unless mentioned otherwise, all the data are from the NIST database of critically selected stability constants of metal complexes at 25 ± 0.1 °C ($\mu = 0.1$ M) (Martell et al. 2004)
[b]A partial adaptation from Begum et al. (2012a) and Pinto et al. (2014)
[c]$\mu = 1$ M; [d]At 20 °C; [e]$\mu = 0.5$ M
[d]At 20 °C
[e]$\mu = 0.5$ M
[f]Data source: BASF (2007)
[g]Data source: Begum et al. (2012a)

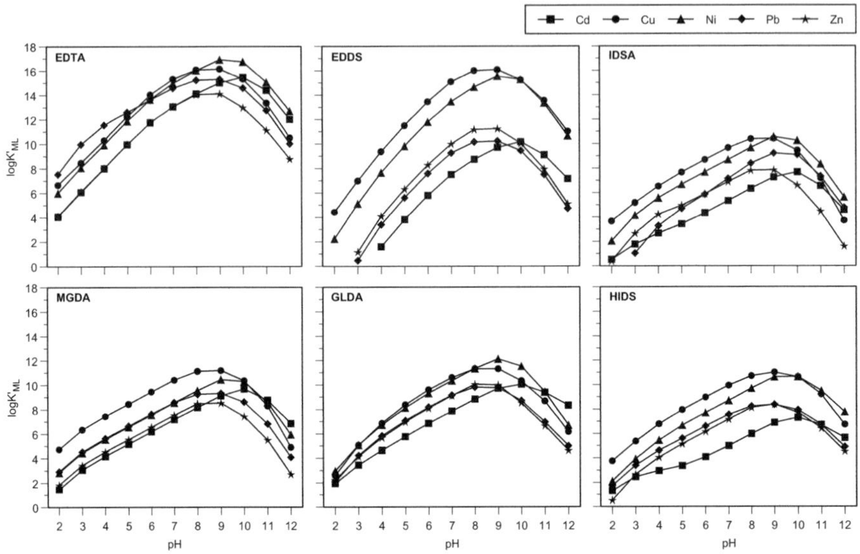

Fig. 9.2 The changes in the conditional stability constants ($\log K'_{ML}$) of the corresponding metal–chelant (ML) complexes as a function of pH (M = Cd, Cu, Ni, Pb, and Zn; L = EDTA, EDDS, IDSA, MGDA, GLDA, and HIDS). The calculation was performed with the aid of the computer program HySS2009 (Alderighi et al. 1999) using the values mentioned in Table 9.1 (Adapted from Begum et al. 2012b)

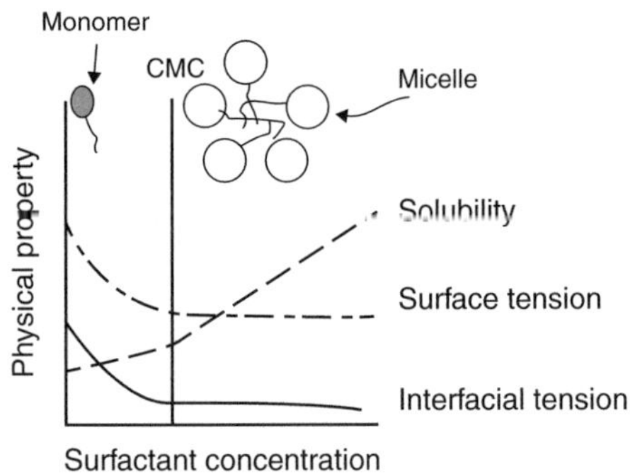

Fig. 9.3 A schematic view of the variation of surface tension, interfacial, and contaminant solubility with surfactant concentration (Adapted from Mulligan et al. 2001a)

an additive to the solvent mixtures has gained advanced research focus from 1990s, and continued thereafter (Herman et al. 1995; Mulligan et al. 1999a, b, 2001a, b, c; Mulligan and Wang 2006; Dahrazma and Mulligan 2007; Song et al. 2008; Wang and Mulligan 2009a, b). The findings conclude that the acidic bio-surfactant performed better in extracting the metals bound to carbonate and oxide, while the alkaline bio-surfactant expedites the release of the organically associated metals

Table 9.3 Basic information about the surfactants used for the washing remediation of contaminated soils[a]

Surfactant	Name/Components[b]	Ionic nature[b]	Mol weight ($g\ mol^{-1}$)[b]
DPC	1-Dodecylpyridinium chloride	Cationic	283.88
TX-100	*p*-tertiary-Octylphenoxy polyethyl alcohol	Nonionic	628
PFOA	Perfluorooctanoic acid	Anionic	414.07
NINOL 40-CO	Cocamide DEA	Nonionic	287.44
CAPB	Cocoanut amide propyl betaine	Zwitterionic	342.52
DDAC	Didecyl dimethyl ammonium chloride	Cationic	362.08
SLES	Sodium laureth sulfate	Anionic	NR
SDS	Sodium dodecyl sulfate	Anionic	288.38
SDHS	Sodium dihexyl sulfosuccinate	Anionic	388.45
JBR-425	Rhamnolipid	Nonionic	504.6/650.8
Ammonyx KP	Oleyl dimethyl benzyl ammonium chloride	Cationic	436.11
CTAB	Cetyltrialkyl Ammonium Bromide	Cationic	364.45/406.53
SDBS	Sodium dodecyl benzene sulfonate	Anionic	348.48
Texapon-40	Sodium lauryl ether sulfate	Anionic	376.48
AOT	Bis(2-ethylhexyl) sulfosuccinate sodium	Anionic	444.56
Brij-35	Poly(oxyethylene)$_{23}$ dodecyl ether	Nonionic	1198
Tween 80	Polyoxyethylene sorbitan monooleate	Nonionic	1310
Empilan KR6	Alcohols, C9–C11, ethoxylated	Nonionic	NR
Tergitol NP-10	Polyoxyethylene nonyl phenyl ether	Nonionic	NR
Sophorolipid	Sophorolipid	Nonionic	NR
Surfactin	Cyclic lipopeptide	Zwitterionic	NR
Guar gam	Galactomannan	Nonionic	NR
TX-405	Polyoxyethylene (40) isooctylphenyl ether	Nonionic	NR
Brij-58	Polyoxyethylene (20) cetyl ether	Nonionic	1123.5
Brij-98	Polyoxyethylene (20) oleyl ether	Nonionic	1149.5
Saponin	Pentacyclic triterpene saponin	Nonionic	NR
CAS	Cocamydopropyl hydroxysultaine	Zwitterionic	452.69
Emulgin W600	Nonyl phenol	Nonionic	483
Canarcel 20	Sorbitan monolaureate	Nonionic	NR
Canasol BJ35	Lautyl alcohol ether	Nonionic	NR
Surfacpol 203	NR	NR	NR
Surfacpol G	NR	Anionic	NR
Surfacpol 14104	NR	NR	NR
Polafix LO	Propyl-cocoamide betaine	Zwitterionic	NR
Maranil Lab	Sodium dodecyl bencen sulfonate	Anionic	NR
Texapon N-40	Sodium lauryl ether sulfate	Anionic	NR

[a]A compilation from the work of Torres et al. (2012), Zacarias-Salinas et al. (2013), and Mao et al. (2015)
[b]'NR' stands for 'Not Reported'

(Mulligan et al. 1999a). The release of the cationic forms of metals from contaminated soil occurs easily with anionic bio-surfactant solutions, e.g., the remediation of Cd, Zn, and Pb-contaminated soil is reported with the use of rhamnolipid (Herman et al. 1995). In addition, bio-surfactants are found to be able to remove chromium and arsenic from contaminated soils (Li et al. 2002; Ozturk et al. 2012; Maity et al. 2013; Mukhopadhyay et al. 2013). The common forms of chromium and arsenic are negatively charged anionic complexes, which facilitate the cationic surfactant-assisted fixation of those species (Li et al. 2002). However, mobilization of arsenic/chromium oxyanions by negatively charged bio-surfactant (e.g., rhamnolipids) might be due to any of following mechanisms: (a) the competition between the arsenic oxyanions and rhamnolipids for the adsorption sites on soil particles; (b) anion exchange reactions among arsenic anions and rhamnolipids; and (c) electrostatic repulsive interactions because of the increase in the negative zeta potential of the soil particles through the adsorption of rhamnolipids (Wang and Mulligan 2009a, b). Song et al. (2008), while investigating the performance of saponin bio-surfactant for the simultaneous removal of cadmium and phenanthrene, concluded that the external carboxyl groups of saponin micelles might have coordinated with cadmium and improved the mobilization rate. Mulligan et al. (1999a) suggested that the metal removal rate with the bio-surfactant can further be enhanced after consecutive washing. However, the surfactant-enhanced washing remediation of soils can be ineffective if the soil has a silt and clay content more than 20–30 % or have substantial quantities of organic matter (Riser-Roberts 1998; Mulligan et al. 2001c; Wen and Marshall 2011).

The surfactants are also used in combination with other extractants for the enhancement of metal removal rate from contaminated soils. The elimination of both heavy metals (Cd, Cr, Mn, Ni, Pb, and Zn) and the organic pollutants from soil were observed by the combined use of surfactant (e.g., Tween 80, Brij-98, saponin, CAS) and aminopolycarboxylate chelants (e.g., EDTA, EDDS) (Ehsan et al. 2006a; Mouton et al. 2009; Wen and Marshall 2011, Alcántara et al. 2012; Cao et al. 2013). Surfactants are also exploited in conjunction with some ligand ions to achieve an enhanced removal rate of metals from soil (Lima et al. 2011) and bio-extraction of soil heavy-metals (Ernst 1996; Langley and Beveridge 1999; Almeida et al. 2009).

9.9 Conclusion

The principles, methodologies, and features of the extractant as adopted during the chemical-assisted soil washing approaches have been discussed briefly within the scope of this chapter. Although a varying range of extractant is available, we have limited our discussion on the use of acid, chelant and surfactant considering the overall frequency of extractant recommendation trend by the researchers. It should be noted that chemical-induced washing remediation of metal-contaminated soils have been often studied at laboratory scale but moderately used at field-scale or full/commercial-scale, mostly due to the higher reagent cost and treatment-issues of the spent washing liquids. Hence, there has been increasing research focus on the formation of a treatment-scheme, which consists of recycling or recovery of the washing additives.

Acknowledgment The research has partially been supported by the Grants-in-Aid for Scientific Research (15H05118 and 25 · 5863) from the Japan Society for the Promotion of Science. One of the authors, Ismail M.M. Rahman, acknowledges the financial grant from the 'The Public Foundation of Chubu Science and Technology Center, Japan' to support his research in the Kanazawa University, Japan.

References

Abumaizar RJ, Smith EH (1999) Heavy metal contaminants removal by soil washing. J Hazard Mater 70:71–86

Alcántara MT, Gómez J, Pazos M, Sanromán MA (2012) Electrokinetic remediation of lead and phenanthrene polluted soils. Geoderma 173–174:128–133

Alderighi L, Gans P, Ienco A, Peters D, Sabatini A, Vacca A (1999) Hyperquad simulation and speciation (HySS): a utility program for the investigation of equilibria involving soluble and partially soluble species. Coord Chem Rev 184:311–318

Almeida CMR, Dias AC, Mucha AP, Bordalo AA, Vasconcelos MTSD (2009) Influence of surfactants on the Cu phytoremediation potential of a salt marsh plant. Chemosphere 75:135–140

Arwidsson Z, Elgh-Dalgren K, von Kronhelm T, Sjoberg R, Allard B, van Hees P (2010) Remediation of heavy metal contaminated soil washing residues with amino polycarboxylic acids. J Hazard Mater 173:697–704

BASF (2007) Technical Information (Ti/EVD 1418 e): Trilon® M types. BASF Aktiengesellschaft, Ludwigshafen

Basta NT, Ryan JA, Chaney RL (2005) Trace element chemistry in residual-treated soil: key concepts and metal bioavailability. J Environ Qual 34:49–63

Begum ZA, Rahman IMM, Tate Y, Egawa Y, Maki T, Hasegawa H (2012a) Formation and stability of binary complexes of divalent ecotoxic ions (Ni, Cu, Zn, Cd, Pb) with biodegradable aminopolycarboxylate chelants (DL-2-(2- carboxymethyl)nitrilotriacetic acid, GLDA, and 3-hydroxy-2,2′-iminodisuccinic acid, HIDS) in aqueous solutions. J Solut Chem 41:1713–1728

Begum ZA, Rahman IMM, Tate Y, Sawai H, Maki T, Hasegawa H (2012b) Remediation of toxic metal contaminated soil by washing with biodegradable aminopolycarboxylate chelants. Chemosphere 87:1161–1170

Begum ZA, Rahman IMM, Sawai H, Tate Y, Maki T, Hasegawa H (2012c) Stability constants of Fe(III) and Cr(III) complexes with dl-2-(2-carboxymethyl)nitrilotriacetic acid (GLDA) and 3-hydroxy-2,2′-iminodisuccinic acid (HIDS) in aqueous solution. J Chem Eng Data 57:2723–2732

Begum ZA, Rahman IMM, Hasegawa H (2013a) Management of EDTA-containing aqueous effluent: environmental concerns and remedies. In: Molnar A (ed) EDTA: synthesis, uses and environmental concerns. Nova, Hauppauge, pp 163–177

Begum ZA, Rahman IMM, Sawai H, Mizutani S, Maki T, Hasegawa H (2013b) Effect of extraction variables on the biodegradable chelant-assisted removal of toxic metals from artificially contaminated European reference soils. Water Air Soil Pollut 224:1381

Bolton H Jr, Li SW, Workman DJ, Girvin DC (1993) Biodegradation of synthetic chelates in subsurface sediments from the southeast coastal plain. J Environ Qual 22:125–132

Bricka RM, Williford CW, Jones LW (1993) Technology assessment of currently available and developmental technique for heavy metals-contaminated soils treatment, Technical report IRRP-93-4. U.S. Army Corps of Engineers, Waterways Experiment Station, Vicksburg

Cao M, Hu Y, Sun Q, Wang L, Chen J, Lu X (2013) Enhanced desorption of PCB and trace metal elements (Pb and Cu) from contaminated soils by saponin and EDDS mixed solution. Environ Pollut 174:93–99

Chu W, Chan KH (2003) The mechanism of the surfactant-aided soil washing system for hydrophobic and partial hydrophobic organics. Sci Total Environ 307:83–92

Cline SR, Reed BE (1995) Lead removal from soils via bench-scale soil washing techniques. J Environ Eng ASCE 121:700–705

Conte P, Agretto A, Spaccini R, Piccolo A (2005) Soil remediation: humic acids as natural surfactants in the washings of highly contaminated soils. Environ Pollut 135:515–522

Dahrazma B, Mulligan CN (2007) Investigation of the removal of heavy metals from sediments using rhamnolipid in a continuous flow configuration. Chemosphere 69:705–711

Davis AP, Hotha BV (1998) Washing of various lead compounds from a contaminated soil column. J Environ Eng ASCE 124:1066–1075

Davis AP, Singh I (1995) Washing of zinc(II) from contaminated soil column. J Environ Eng ASCE 121:174–185

Dermont G, Bergeron M, Mercier G, Richer-Lafleche M (2008) Soil washing for metal removal: a review of physical/chemical technologies and field applications. J Hazard Mater 152:1–31

Deshpande S, Shiau BJ, Wade D, Sabatini DA, Harwell JH (1999) Surfactant selection for enhancing ex situ soil washing. Water Res 33:351–360

Egli T (2001) Biodegradation of metal-complexing aminopolycarboxylic acids. J Biosci Bioeng 92:89–97

Ehsan S, Prasher SO, Marshall WD (2006a) A washing procedure to mobilize mixed contaminants from soil: II. Heavy metals. J Environ Qual 35:2084–2091

Ehsan S, Prasher SO, Marshall WD (2006b) A washing procedure to mobilize mixed contaminants from soil. I. Polychlorinated biphenyl compounds. J Environ Qual 35:2146–2153

Elliott HA, Brown GA (1989) Comparative evaluation of NTA and EDTA for extractive decontamination of Pb-polluted soils. Water Air Soil Pollut 45:361–369

Ernst WHO (1996) Bioavailability of heavy metals and decontamination of soils by plants. Appl Geochem 11:163–167

ESTCP (1997) Joint small-arms range remediation: cost and performance report. Environmental Security Technology Certification Program (ESTCP). U.S. Department of Defense, Arlington

Evanko CR, Dzombak DA (1997) Remediation of metals-contaminated soils and groundwater (TE-97-01). Ground-Water Remediation Technologies Analysis Center (GWRTAC), Pittsburgh

Gao Y, He J, Ling W, Hu H, Liu F (2003) Effects of organic acids on copper and cadmium desorption from contaminated soils. Environ Int 29:613–618

Griffiths RA (1995) Soil-washing technology and practice. J Hazard Mater 40:175–189

Grundler OJ, van der Steen ATM, Wilmot J (2005) Overview of the European risk assessment on EDTA. In: Nowack B, VanBriesen JM (eds) Biogeochemistry of chelating agents, ACS symposium series. American Chemical Society, Washington, DC, pp 336–347

Hasegawa H, Rahman IMM, Kinoshita S, Maki T, Furusho Y (2010) Non-destructive separation of metal ions from wastewater containing excess aminopolycarboxylate chelant in solution with an ion-selective immobilized macrocyclic material. Chemosphere 79:193–198

Hasegawa H, Rahman IMM, Nakano M, Begum ZA, Egawa Y, Maki T, Furusho Y, Mizutani S (2011) Recovery of toxic metal ions from washing effluent containing excess aminopolycarboxylate chelant in solution. Water Res 45:4844–4854

Herman DC, Artiola JF, Miller RM (1995) Removal of cadmium, lead, and zinc from soil by a rhamnolipid biosurfactant. Environ Sci Technol 29:2280–2285

Hester RE, Harrison RM (eds) (1997) Contaminated land and its reclamation. Royal Society of Chemistry, Cambridge

Hong PKA, Jiang W (2005) Factors in the selection of chelating agents for extraction of lead from contaminated soil: effectiveness, selectivity, and recoverability. In: Nowack B, VanBriesen JM (eds) Biogeochemistry of chelating agents, ACS symposium series. American Chemical Society, Washington, DC, pp 421–432

Isoyama M, Wada SI (2007) Remediation of Pb-contaminated soils by washing with hydrochloric acid and subsequent immobilization with calcite and allophanic soil. J Hazard Mater 143:636–642

ITRC (1997) Technical and regulatory guidelines for soil washing. Interstate Technology & Regulatory Council (ITRC), Washington, DC

Kari FG, Giger W (1995) Modeling the photochemical degradation of ethylenediaminetetraacetate in the river Glatt. Environ Sci Technol 29:2814–2827

Kari FG, Hilger S, Canonica S (1995) Determination of the reaction quantum yield for the photochemical degradation of Fe(III)-EDTA: implications for the environmental fate of EDTA in surface waters. Environ Sci Technol 29:1008–1017

Khan S, Cao Q, Zheng YM, Huang YZ, Zhu YG (2008) Health risks of heavy metals in contaminated soils and food crops irrigated with wastewater in Beijing, China. Environ Pollut 152:686–692

Ko I, Chang YY, Lee CH, Kim KW (2005) Assessment of pilot-scale acid washing of soil contaminated with As, Zn and Ni using the BCR three-step sequential extraction. J Hazard Mater 127:1–13

Ko I, Lee CH, Lee KP, Lee SW, Kim KW (2006) Remediation of soil contaminated with arsenic, zinc, and nickel by pilot-scale soil washing. Environ Prog 25:39–48

Kuo S, Lai MS, Lin CW (2006) Influence of solution acidity and $CaCl_2$ concentration on the removal of heavy metals from metal-contaminated rice soils. Environ Pollut 144:918–925

Langley S, Beveridge TJ (1999) Effect of O-side-chain-lipopolysaccharide chemistry on metal binding. Appl Environ Microbiol 65:489–498

Leštan D, C-l L, X-d L (2008) The use of chelating agents in the remediation of metal-contaminated soils: a review. Environ Pollut 153:3–13

Li Z, Alessi D, Zhang P, Bowman R (2002) Organo-illite as a low permeability sorbent to retard migration of anionic contaminants. J Environ Eng 128:583–587

Lima TMS, Procópio LC, Brandão FD, Carvalho AMX, Tótola MR, Borges AC (2011) Simultaneous phenanthrene and cadmium removal from contaminated soil by a ligand/biosurfactant solution. Biodegradation 22:1007–1015

Lin HK, Man XD, Walsh DE (2001) Lead removal via soil washing and leaching. J Miner Met Mater Soc 53:22–25

Maity JP, Huang YM, Fan C-W, Chen C-C, Li C-Y, Hsu C-M, Chang Y-F, Wu C-I, Chen C-Y, Jean J-S (2013) Evaluation of remediation process with soapberry derived saponin for removal of heavy metals from contaminated soils in Hai-Pu, Taiwan. J Environ Sci China 25:1180–1185

Makino T, Kamiya T, Takano H, Itou T, Sekiya N, Sasaki K, Maejima Y (2007) Remediation of cadmium-contaminated paddy soils by washing with calcium chloride: verification of on-site washing. Environ Pollut 147:112–119

Mann MJ (1999) Full-scale and pilot-scale soil washing. J Hazard Mater 66:119–136

Mao X, Jiang R, Xiao W, Yu J (2015) Use of surfactants for the remediation of contaminated soils: a review. J Hazard Mater 285:419–435

Martell AE, Smith RM (1974) Critical stability constants, vol 1: amino acids. Plenum Press, New York

Martell AE, Smith RM, Motekaitis RJ (2004) NIST standard reference database 46: NIST critically selected stability constants of metal complexes database (Version 8.0 For Windows). Texas A&M University, College Station, TX

Maturi K, Reddy KR (2008) Extractants for the removal of mixed contaminants from soils. Soil Sediment Contam 17:586–608

Means JL, Kucak T, Crerar DA (1980) Relative degradation rates of NTA, EDTA and DTPA and environmental implications. Environ Pollut Ser B Chem Phys 1:45–60

Mouton J, Mercier G, Blais JF (2009) Amphoteric surfactants for PAH and lead polluted-soil treatment using flotation. Water Air Soil Pollut 197:381–393
Moutsatsou A, Gregou M, Matsas D, Protonotarios V (2006) Washing as a remediation technology applicable in soils heavily polluted by mining-metallurgical activities. Chemosphere 63:1632–1640
Mukhopadhyay S, Hashim MA, Sahu JN, Yusoff I, Sen Gupta B (2013) Comparison of a plant based natural surfactant with SDS for washing of As(V) from Fe rich soil. J Environ Sci China 25:2247–2256
Mulligan CN, Wang SL (2006) Remediation of a heavy metal-contaminated soil by a rhamnolipid foam. Eng Geol 85:75–81
Mulligan CN, Yong RN, Gibbs BF (1999a) On the use of biosurfactants for the removal of heavy metals from oil-contaminated soil. Environ Prog 18:50–54
Mulligan CN, Yong RN, Gibbs BF (1999b) Removal of heavy metals from contaminated soil and sediments using the biosurfactant surfactin. J Soil Contam 8:231–254
Mulligan CN, Yong RN, Gibbs BF (2001a) Remediation technologies for metal-contaminated soils and groundwater: an evaluation. Eng Geol 60:193–207
Mulligan CN, Yong RN, Gibbs BF (2001b) Heavy metal removal from sediments by biosurfactants. J Hazard Mater 85:111–125
Mulligan CN, Yong RN, Gibbs BF (2001c) Surfactant-enhanced remediation of contaminated soil: a review. Eng Geol 60:371–380
Nedwed T, Clifford DA (2000) Feasibility of extracting lead from lead battery recycling site soil using high-concentration chloride solutions. Environ Prog 19:197–206
Nörtemann B (2005) Biodegradation of chelating agents: EDTA, DTPA, PDTA, NTA, and EDDS. In: Nowack B, VanBriesen JM (eds) Biogeochemistry of chelating agents, ACS symposium series. American Chemical Society, Washington, DC, pp 150–170
Nowack B (2002) Environmental chemistry of aminopolycarboxylate chelating agents. Environ Sci Technol 36:4009–4016
Ochoa-Loza FJ, Artiola JF, Maier RM (2001) Stability constants for the complexation of various metals with a rhamnolipid biosurfactant. J Environ Qual 30:479–485
Ozturk S, Kaya T, Aslim B, Tan S (2012) Removal and reduction of chromium by *Pseudomonas* spp. and their correlation to rhamnolipid production. J Hazard Mater 231–232:64–69
Paria S (2008) Surfactant-enhanced remediation of organic contaminated soil and water. Adv Colloid Interface Sci 138:24–58
Pavel LV, Gavrilescu M (2008) Overview of *ex situ* decontamination techniques for soil cleanup. Environ Eng Manage J 7:815–834
Peters RW (1999) Chelant extraction of heavy metals from contaminated soils. J Hazard Mater 66:151–210
Pichtel J, Pichtel TM (1997) Comparison of solvents for ex situ removal of chromium and lead from contaminated soil. Environ Eng Sci 14:97–104
Pickering WF (1986) Metal ion speciation – soils and sediments (a review). Ore Geol Rev 1:83–146
Pinto ISS, Neto IFF, Soares HMVM (2014) Biodegradable chelating agents for industrial, domestic, and agricultural applications—a review. Environ Sci Pollut Res 21:11893–11906
Polettini A, Pomi R, Rolle E, Ceremigna D, De Propris L, Gabellini M, Tornato A (2006) A kinetic study of chelant-assisted remediation of contaminated dredged sediment. J Hazard Mater 137:1458–1465
Ramamurthy AS, Vo D, Li XJ, Qu J (2008) Surfactant-enhanced removal of Cu (II) and Zn (II) from a contaminated sandy soil. Water Air Soil Pollut 190:197–207
Rampley CG, Ogden KL (1998) Preliminary studies for removal of lead from surrogate and real soils using a water soluble chelator: adsorption and batch extraction. Environ Sci Technol 32:987–993
Reed B, Carriere P, Moore R (1996) Flushing of a Pb(II) contaminated soil using HCl, EDTA, and $CaCl_2$. J Environ Eng ASCE 122:48–50

Riser-Roberts E (1998) Remediation of petroleum contaminated soils : biological, physical, and chemical processes. Lewis Publishers, Boca Raton

Rosen MJ, Kunjappu JT (2012) Surfactants and interfacial phenomena. Wiley, Hoboken

Roundhill DM (2001) Extraction of metals from soils and waters. Kluwer Academic/Plenum Publishers, New York

Sachdev DP, Cameotra SS (2013) Biosurfactants in agriculture. Appl Microbiol Biotechnol 97:1005–1016

Sillanpää MET (2005) Distribution and fate of chelating agents in the environment. In: Nowack B, VanBriesen JM (eds) Biogeochemistry of chelating agents, ACS symposium series. American Chemical Society, Washington, DC, pp 226–233

Sillanpää M, Oikari A (1996) Assessing the impact of complexation by EDTA and DTPA on heavy metal toxicity using microtox bioassay. Chemosphere 32:1485–1497

Slizovskiy IB, Kelsey JW, Hatzinger PB (2011) Surfactant-facilitated remediation of metal-contaminated soils efficacy and toxicological consequences to earthworms. Environ Toxicol Chem 30:112–123

Song S, Zhu L, Zhou W (2008) Simultaneous removal of phenanthrene and cadmium from contaminated soils by saponin, a plant-derived biosurfactant. Environ Pollut 156:1368–1370

Sorvari J, Sillanpää M (1996) Influence of metal complex formation on heavy metal and free EDTA and DTPA acute toxicity determined by *Daphnia magna*. Chemosphere 33:1119–1127

Sposito G, Page AL (1984) Cycling of metal ions in the soil environment. In: Sigel H, Sigel A (eds) Metal ions in biological systems, vol 18, Circulation of metals in the environment. Marcel Dekker, New York, pp 287–332

Steele M, Pichtel J (1998) Ex-situ remediation of a metal-contaminated superfund soil using selective extractants. J Environ Eng 124:639–645

Swarnkar V, Agrawal N, Tomar R (2011) Sorption of chromate and arsenate by surfactant modified Erionite (E-SMZ). J Dispers Sci Technol 33:919–927

Tampouris S, Papassiopi N, Paspaliaris I (2001) Removal of contaminant metals from fine grained soils, using agglomeration, chloride solutions and pile leaching techniques. J Hazard Mater 84:297–319

Tandy S, Bossart K, Mueller R, Ritschel J, Hauser L, Schulin R, Nowack B (2004) Extraction of heavy metals from soils using biodegradable chelating agents. Environ Sci Technol 38:937–944

Tejowulan RS, Hendershot WH (1998) Removal of trace metals from contaminated soils using EDTA incorporating resin trapping techniques. Environ Pollut 103:135–142

Torres LG, Lopez RB, Beltran M (2012) Removal of As, Cd, Cu, Ni, Pb, and Zn from a highly contaminated industrial soil using surfactant enhanced soil washing. Phys Chem Earth 37–39:30–36

US EPA (1995) Contaminants and remedial options at selected metal-contaminated sites (EPA/540/R-95/512). . United States Environmental Protection Agency, Washington, DC

US EPA (1996) A citizen's guide to soil washing (EPA 542-F-96-002). . United States Environmental Protection Agency, Cincinnati

US EPA (1997) Recent developments for in situ treatment of metal contaminated soils. United States Environmental Protection Agency, Washington, DC

VanBenschoten JE, Matsumoto MR, Young WH (1997) Evaluation and analysis of soil washing for seven lead-contaminated soils. J Environ Eng ASCE 123:217–224

Vandevivere P, Hammes F, Verstraete W, Feijtel T, Schowanek D (2001) Metal decontamination of soil, sediment, and sewage sludge by means of transition metal chelant [S, S]-EDDS. J Environ Eng ASCE 127:802–811

Wang S, Mulligan CN (2004) An evaluation of surfactant foam technology in remediation of contaminated soil. Chemosphere 57:1079–1089

Wang S, Mulligan CN (2009a) Rhamnolipid biosurfactant-enhanced soil flushing for the removal of arsenic and heavy metals from mine tailings. Process Biochem 44:296–301

Wang SL, Mulligan CN (2009b) Arsenic mobilization from mine tailings in the presence of a biosurfactant. Appl Geochem 24:928–935

Wasay SA, Parker WJ, VanGeel PJ (2002) Removal of lead from a calcareous soil by chloride complexation. Soil Sediment Contam 11:841–859

Wen Y, Marshall WD (2011) Simultaneous mobilization of trace elements and polycyclic aromatic hydrocarbon (PAH) compounds from soil with a nonionic surfactant and [S, S]-EDDS in admixture: Metals. J Hazard Mater 197:361–368

Wood AL, Bouchard DC, Brusseau ML, Rao PSC (1990) Cosolvent effects on sorption and mobility of organic contaminants in soils. Chemosphere 21:575–587

Yang R, Luo C, Zhang G, Li X, Shen Z (2012) Extraction of heavy metals from e-waste contaminated soils using EDDS. J Environ Sci China 24:1985–1994

Yu J, Klarup D (1994) Extraction kinetics of copper, zinc, iron, and manganese from contaminated sediment using disodium ethylenediaminetetraacetate. Water Air Soil Pollut 75:205–225

Zacarias-Salinas M, Vaca M, Flores MA, Bandala ER, Torres LG (2013) Surfactant-enhanced washing of soils contaminated with wasted-automotive oils and the quality of the produced wastewater. J Environ Prot 4:1495–1501

Chapter 10
Application of Nanotechnology to Remediate Contaminated Soils

Mohammad Mahbub Rabbani, Imteaz Ahmed, and Soo-Jin Park

Abstract Soil is an important part of environment which is under threat due to various types of contaminations happening since last few decades. Recovery and regeneration of soil have become a global problem. Recently nanotechnology has emerged as an efficient, cost effective, environment friendly and promising technology for soil remediation. This technology has significant potentiality to remove contaminants from environment by various approaches like adsorption, redox reaction, conversion, stabilization, etc. Various types of nanomaterials and devices are used to remove contaminants from soil. Therefore, soil could be remediated effectively by utilizing nanotechnology based concepts, processes and products which cannot be achieved from conventional methods. One of the most hazardous soil contaminants is heavy metal. Like other contaminants, heavy metals could be removed from soil using nanotechnology based approaches. The applications of nanotechnology to remove heavy metals from the contaminated soil are discussed in this chapter.

Keywords Environment • Soil • Contamination • Heavy metal • Remediation • Nanotechnology • Application

M.M. Rabbani (✉)
Department of Chemistry, Inha University, 100 Inharo, Incheon 402-751, Republic of Korea

Department of Chemistry, American International University-Bangladesh, Banani, Dhaka, 1213, Bangladesh
e-mail: rabbani_che@yahoo.com

I. Ahmed
Department of Chemistry and Green-Nano Materials Research Center, Kyungpook National University, Daegu 702-701, Republic of Korea

S.-J. Park (✉)
Department of Chemistry, Inha University, 100 Inharo, Incheon 402-751, Republic of Korea
e-mail: sjpark@inha.ac.kr

H. Hasegawa et al. (eds.), *Environmental Remediation Technologies for Metal-Contaminated Soils*, DOI 10.1007/978-4-431-55759-3_10

10.1 Introduction

Soil, one of the most valuable parts of environment, is being contaminated rapidly for last few decades due to the side effect of industrialization. Industrial waste chemical disposal to soil, uncontrolled use of fertilizer, pesticides and other agro-chemicals cause the soil contamination. The prominent soil contaminants are heavy metals and metalloids, petroleum, solvents, pesticides and herbicides (Caliman et al. 2011). An overview of soil and water contaminants in Europe as reported in 2011 could be found in Fig. 10.1 (van Liedekerke et al. 2014). The report demonstrates that 31 % and 35 % of total contaminants in water and soil, respectively, are heavy metals. Contaminated soil affects human health, ecosystem and agriculture and thus threatens the entire environment. Therefore, the remediation of contaminated soil has become a major environmental issue (Cundy et al. 2008; Li et al. 2006b; Science Communication Unit, University of the West of England, Bristol 2013; Wuana and Okieimen 2011).

Nanoscience and nanotechnology are the terms used for the dealing of the science and technology which discuss about the materials having sizes within the range of 1–100 nanometers (nm). Nanomaterials have small size, different shapes, very large surface area, many reactive sites and show extraordinary reactivity. The different shapes of nanomaterials are shown in Fig. 10.2. Recently, nanotechnology has emerged as an invaluable technology and attracted much attention among the environmental scientists for removal of pollutants from water, air and soil. Nanotechnology based remediation techniques are effective, low cost and environmentally friendly (Caliman et al. 2011; National nanotechnology initiative 2014; Nowack 2008; Re et al. 2012; Tratnyek and Johnson 2006; Watlington 2005; Wei et al. 2013; Zhang 2003).

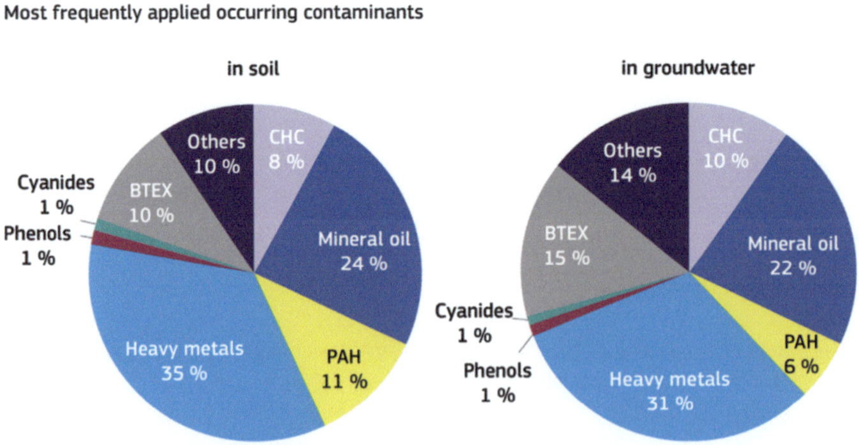

Fig. 10.1 Overview of contaminants affecting soil and groundwater in Europe as reported in 2011 (Reprinted from reference: van Liedekerke et al. 2014. Copyright 2014 European Union)

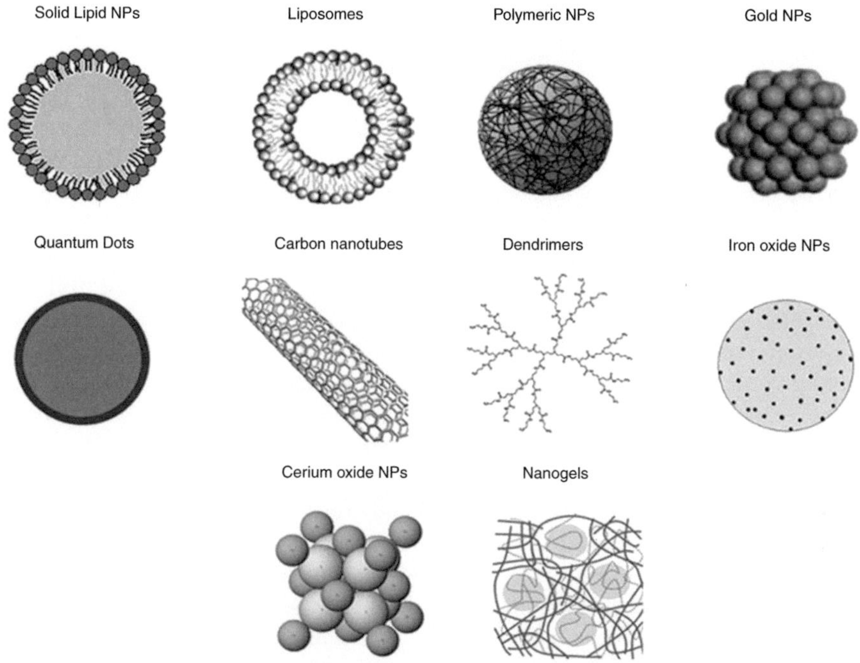

Fig. 10.2 Nanomaterials with different shapes (Reprinted with permission from reference: Re et al. 2012. Copyright 2012 Elsevier Inc.)

During decontamination process, most of the interactions occur at the interface between the nanomaterials and pollutants. As the nanomaterials have very large surface area so these interactions become faster. Active sites of the nanomaterials increase their decontamination efficiency significantly. The small size of the nanoparticles is favorable for their delivery into the contaminated sites. Nanomaterials can be modified depending on the application requirements, and generally, they are non-toxic. These nanomaterials are used not only for decontamination but also for detection of the pollutants. All these properties of the nanomaterials have emerged themselves a promising candidate for environmental remediation (Caliman et al. 2011; EPA 2008; Noubactep et al. 2012; Wei et al. 2013).

Heavy metals are major contaminants in soil and hence removal of heavy metals is of great importance. A wide range of the metals and even some of the metalloids are termed as heavy metals based on environmental concern. The prominent heavy metals which are responsible and considered as threatening for environment and health issue are: Arsenic (As), Lead (Pb), Cadmium (Cd), Chromium (Cr), Copper (Cu), Mercury (Hg), Nickel (Ni), etc. These heavy metals can be removed effectively from soil by applying nanotechnology (Al-Rashdi et al. 2011; Li et al. 2006a; Mulligan et al. 2001; Wei et al. 2013; Wuana and Okieimen 2011). In this chapter, we have discussed the applications of nanotechnology to remediate heavy metals contaminated soil. The applied methods, nanomaterials and the removal mechanisms are discussed as well.

10.2 Remediation

In situ, on site and ex situ techniques are globally applied for soil remediation (Caliman et al. 2011). A colloidal solution or aqueous slurry of nanoparticles can be injected into or sprayed on the contaminated soil using pressure or gravity. While nanoparticles are injected in soil, they form a treatment zone and remain in suspension (Noubactep et al. 2012; Zhang 2003; Watlington 2005). The widely investigated biocompatible technologies for removal of heavy metals from soil are adsorption and stabilization or immobilization of heavy metals in soil by nanoparticles (Almaroai et al. 2014; Chen et al. 2010; Nasiri et al. 2013). Materials which are used for the remediation of contaminated soil are broadly classified into two categories such as adsorptive and reactive materials depending on the remediation mechanism. Adsorptive materials remove contaminants by adsorption on both surface and internal structure (Caliman et al. 2011; Wei et al. 2013). And, adsorption is one of the most promising and effective techniques for heavy metals removal in soils. It is simple, relatively of low cost and effective in removing heavy metal ions such as Cu, Hg, Cd and Cr. The adsorption capacities depend on the properties of adsorbent. The extremely small size and high surface area of the nanomaterials help them to achieve better kinetics for the adsorption of heavy metal (Shan et al. 2009). On the other hand, reactive materials involve chemical reactions such as acid-base, redox, precipitation/dissolution, ion exchange and photocatalysis reactions (Caliman et al. 2011; Wei et al. 2013). During remediation processes, toxic contaminants are rapidly converted into significantly less toxic products (Wei et al. 2013). There are various nanomaterials which can be utilized for the removal of metallic contaminants from waste water (Al-Rashdi et al. 2011; Kumar et al. 2014; Mahdavi et al. 2013) or aquifer (Shan et al. 2009), but reports on the removal of heavy metals from contaminated soil are very limited. Moreover, most of the reports describe the heavy metal removal process in laboratory scale but field scale reports are very rare. The widely tested nanomaterials for heavy metal removal from contaminated soil are discussed below.

10.2.1 Zero-Valent Iron Nanoparticles

Iron-based remediation technologies are being developed rapidly and tested in different scales both in laboratory and field for environmental remediation (O'Carroll et al. 2013). The main advantage of iron is, it is the most abundant metal of the earth and hence the adsorbents prepared from this metal will be very cost effective. Additionally iron metal is non-toxic in their elemental form and hence they are environmentally friendly, and no special account has to be taken for their application in the soil contaminated site itself. Zero-valent iron (ZVI) nanoparticles are very small in size, which allows them to penetrate into the contaminated sites; they have very large surface area which facilitates close contact with pollutant and thus improves decontamination efficiency and they contain many reactive sites which enhances their reactivity compared with their bulk counterpart. Nanoscale iron particles remain suspended in their colloidal solution and so they can be injected straight into contaminated soils, sediments and aquifers. They have very good adsorption and reducing properties which enable them to react with heavy metals such as mercury (Hg), nickel (Ni) and cadmium (Cd). The extraordinary properties and feasible applications made zero-valent iron nanoparticles a promising candidate for soil remediation (Caliman et al. 2011; Cundy et al. 2008; Deliyanni, et al. 2004; Li et al. 2006a, b; Noubactep et al. 2012; USEPA 2007; Wei et al. 2013).

Zero-valent iron nanoparticles are strong reducing agent and very good adsorbent for various pollutants (Cundy et al. 2008). They are widely used to remove heavy metal such as Mercury (Hg^{2+}), Nickel (Ni^{2+}), Cadium (Cd^{2+}), Lead (Pb^{2+}) and Chromium (Cr^{6+}) from contaminated soil (Caliman et al. 2011; Li et al. 2006a; Zhang 2003). The key mechanisms involved in heavy metal removal are adsorption and reduction through which heavy metals are converted to less toxic species, immobilized or become less available in soil (Cundy et al. 2008; Li et al. 2006b). The reaction mechanism during decontamination process occurred in between metal contaminant and zero-valent iron depends on the standard redox potential (E^0) of the heavy metal. Heavy metals such as Cd and Zn which have more negative E^0 value than ZVI nanoparticles or similar E^0 value to ZVI nanoparticles are being adsorbed on iron shell. On the other hand, heavy metals (Cr, As, Cu) with much more positive E^0 value than ZVI nanoparticles are reduced and precipitated in soil. Heavy metals (Pb, Ni) with slightly positive E^0 value than ZVI nanoparticles are reduced and also adsorbed on ZVI nanoparticles. A core-shell structure of zero-valent iron nanoparticles is presented in Fig. 10.3, depicting the mechanisms involved in pollutant removal (Li et al. 2006a; Li and Zhang 2007; O'Carroll et al. 2013). There is a drawback associated with ZVI nanoparticles and that is they are rapidly oxidized in air due to their high surface area and reactivity. To reduce this oxidation rate, ZVI nanoparticles are passivated with some metal oxides or noble metals such as Pd and Pt which is then referred as bimetallic iron (Cundy et al. 2008; Li et al. 2006b). ZVI nanoparticles exhibit a strong tendency to aggregate which restrict their application. To overcome the aggregation tendency and to improve their mobility and deliverability into soil, various kind of modifications are made on the ZVI nanoparticles. Generally modified nanoparticles provide some additional advantage in removal of contaminants compared with bare nanoparticles (Li et al. 2006a, b).

Nasiri et al. effectively removed cadmium (Cd) from aqueous soil solution in laboratory through adsorption technique using colloidal zero valent iron (ZVI) nanoparticles (Nasiri et al. 2013). They have stabilized ZVI nanoparticles with carboxymethyl cellulose (CMC), which improves colloidal stability and diffusion of nanoparticles suspensions and consequently accelerates the cadmium adsorption rate from aqueous soil solution compared with uncoated ZVI nanoparticles. The concentration of ZVI nanoparticles has direct influence on the cadmium removal process and higher concentration of nanoparticles showed improved efficiency.

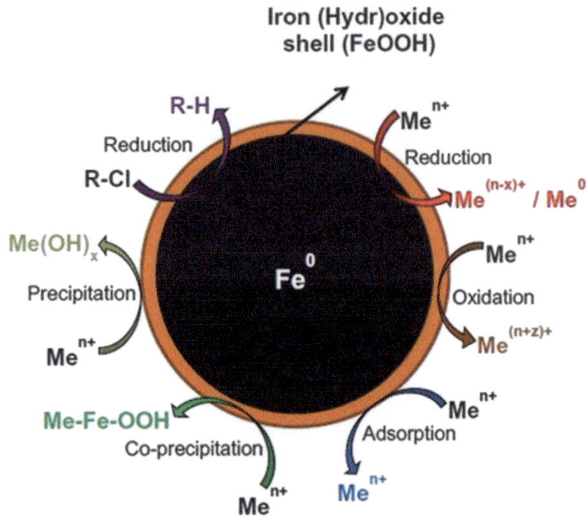

Fig. 10.3 The core–shell structure of a ZVI nanoparticle depicting various mechanisms involved in pollutant removal (Reprinted with permission from reference: O'Carroll et al. 2013. Copyright 2012 Elsevier Inc.)

Shariatmadari et al. remediates Cr(VI) contaminated clayey soil experimentally using nano-scaled zero valent iron (nZVI) (Shariatmadari et al. 2009). They have applied electrokinetics (EK) process with a permeable reactive barrier (PRB) filled with nZVI as a reactive materials and achieved a maximum of 19 % total Cr removal, whereas applying EK process without nZVI barrier only 14.78 % Cr(VI) removal was achieved. Combined EK/PRB process significantly enhances Cr(VI) reduction to less reactive Cr(III) very effectively up to 88 %. The EK/PRB process involves redox and adsorption/desorption reactions to reduce Cr(VI) to Cr(III) and total Cr removal. Addition of nZVI barrier in the EK process improves both Cr(VI) reduction and total Cr removal efficiency. Experimental conditions greatly influence the efficiencies of Cr(VI) reduction and Cr removal using nZVI from soil. Xu and Zhao reported the reductive immobilization of chromate in a sandy loam soil using stabilized ZVI nanoparticles (Xu and Zhao 2007). The ZVI nanoparticles were highly deliverable in the soil and they have found that only 0.08 g/L ZVI nanoparticles can reduce Cr leachability by 50 % at a soil-to-solution ratio of 1 g: 10 mL. The TCLP (toxicity characteristic leaching procedure) leachability of Cr in soil was reduced by 90 %. Kumpiene et al. reported the reduction of mobility and bioavailability of Cr, Cu, As and Zn in a chromated copper arsenate (CCA)-contaminated soil using ZVI nanoparticles. They have shown that it is possible to decrease As and Cr concentrations by 98 % and 45 % respectively in soil leachates by using ZVI nanoparticles (Kumpiene et al. 2006).

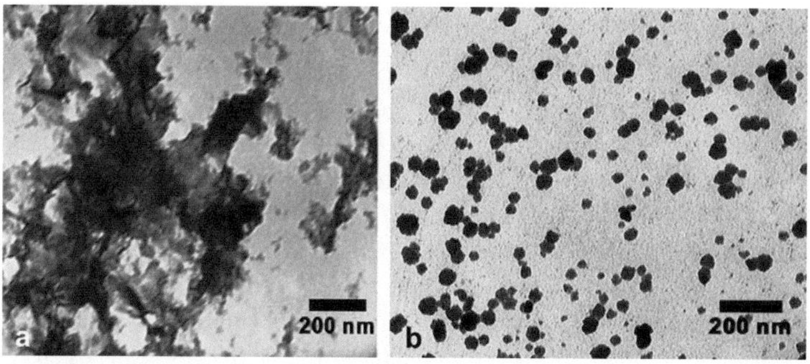

Fig. 10.4 TEM images of (**a**) FeS and (**b**) CMC-FeS nanoparticles (Reprinted with permission from reference: Xiong et al. 2009. Copyright 2012 Elsevier Inc.)

10.2.2 Iron Sulfide Nanoparticles

Xiong et al. and Gong et al. stabilized iron sulfide (FeS) nanoparticles with carboxymethyl cellulose (CMC) to immobilize mercury (Hg) and to improve deliverability of nanoparticles in field soil and sediment (Gong et al. 2012; Xiong et al. 2009). For in situ remediation, the delivery of nanoparticles to the contaminated soil is very important. CMC reduces the inherent aggregation tendency of FeS nanoparticles. Figure 10.4 shows the effect of CMC stabilization on the dispersion of FeS nanoparticles (Xiong et al. 2009). During remediation process, mercury might be bound to CMC-FeS nanoparticles through three possible reaction mechanisms such as adsorption, structural incorporation and precipitation and thus immobilized in soil and sediment. CMC-FeS nanoparticles are found to be efficient candidate to immobilize Hg significantly in soil and sediment. CMC-FeS nanoparticles reduced water-leachable Hg by 79–96 % at an FeS-to-Hg molar ratio of 28:1–118:1. The CMC stabilized FeS nanoparticles showed excellent deliverability in soils or sediments under moderate injection pressure, and the deliverability of nanoparticles can be controlled by varying injection pressure in field media. Even after the withdrawal of injection pressure, the delivered nanoparticles continue their mobility due to groundwater flow. CMC-FeS nanoparticles are highly capable of adsorbing mercury and they have very strong affinity for Hg. However, the authors have tested the in situ immobilization of mercury using CMC-FeS nanoparticles in laboratory with naturally contaminated soil, and the deliverability of nanoparticles was examined in a column breakthrough tests with wet-packed contaminated soil. Further research is required for practical use in field media.

10.2.3 Iron Phosphate Nanoparticles

Although the remediation of heavy metal contaminated soil using phosphate materials such as phosphoric acid (Yang et al. 2001), natural phosphate rocks (Basta and McGowen 2004) has been reported in literature but the physical and chemical properties of these phosphate materials restrict their applications in field condition and cause other secondary environmental hazards (Liu and Zhao 2007). Phosphate materials react with heavy metals and form secondary metal phosphate precipitates and thus immobilize heavy metals in soil. Iron is naturally abundant, low cost and less toxic element. So iron phosphate nanoparticles can overcome the problems arisen from the commonly used phosphate materials but can follow the same reaction mechanism. Liu and Zhao reported the in situ remediation of Cu(II) contaminated soils using sodium carboxymethyl cellulose (CMC) stabilized iron phosphate (vivianite) nanoparticles in a laboratory scale (Liu and Zhao 2007). In this process, Cu(II) ions are immobilized through forming less soluble copper phosphate minerals. Precipitation and adsorption mechanism are suggested for copper remediation in soil. Iron phosphate nanoparticles do not cause secondary environmental risk and they are more effective than commonly used phosphate materials even at low dose. As the nanoparticles sizes are very small (~10 nm) so they showed a very good deliverability in soils and they can be injected easily into the contamination soil.

10.2.4 Iron Oxides Nanoparticles

Iron rich nanomaterials are also used for heavy metal removal from contaminated soil. Iron oxide (FeO) has a very good adsorption capacity of heavy metals. Zirconium-iron oxide (Zr-FeO) is chemically more stable than FeO, has better binding affinity and show high adsorption capacity over a wide range of pH. Almaroai et al. reported the remediation of arsenic (As) and lead (Pb) contaminated agricultural soil using iron rich nanomaterials such as iron oxide (FeO) and Zirconium-iron oxide (Zr-FeO) in a laboratory scale (Almaroai et al. 2014). In this process, the possible mechanism of As and Pb immobilization is adsorption. The increased amount of iron oxide nanoparticles enhances the heavy metal immobilization rate in soil. They have demonstrated that the addition of Fe and Zr–Fe oxides (5 %) decreased ammonium acetate (NHO_4Ac)-extractable lead (Pb) by up to 83 % and 65 %, respectively. On the other hand, NHO_4Ac-extractable As in soil was decreased by an average of 77.3 % and 64.7 %, respectively, with the addition of FeO and Zr-FeO nanoparticles. These results suggest that iron rich nanomaterials have a very good capacity of As and Pb immobilization in soil.

10.2.5 Allophane

Allophane is a nano-scale hydrous aluminosilicate and an effective sorbent of copper. G. Yuan demonstrated the adsorptive removal of Cu from soil using a natural nanomaterial, allophane, in laboratory scale (Yuan 2004). The adsorption of copper to allophane occurs through cation-exchange reaction and specific complexation between copper ions and the $(OH)Al(OH_2)$ groups of allophane. The pH value of contaminated soil has an effect on the adsorption of Cu, and the adsorption is enhanced at a higher pH. Allophane can take up 4448 mg Cu/kg at the equilibrium concentration of 10 mg Cu/L at pH 5.5. As an environment friendly material, allophane can be used in remediating copper from contaminated soil.

10.2.6 Carbon Black

Surface-modified carbon black is used for removal of heavy metals through adsorption because it has very good affinity to heavy metals (Zhou et al. 2010). Cheng et al. reported the remediation of soil polluted by Cu and Zn using surface-modified nanoscale carbon black in a laboratory scale (Cheng et al. 2014). The suggested mechanisms involved in removal of these metals are adsorption and complexation. They oxidized carbon black with HNO_3 to introduce functional groups which increases surface cation exchange and complexation capacity of carbon black. The experiment of heavy metals removal was conducted in a green house using ryegrass cultivation. They have found that the amount of Cu and Zn accumulated in ryegrass shoot and root was significantly decreased with the increase of carbon black added to the contaminated soil. This result suggests that Cu and Zn were adsorbed and complexed by the modified carbon black and thus the transfer of these metals from soil to the ryegrass shoot and root was prohibited. Modified carbon black could be applied effectively for the remediation of soils contaminated by Cu and Zn.

10.3 Conclusion

Nanotechnology opens the opportunity to tune the properties of materials in a very specific way for one's specific purpose and therefore can be more efficient compared with the conventional materials. Nanoparticles are much more efficient and highly selective for the heavy metal contaminants. Iron-based nanoparticles are widely used for the heavy metal removal from soil. They have very strong adsorption property and excellent reducing property. Nanoparticles are generally injected into the soil and they remain suspended making a decontamination zone. The small size of the nanomaterials increases their mobility and deliverability in soil, and the heavy metals are stabilized or converted to less toxic species in soil. Nanotechnology has become a reliable means to remediate heavy metal contaminated soil. Most of the studies are conducted in laboratory scale, and therefore, much effort should be given to the field-scale remediation.

References

Almaroai YA, Vithanage M, Rajapaksha AU, Lee SS, Dou Z, Lee YH, Sung JK, Ok YK (2014) Natural and synthesized iron-rich amendments for As and Pb immobilization in agricultural soil. Chem Ecol 30:267–279

Al-Rashdi B, Somerfield C, Hilal N (2011) Heavy metals removal using adsorption and nanofiltration techniques. Sep Purif Rev 40:209–259

Basta NT, McGowen SL (2004) Evaluation of chemical immobilization treatments for reducing heavy metal transport in a smelter contaminated soil. Environ Pollut 127:73–82

Caliman FA, Robu BM, Smaranda C, Pavel VL, Gavrilescu M (2011) Soil and groundwater cleanup: benefits and limits of emerging technologies. Clean Techn Environ Policy 13:241–268

Chen G, Zeng G, Chunyan D, Huang D, Lin T, Wang L, Guoli S (2010) Transfer of heavy metals from compost to red soil and groundwater under simulated rainfall conditions. J Hazard Mater 181:211–216

Cheng JM, Liu YZ, Wang HW (2014) Effects of surface-modified nano-scale carbon black on Cu and Zn fractionations in contaminated Soil. Int J Phytoremediation 16:86–94

Cundy AB, Hopkinson L, Whitby RLD (2008) Use of iron-based technologies in contaminated land and groundwater remediation: a review. Sci Total Environ 400:42–51

Deliyanni EA, Lazaridis NK, Peleka EN, Matis KA (2004) Metals removal from aqueous solution by iron-based bonding agents. Environ Sci Pollut Res 11:18–21

EPA (2008) Nanotechnology for site remediation fact sheet. EPA, Cincinnati

Gong Y, Liu Y, Xiong Z, Kaback D, Zhao D (2012) Immobilization of mercury in field soil and sediment using carboxymethyl cellulose stabilized iron sulfide nanoparticles. Nanotechnology 23:294007 (13pgs)

Kumar R, Khan MA, Haq N (2014) Application of carbon nanotubes in heavy metals remediation. Crit Rev Environ Sci Technol 44:1000–1035

Kumpiene J, Ore S, Renella G, Mench M, Lagerkvist A, Maurice C (2006) Assessment of zerovalent iron for stabilization of chromium, copper, and arsenic in soil. Environ Pollut 144:62–69

Li XQ, Zhang WX (2007) Sequestration of metal cations with zerovalent iron nanoparticles – a study with high resolution X-ray photoelectron spectroscopy (HR-XPS). J Phys Chem C 111:6939–6946

Li XQ, Elliott DW, Zhang WX (2006a) Zero-valent iron nanoparticles for abatement of environmental pollutants: materials and engineering aspects. Crit Rev Solid State Mater Sci 31:111–122

Li L, Fan M, Brown RC, Leeuwen JHV, Wang J, Wang W, Song Y, Zhang P (2006b) Synthesis, properties, and environmental applications of nanoscale iron-based materials: a review. Crit Rev Environ Sci Technol 36:405–431

Liu R, Zhao D (2007) In situ immobilization of Cu(II) in soils using a new class of iron phosphate nanoparticles. Chemosphere 68:1867–1876

Mahdavi S, Jalali M, Afkhami A (2013) Heavy metals temoval from aqueous solutions using TiO_2, MgO, and Al_2O_3 nanoparticles. Chem Eng Commun 200:448–470

Mulligan CN, Yong RN, Gibbs BF (2001) Remediation technologies for metal-contaminated soils and groundwater: an evaluation. Eng Geol 60:193–207

Nasiri J, Gholami A, Panahpour E (2013) Removal of cadmium from soil resources using stabilized zero-valent iron nanoparticles. J Civil Eng Urban 3:338–341
National nanotechnology initiative (2014) What is nanotechnology? http://www.nano.gov/nanotech-101/what/definition. Accessed 22 Sept 2014
Noubactep C, Caré S, Crane R (2012) Nanoscale metallic iron for environmental remediation: prospects and limitations. Water Air Soil Pollut 223:1363–1382
Nowack B (2008) Pollution prevention and treatment using nanotechnology. In: Krug H (ed) Nanotechnology, volume 2: environmental aspects. WILEY-VCH Verlag GmbH & Co. KGaA, Weinheim, pp 1–15
O'Carroll D, Sleep B, Krol M, Boparai H, Kocur C (2013) Nanoscale zero valent iron and bimetallic particles for contaminated site remediation. Adv Water Resour 51:104–122
Re F, Gregori M, Masserini M (2012) Nanotechnology for neurodegenerative disorders. Nanomedicine: NBM 8:S51–S58
Science Communication Unit, University of the West of England, Bristol (2013) Science for environment policy in-depth report: soil contamination: impacts on human health. Report produced for the European Commission DG Environment, September 2013
Shan G, Yan S, Tyagi RD, Surampalli RY, Zhang TC (2009) Applications of nanomaterials in environmental science and engineering: review. Pract Period Hazard Toxic Radioact Waste Manage 13:110–119
Shariatmadari N, Weng CH, Daryaee H (2009) Enhancement of hexavalent chromium [Cr(VI)] remediation from clayey soils by electrokinetics coupled with a nano-sized zero-valent iron barrier. Environ Eng Sci 26:1071–1079
Tratnyek PG, Johnson RL (2006) Nanotechnologies for environmental cleanup. Nanotoday 1:44–48
USEPA (2007) Nanotechnology white paper. EPA 100/B-07/001. Science Policy Council, U.S. Environmental Protection Agency, Washington, DC, p 120
van Liedekerke M, Prokop G, Rabl-Berger S, Kibblewhite M, Louwagie G (2014) Progress in the management of contaminated sites in Europe, Reference Report by the Joint Research Centre of the European Commission
Watlington K (2005) Emerging nanotechnologies for site remediation and wastewater treatment. Report, U.S. Environmental Protection Agency, Office of Solid Waste and Emergency Response, Office of Superfund Remediation and Technology Innovation, Technology Innovation and Field Services Division, Washington, DC
Wei Q, Yang D, Fan M, Harris HG (2013) Applications of nanomaterial-based membranes in pollution control. Crit Rev Environ Sci Technol 43:2389–2438
Wuana RA, Okieimen FE (2011) Heavy metals in contaminated soils: a review of sources, chemistry, risks and best available strategies for remediation. International Scholarly Research Network, ISRN Ecology, Volume 2011, Article ID 402647, doi:10.5402/2011/402647
Xiong Z, He F, Zhao D, Barnett MO (2009) Immobilization of mercury in sediment using stabilized iron sulfide nanoparticles. Water Res 43:5171–5179
Xu YH, Zhao DY (2007) Reductive immobilization of chromate in water and soil using stabilized iron nanoparticles. Water Res 41:2101–2108
Yang J, Mosby DE, Casteel SW, Blanchar RW (2001) Lead immobilization using phosphoric acid in a smelter-contaminated urban soil. Environ Sci Technol 35:3553–3559
Yuan G (2004) Natural and modified nanomaterials as sorbents of environmental contaminants. J Environ Sci Health A 39:2661–2670
Zhang WX (2003) Nanoscale iron particles for environmental remediation: an overview. J Nanoparticle Res 5:323–332
Zhou D, Wang Y, Wang H (2010) Surface-modified nano-scale carbon black used as sorbents for Cu(II) and Cd(II). J Hazard Mater 174:34–39

Chapter 11
Risk Evaluation for Remediation Techniques to Metal-Contaminated Soils

Aiichiro Fujinaga

Abstract In Japan, environmental standards for contaminants in groundwater and in leachate from soil are set with the assumption that they are used for drinking water over a human lifetime. Where there is neither a well nor groundwater used for drinking, the standard is thus too severe. Therefore, remediation based on these standards incurs excessive effort and cost. In contrast, the environmental assessment procedure used in the USA and the Netherlands considers the site conditions (land use, existing wells, etc.); however, a risk assessment is required for each site. This chapter shows a framework for validating contamination by considering the merits of the environmental standards used and a method for risk assessment. The framework involves setting risk-based concentrations (RBCs), which are attainable remediation goals for contaminants in soil and groundwater. The framework was then applied to a model contaminated site for risk management. RBCs of Cr(VI) in a contaminated site were set according to the site conditions. The RBCs of Cr (VI) with/without drinking water in residential area are calculated. Second, an experiment for contaminated soil was introduced by using column equipment. The equipment was designed by applying water permeability test. And then, variation of concentration of the contaminant was simulated using an advection-diffusion model. Simulation by the mathematical model is also useful for monitored natural attenuation or *in situ* treatment, because the simulation can estimate clean-up time at the contaminated site. Even though the estimated clean-up time is not exact time, the cost of *in situ* treatment is not expensive. And then land owners can choose the *in situ* treatment.

Keywords Risk assessment • Metal • Contaminated soil and groundwater • Risk-based concentration • Column experiment

---

A. Fujinaga (✉)
Department of Civil Engineering, Faculty of Engineering, Osaka Sangyo University, 3-1-1 Nakagaito, Daito City, Osaka 574-8530, Japan
e-mail: fujinaga@ce.osaka-sandai.ac.jp

H. Hasegawa et al. (eds.), *Environmental Remediation Technologies for Metal-Contaminated Soils*, DOI 10.1007/978-4-431-55759-3_11

11.1 Introduction

11.1.1 Countermeasure to Soil Contamination

If contamination of soil or groundwater is found in residential or occupational sites, the source of the contamination should be treated and removed before it releases the contamination into the environment. Otherwise, the contamination will spread out widely and the consequences of the contamination will worsen. After the removal of the source of the contaminant, countermeasures of contaminated sites, which have relatively low concentration of the contaminant, should be planned. The remediation plan has to decide target area and depth (soil volume), and the remediation method should be chosen. However, it is difficult to decide contaminated area and un-contaminated area, because border line of the contamination is not clear.

Small volume of contaminated soil in Japan is often treated by excavation and disposal. If the contaminated area or volume is large, treatment method is compared among several possible methods. For example, excavation followed by removal and excavation followed by *in situ* washing are often compared for efficiency, cost, time and resident's acceptance. On the other hand, if the contamination is already spread out to vast area and depth, remediation of the soil and groundwater is difficult. Therefore, approximately half of the sites having very high contamination are left abandoned to become brownfields, because the cost of remediation is higher than the land price or the potential benefit to the site (Fig. 11.1, JMOE 2009).

Risk management is very important in order to solve the problem of the brownfield. The risk management is a total solution method of the contaminated site. Setting an acceptable concentration for a land use is one way of the risk management. The risk management method is explained in Section 11.2.

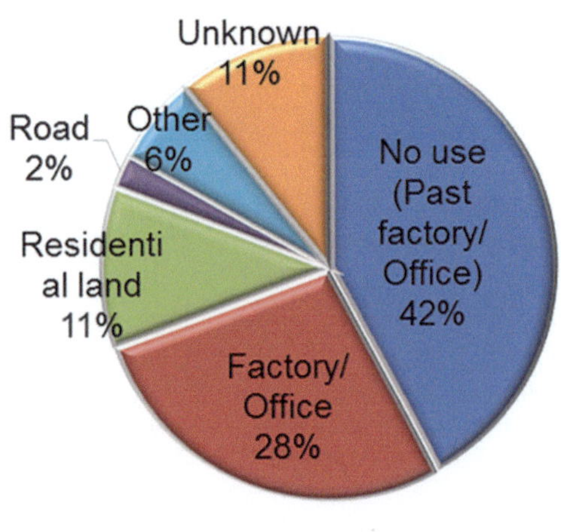

Fig. 11.1 Land use of contaminated sites which are registered according to Soil Contamination Countermeasures Act in Japan (JMOE 2009)

11.1.2 Risk-Based Concentration

There are two methods currently used to assess contamination in soil and groundwater: (1) the use of environmental standards and (2) risk assessment on human health for each contaminated site. With the former method, if the concentration of a contaminant is above the standard, the site is defined as a contaminated site. In Japan, the goal of remediation is to achieve the environmental standards set for drinking water. However, if the groundwater is not used for drinking purpose, remediation to achieve the environmental standard is considered as unnecessary and expensive (Fujinaga et al. 2012).

In contrast, the environmental assessment methods used in the USA and the Netherlands consider site conditions. Here, the site conditions include how the site and groundwater are used, the conditions of the ground surface (soil covered or uncovered), and whether there is surface water (such as a lake, pond or river) at or near the site. Therefore, in many cases, required remediation can be performed at a reasonable cost. However, risk assessors must perform risk assessment for remediation of the contaminated sites, and thus the cost of the risk assessment should be included in a risk-management project.

11.2 Methodology for Risk Assessment

11.2.1 Basic Risk Assessment

In this section, the basic risk-assessment procedure is described (Fig. 11.2). First, the human intake of a particular contaminant is estimated from exposures at the contaminated site. The intake is calculated from concentrations of the contaminants in the groundwater, soil and surface water. The intake calculation also includes ingestion rate, exposure frequency, exposure duration, body weight, averaging exposure time and other parameters. The risk of the contaminants to human health is then calculated based on their toxicity.

Non-cancer and cancer risk are estimated separately. The non-cancer hazard quotient (*HQ*) is calculated as the intake (*I*) divided by the reference dose (*RfD*). *HQ* is expressed as Eq. 11.1.

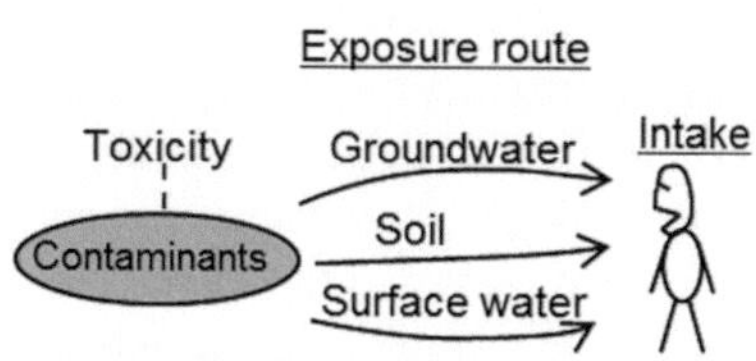

Fig. 11.2 Image of intake of contaminants by exposure routes

$$HQ = \frac{I}{RfD} \tag{11.1}$$

If $HQ < 1$, it presents a strong likelihood that a toxic effect of concern will not occur.

The mechanism of cancer development is complicated. Contaminants may attack genes directly or indirectly, and cancer may arise in one step or in several steps. Therefore, cancer risk cannot be expressed as a simple dose-response reaction (Kammen and Hassenzahl 1999). However, to calculate the risk, cancer risk was assumed here to be proportional to the dose of a carcinogen. Risk is calculated as "*I* (mg/kg·day)" times "the cancer-slope factor of a contaminant (*Sf*) $(\text{mg}/(\text{kg}\cdot\text{day}))^{-1}$" (Eq. 11.2).

Acceptable cancer risk over a lifetime is set as 10^{-5} for individual contaminants by the World Health Organization (WHO) for drinking water (WHO 1993, 2001). Japanese standards also use a value of 10^{-5} as an acceptable risk (Hayamizu 1993). For remediation of Superfund sites in the USA, the acceptable risk can be up to 10^{-4} (USEPA 1989).

$$\text{Risk} = I \times Sf \tag{11.2}$$

where Risk is the cancer risk of a contaminant.

11.2.2 Collection and Selection of Toxicity Values

Toxicity values are necessary for estimating risk for non-cancer and cancer. The toxicity values are expressed by the *RfD* and *Sf*. Both *RfD* and *Sf* have different values for oral intake and inhalation. Toxicity values (*RfD* and *Sf*) are generally set from animal experiments or epidemiological studies. Toxicity experiments vary in regards to species of experimental animals, intake duration, etc. Therefore, it is difficult to compare toxicity values across studies and select a representative value. The selection often depends on the organization conducting the risk evaluation.

Table 11.1 shows the toxicity values of Cr(VI) set by the United States Environmental Protection Agency (USEPA) and WHO. A risk assessor must choose a toxicity value to evaluate risk. In this study, the *RfD* and *Sf* of the WHO or the Integrated Risk Information System (IRIS) of the USEPA were chosen first. Where there was no value available from either of these sources, the value was taken from other agencies or institutes, such as the National Center for Environmental Assessment of USEPA (NCEA), the California EPA, the Agency for Toxic Substances and Disease Registry (ATSDR) or a state. When there were more than two values, the most conservative value (smaller value for *RfD* and larger value for *Sf*) was chosen.

Choosing toxicity values has some difficulties and needs further discussion. The differences between the values chosen by the two agencies may affect risk assessment. For example, for the inhalation *Sf*, the calculated USEPA IRIS's value is

Table 11.1 Toxicities of Cr(VI) set by USEPA (2014) and WHO (2013)

	Sfo (mg/kg-d)$^{-1}$	*Sfi* (mg/kg-d)$^{-1}$	*RfDo* (mg/kg-d)	*RfDi* (mg/kg-d)
USEPA (2014)	**5.0 × 10^{-1}** (1-a)	**294** (2-a)	3.0 × 10^{-3} (1-b)	**2.9 × 10^{-5}** (1-b,2-b)
WHO (2013)	–	140 (2-a)	9.0 × 10^{-4}	–

For *Sfi*, the bigger value (294 (mg/kg-d)$^{-1}$) is selected, and for *RfDo*, the smaller value (9.0 × 10^{-4} (mg/kg-d)) is selected, because of conservative choice
(1-a) The New Jersey Department of Environmental Protection (NJDEP)
(1-b) USEPA, integrated risk information system
(2-a) Conversion of inhalation unit risk to *Sf*

$$\text{Risk} = Sfi \times \text{Intake} = Sf \times C \times Ih/W$$

$$\begin{aligned}\text{Unit Risk } (\text{ug/m}^3)^{-1} &= Sfi\ (\text{mg/kg-day})^{-1} \times 1\ (\text{ug/m}^3) \times 20(\text{m}^3/\text{d})/70\,\text{kg}\\ &= Sfi(\text{mg/kg-day})^{-1} \times 0.286\ (\text{ug/kg}\cdot\text{d})\end{aligned}$$

$$\text{Therefore, } Sf(\text{mg}/(\text{kg}\cdot\text{d}))^{-1} = \text{Unit Risk}/\left(2.86 \times 10^{-4}\right)$$

(2-b) Conversion of inhalation reference concentration (*RfCi*) to *RfDi*

$$RfDi\,(\text{mg}/(\text{kg}\cdot\text{d})) = RfCi\ (\text{mg/m}^3) \times 20(\text{m}^3/\text{d})/70(\text{kg}) = 0.286\,RfCi$$

294 (mg/(kg·d))$^{-1}$ and the WHO's value is 140 (mg/(kg·d))$^{-1}$ (Table 11.1). The value given by USEPA is about two times higher than that given by WHO. This difference of *Sfi* means that risk assessed using the USEPA value will be more severe than that of using the WHO value. When choosing toxicity values, the true values should be sought scientifically, but it is often difficult to choose which value is close to the true value. Therefore, the most conservative value is often chosen because it is better to estimate risk severely for managing human health risk.

11.2.3 Flow Diagram for Setting RBCs

A flow diagram for setting RBCs at the contaminated site is shown in Fig. 11.3. First, the exposure scenarios are set by choosing exposure routes, and intake is then calculated. Then, the toxicity values are selected. Equations of *HQ* and *Risk* are formulated with groundwater and soil. Finally, the concentrations in groundwater and soil are calculated for acceptable *HQ* (<1) and/or *Risk* (<10^{-5}). These concentrations are termed as RBCs.

11.2.4 Exposure Route and Scenario

First, the exposure route is set according to the site conditions of the groundwater and soil (Fig. 11.4). For drinking groundwater, the exposure routes are oral intake (W1), skin contact (W2) and inhalation during bathing (W3). For bathing without drinking groundwater, W2 and W3 are applicable. For volatilization from groundwater, exposure routes are inhalation inside buildings (W4) and outside (W5).

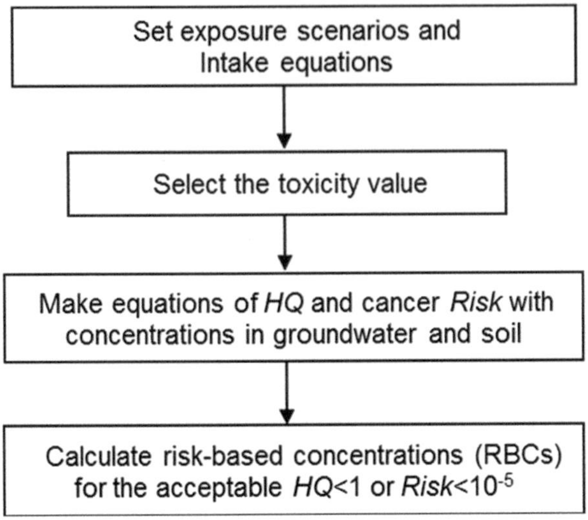

Fig. 11.3 Flow diagram for non-cancer and cancer risk-based concentrations (RBCs)

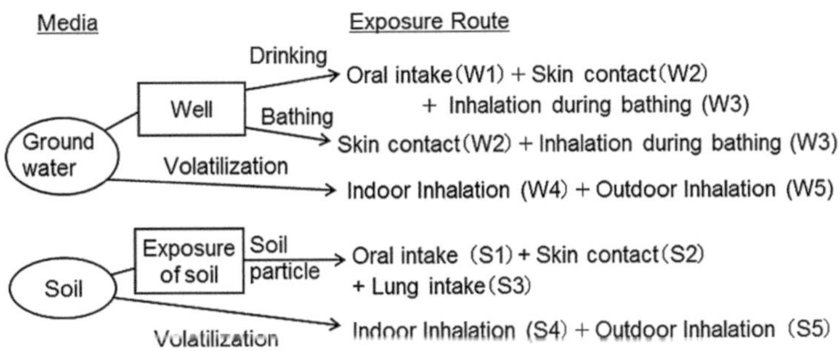

Fig. 11.4 Exposure routes of contaminants at a contaminated site

For soil, the exposure routes of soil particles are oral intake (S1), skin contact (S2), lung intake (S3) and volatilization from the contaminated soil inside buildings (S4) and outside (S5).

Exposure scenarios are set by the combination of the exposure routes (Table 11.2). In a residential area and industrial area, 11 scenarios are set according to groundwater usage, the existence of surface water and soil, respectively (Res. 1 to 11 and Occ. 1 to 11). Thus, a total of 22 scenarios are set, and intakes for each exposure scenario are then calculated.

Intake by swimming and consuming fish should be considered if surface water is present (lake, pond, or river). However, it is not a general scenario in Japan. Therefore, intake from surface water is not considered in this review. If there is background contaminant in any exposure routes, the concentration in the background should be added to source of the contaminated site. Nakanishi and Ono (2008) reported that Cr(VI) is existed in air. However, it is not counted in this review.

Table 11.2 Exposure scenarios consisting of a combination of exposure routes for metals

Land use	Media	Exposure conditions	Exposure scenario no.	
Residential /Occupational	GW	Drinking: W1 + W2		1
		Bathing: W2		2
		No use		3
	Soil	Exposure from ground: S1 + S2 + S3		4
		No exposure from ground		5
	GW + Soil	Drinking: W1 + W2	Exposure from ground: S1 + S2 + S3	6
			No exposure from ground	7
		Bathing: W2	Exposure from ground: S1 + S2 + S3	8
			No exposure from ground	9
		No use	Exposure from ground: S1 + S2 + S3	10
			No exposure from ground	11

For groundwater, W1: Oral intake; W2: Skin contact; W3: Inhalation during bathing
For soil, S1: Oral intake; S2: Skin contact; S3: Lung intake
In these scenarios, inhalation of metal is not counted because contaminated media is only soil of groundwater

11.2.5 Intake Estimation

Intake of groundwater (W1 and W2) and soil (S1 to S5) were calculated. Calculation of intake uses the physical properties of the contaminant and parameters related to soil, body, lifestyle and other factors (ASTM 2000). Parameter of body weight and lifetime can be used from Japanese data (JMHLW 1997, JMIAC 2001). And then, intake of all exposure routes was calculated.

Table 11.3 shows equations of intake for exposure routes chosen in Fig. 11.4, and Table 11.2 shows the parameters for the equations in Table 11.3. There are parameters for the physical and chemical properties of the contaminants and for the exposed human body and lifestyle. These equations and parameters are quoted from USEPA guidelines (USEPA 1989) and the American Society for Testing and Materials (ASTM 2000) and complemented by the guidelines of the Oregon Department of Environmental Quality (ODEQ 2003, 1998).

For the body weight of children, a weighted average by population ratio from 1 year old through 6 years old was used. For adults, a weighted average from 7 years old through 80 years old was used (JMHLW 1997). A lifetime of 80 years was used as the average of the lifetime for men and women (JMHLW 1999). After calculation using intake-equations in Tables 11.3 and 11.4, intake can be expressed simply by concentrations of the contaminants in groundwater and soil in Eq. 11.3.

Table 11.3 Equations describing contaminant intakes for each exposure route (USEPA 1989, ODEQ 2003)

Route	Intake equation (mg/kg-d)
For groundwater	
W1	$\frac{C_{\mathrm{GW}} \cdot Ef}{At} \cdot \left(\frac{Ed_C \cdot Iw_C}{Bw_C} + \frac{Ed_A \cdot Iw_A}{Bw_A}\right)$
W2	$\frac{C_{\mathrm{GW}} \cdot Pc \cdot Et_{\mathrm{Bath}} \cdot Ef \cdot 10^{-6}}{At} \cdot \left(\frac{Ed_C \cdot Ss_C}{Bw_C} + \frac{Ed_A \cdot Ss_A}{Bw_A}\right)$
For soil	
S1	$\frac{C_{\mathrm{S}} \cdot Ef \cdot 10^{-6}}{At} \cdot \left(\frac{Ed_C \cdot Iw_C}{Bw_C} + \frac{Ed_A \cdot Iw_A}{Bw_A}\right)$
S2	$\frac{C_{\mathrm{S}} \cdot Ef \cdot Df \cdot 10^{-6}}{At}$ \$\$ $\cdot \left(\frac{Ed_C \cdot Af_C \cdot Sa_C}{Bw_C} + \frac{Ed_A \cdot Af_A \cdot Sa_A}{Bw_A}\right)$
S3	$\frac{C_P \cdot Ef}{At} \cdot \left(\frac{Ed_C \cdot Ih_C}{Bw_C} + \frac{Ed_A \cdot Ih_A}{Bw_A}\right)$

C_{GW}:Concentration in groundwater (mg/L)
C_{S}:Concentration in soil (mg/kg)

$$I = k_{\mathrm{GW}} \times C_{\mathrm{GW}} + k_{\mathrm{S}} \times C_{\mathrm{S}} \tag{11.3}$$

Where

C_{GW}, C_{S}: Concentrations of a contaminant in groundwater (mg/L) and soil (mg/kg)
k_{GW}, k_{S}: Coefficients of C_{GW} (L/(kg·d)) and C_{S} (kg/(kg·d)).

11.3 Risk-Based Concentrations

11.3.1 Equations for Calculating RBCs

Equation 11.3 is substituted into Eq. 11.1, and then *HQ* for each concentration of contaminants in groundwater or soil is shown in Eq. 11.4. Cancer *Risk* can also be expressed by concentrations of contaminants in groundwater and soil and the coefficients by Eqs 11.3 and 11.2.

$$HQ = \frac{k_{\mathrm{GW}}}{RfD} \times C_{\mathrm{GW}} + \frac{k_S}{RfD} \times C_{\mathrm{S}} \tag{11.4}$$

Table 11.4 Parameters of exposure equations

Parameters				Values	
	Term			Residence	Occupation
Ef	Exposure frequency (day/year)			350 (ASTM 2000)	250 (ASTM 2000)
Ed_A	Exposure duration (year)	Adult		24 (Berg 1994)	25(ASTM 2000)
Ed_C		Child		6 (ODEQ 1998)	–
Iw_A	Ingestion of water (L/day)	Adult		2.3 (USEPA 1997)	
Iw_C		Child		1.5 (USEPA 1997)	–
Bw_A	Body weight (kg)	Adult		57 (JMHLW 1997, JMIAC 2001)	
Bw_C		Child		14 (USEPA 1997)	–
At	Exposure averaging time (day)		Non-cancer : At_{Non}	10,950 (30 year × 365 day) (ODEQ 1998)	9125 (25 year × 365 day) (ODEQ 1998)
			Cancer: At_{Can}	29,200 (lifetime: 80year) (JMHLW 1999)	
Pc	Dermal permeability constant (cm/hr)			0.002 (USEPA 2014)	
Et_{Bath}	Exposure time for bathing (hr/day)			0.25 (ODEQ 1998)	–
Ss_A	Skin surface area (cm^2)	Adult		22,000 (USEPA 1997)	–
Ss_C		Child		7300 (USEPA 1997)	–
Is_A	Ingestion of soil particles (mg/day)	Adult		100 (Hayamizu 1993, JMOE 2001)	
Is_C		Child		200 (Hayamizu 1993, JMOE 2001)	–
Df	Dermal absorption factor			1 (USEPA 1989)	
Af_A	Soil-to-skin adherence factor (mg/cm^2)	Adult		0.08 (ODEQ 1998)	
Af_C		Child		1 (ODEQ 1998)	–
Sa_A	Exposed skin surface area (cm^2)	Adult		6900 (USEPA 1997)	4100 (USEPA 1997)
Sa_C		Child		5000 (USEPA 1997)	–
Cp	Conc. in air from soil particles			C_S/Pe (mg/m^3) (ODEQ 1998)	
Pe	Particle-emission factor			$5.6 \times 10^7 (m^3/kg)$ (MDEQ 2012)	$4.7 \times 10^7 (m^3/kg)$ (MDEQ 2012)

11.3.2 Relations Among Groundwater and Soil Concentrations

The intake equation includes concentrations of contaminants in groundwater and soil (C_{GW} and C_S) as variables. The risk is calculated from the intake equation and the toxicity of the contaminants. Finally, the RBC is calculated using the risk equation and the given acceptable risk. However, the equation relating the RBCs has two unknown variables and, therefore, cannot be solved based on mathematical principles. Therefore, the variables should be combined using the relationship between groundwater and soil. C_S is expressed as Eq. 11.5 using C_{GW} by considering the process of contaminant sorption onto and desorption from soil as a linear equilibrium (USEPA 2005).

$$C_S(\text{mg/kg}) = Kd \times C_{GW} \quad (11.5)$$

Where

Kd: Soil-water partition coefficient (L/kg).

Kd depends on the type of soil and the containing material in soil. Therefore, the data from references have a wide range. In this chapter, a mean of Kd values in U.S. EPA report (2005) is used practically. Equation 11.5 is substituted into Eq. 11.4, yielding Eq. 11.6. This equation gives values of C_{GW}, which is the RBC of the contamination in a site. The RBC of cancer risk can be calculated using the same procedure.

Total *HQ* is then expressed as follows;

$$HQ = (k_{GW} + k_S \times K_d) \times C_{GW} \quad (11.6)$$

11.3.3 Calculated RBCs

Table 11.5 shows the RBCs of groundwater calculated for a site contaminated with Cr(IV). Two types of RBCs were calculated from *HQ* and *Risk*. The smaller value was chosen as the representative RBC because it is the most conservative. Bold numbers in Table 11.5 show the chosen RBCs. All RBCs were calculated using the cancer risk.

Comparing the RBC values of scenarios Res. 1 through 11, it is clear that different uses of groundwater and land use give different RBC values. The most important factor is drinking groundwater (Res./Occ. 1, 6 and 7). If residents do not drink groundwater, the RBC becomes about 250 times higher (0.28 mg/L/0.0011 mg/L), which are much smaller than Japanese environmental standard, 0.05 mg/L. This is because cancer risk is not counted severely on Japanese standard (RIEMAM 2009).

The second impacted factor is bathing (Res./Occ. 2, 8 and 9). Other impacted factors are no use of groundwater and only soil contamination (Res./Occ. 4 and 10).

Table 11.5 RBCs of groundwater calculated for a site contaminated with Chromium(VI) (Scenarios Res. 1 to 11 are for residential areas, and Occ. 1 to 11 are for occupational areas)

No.	RBC by *HQ* (mg/L)	RBC by $Risk_T$ (mg/L)
Res.1	0.017	**0.0011**
Res.2	4.5	**0.28**
Res.3	–	–
Res.4	13	**0.6**
Res.5	–	–
Res.6	0.017	**0.0011**
Res.7	0.017	**0.0011**
Res.8	3.3	**0.19**
Res.9	4.5	**0.28**
Res.10	13	**0.60**
Res.11	–	–
Occ.1	0.032	**0.0024**
Occ.2	8.3	**0.53**
Occ.3	–	–
Occ.4	36	**2.6**
Occ.5	–	–
Occ.6	0.032	**0.0024**
Occ.7	0.032	**0.0024**
Occ.8	6.8	**0.44**
Occ.9	8.3	**0.53**
Occ.10	36	**2.6**
Occ.11	–	–
(a)	0.05	

Bold numbers are selected as RBC
The values are two significant digits
(a) Japanese tap water and environmental standards

Here, scenarios of Res./Occ. 3, 5, 11 have no risk because volatilization of Cr(VI) is not counted in this occasion. RBCs of occupational areas are several times higher than those of residential areas because the exposure time in commercial areas is shorter than that in residential areas. In addition, the calculated RBCs are for GW (such as in Table 11.5). The RBCs for soil can be calculated by Eq. 11.5 under the condition of equilibrium.

11.4 Advection and Dispersion of Metal

11.4.1 Transportation of Metal

After contaminants (metals) are released into soil or groundwater, feature of the contamination depends on the type of the metals. The transformation characteristic is very important to choose remediation or risk management of the contaminated site. GEPC (2003) categorized metals into three groups according to their mobility characteristic. From the most transporting metal or ion groups; Group 1 through Group 3.

Group 1: Cr(VI). Highly mobile. There are many groundwater contaminations in Japan.
Group 2: As, B, F. Relatively less mobile than Group 1 metals. There are many groundwater contaminations in Japan.
Group 3: Pb, Hg, CN. Transport relatively little.

A group of harmful and highly mobile metals is called as Group 1. Cr(VI) is the Group 1 metal. Cr(VI) is dissolved in water as Cr^{6+} ion, which can easily be transported in groundwater. Therefore, Cr(VI)-contaminated groundwater should be treated or controlled before it is transported far from the source of the contaminant. The most important countermeasure for groundwater contamination with Cr (VI) is removal of the sources of contaminant. If the sources remain at the site, any treatment will be useless. After the removal of the sources, groundwater treatment such as pump and treatment can be used for decreasing concentration of Cr(VI). Because soluble contaminants such as Cr(VI) transport easily, the concentration of Cr(VI) decreases according to time. Therefore, natural attenuation (dilution) can be applied under the condition. And then, risk management such as inhibit of using groundwater may be needed until the concentration decreases as low as safety level.

The natural attenuation is often rational and effective for widespread contamination. After excavation, countermeasures need enormous expenditure, which is not economically feasible. Stopping drinking groundwater around the contaminated site can also be a practical countermeasure for risk management. For example, when Trichloroethylene (TCE) contamination was found in groundwater in Hyogo prefecture, Japan in 1987 (Kobayashi 1987), the city government announced the residents not to drink groundwater. When contamination of organic arsine in groundwater was found in Ibaraki prefecture, Japan in 2003 (JMOE 2003), the Japan Ministry of Environment (JMOE) surveyed the reason the contamination and asked the residents not to drink the groundwater.

On the other hand, metals of Group 3 are not easily transported from the source of the contamination. The area of the contaminated site is limited, and the concentration would not be changed rapidly. Therefore, risk management of the site is not difficult. Metals of Group 2 have nature of middle between Groups 1 and 3.

11.4.2 Prediction of Concentrations Using Soil Column Experiment

In order to manage risk of the contaminated site, the prediction of concentration of the contaminant is important. Risk of the contaminated site is managed by the RBCs on each condition at the site (Fig. 11.5). Therefore, future concentrations of the contaminant are important (Fig. 11.6).

The future concentration is predicted by simulating the condition of the contaminated site and/or a soil column test in the laboratory. Surveillance of distribution

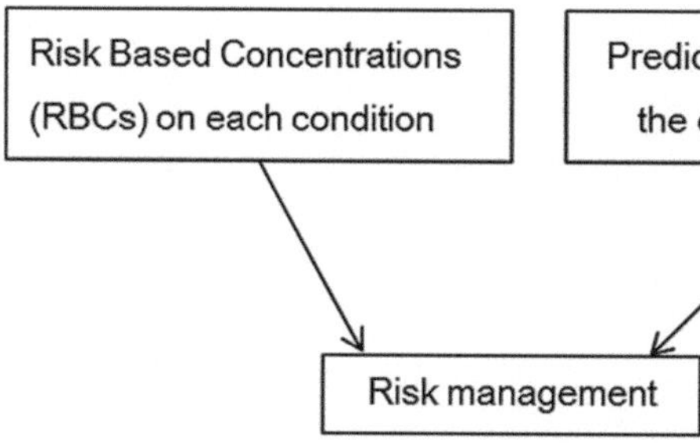

Fig. 11.5 Flow of risk management at the contaminated site

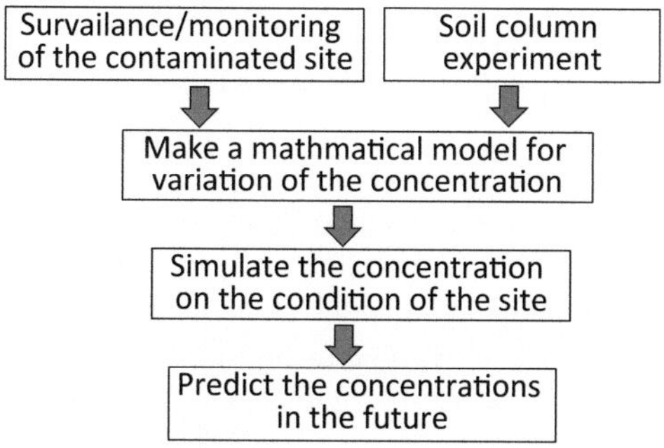

Fig. 11.6 Flow of prediction of the concentration in the future

(area and depth) of the contamination is necessary. In addition, information of groundwater conditions such as velocity, coefficient of permeability, hydraulic gradient, variation of water level, etc. are necessary for the simulation. Geography and soil property are also important for predicting future concentrations.

For simulating and predicting future concentrations, feasibility test of a countermeasure at a contaminated site gives useful data. However, the feasibility test often takes long time (several months up to several years). Therefore, soil column experiment can compensate data of advection and dispersion at the site.

11.4.3 Soil Column Experiment

For soil column experiment, it is better to use the same soil from the site and the same groundwater velocity at the site. If the soil and groundwater cannot be collected from the contaminated site, alternative soil and groundwater can be used. However, the effect of the difference in soil properties and groundwater velocity should be considered in the column experiment.

Soil column experiment is conducted by making simple equipment based on a groundwater permeability test; and transport of benzene is analyzed by an advection-dispersion model in this study. Figure 11.7 shows three kinds of column

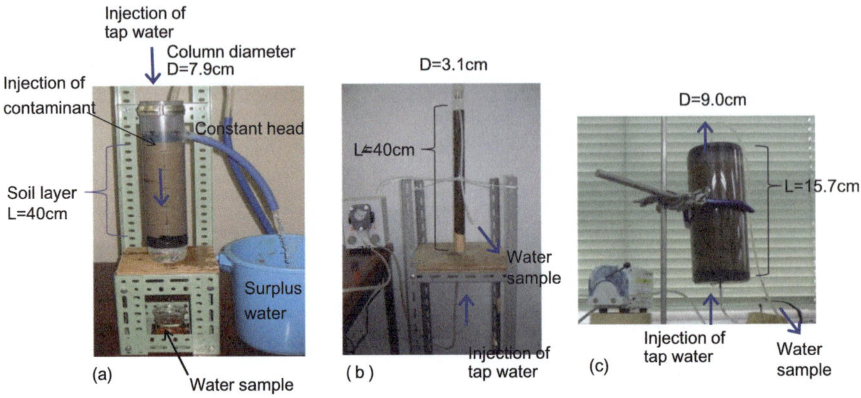

Fig. 11.7 Soil column experiment (**a**) down flow type; (**b**) and (**c**) upper flow type

tests. The column (a) is down flow type. It compacted with sand (Toyoura standard sand), and the length of soil layer is 40 cm. Tap water was input from upper side. Extra water was drained out of the column to maintain the water level. Then, high concentration of contaminant is injected from upper side of the column. The columns (b) and (c) are upper flow type using mini-pump in order to control flow rate, which is slow and same to the groundwater velocity of the site.

11.4.4 Porosity and Velocity of Water

Figure 11.8 shows flow rate and porosity of the soil column experiment. There are two kinds of velocities for groundwater. One is 'Darcy velocity' or 'superficial velocity' v (cm/sec) from macro point of view.

Darcy or superficial velocity, v (cm/sec) is expressed in Eq. 11.7.

$$v = \frac{Q}{A} = k \times \frac{\Delta H}{L} \tag{11.7}$$

k: hydraulic conductivity (cm/sec)
ΔH: Hydraulic head (cm)
L: Soil layer length (cm)

The other velocity for groundwater is average linear velocity, v' (cm/sec) from micro point of viewing. v' is a speed in porosity of sand in soil column or groundwater. v' is expressed in Eq. 11.8.

$$v' = \frac{v}{n} \tag{11.8}$$

where

n: porosity

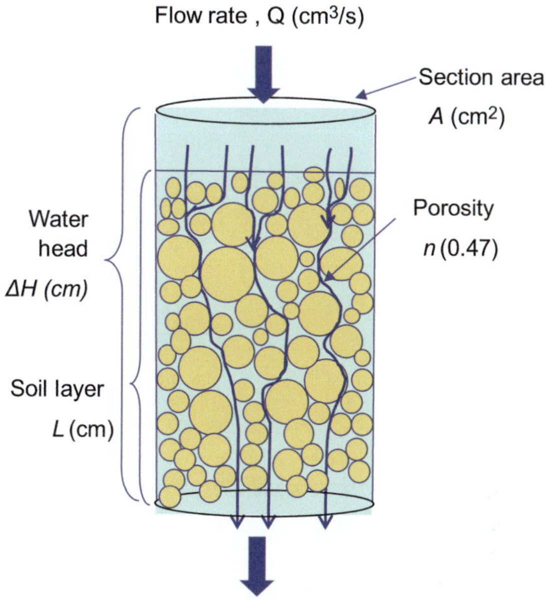

Fig. 11.8 Flow rate and porosity

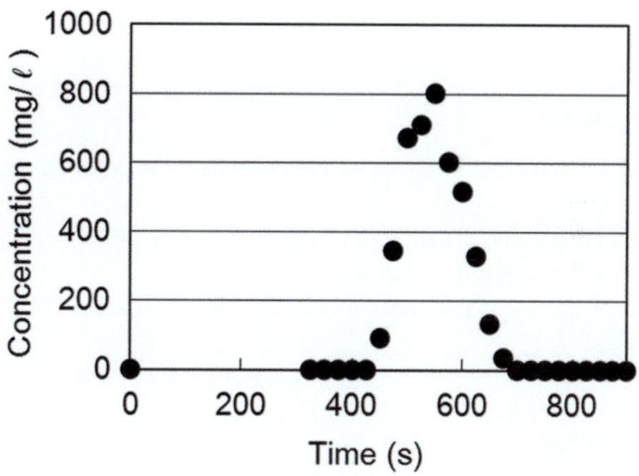

Fig. 11.9 Variation of contaminant concentration

11.4.5 Result of the Soil Column Experiment

Groundwater contamination is transported by advection and dispersion. For example, Fig. 11.9 shows transportation of contaminant for benzene. X-axis is time after injection of contaminant.

Fig. 11.10 shows calculated time of velocity (v') for groundwater as a bold line in Fig. 11.9. Figure 11.10 put a thin line in the centre of the time distribution of contaminant velocity, u (cm/s). The difference of time is called 'retardation', which is caused by sorption and desorption.

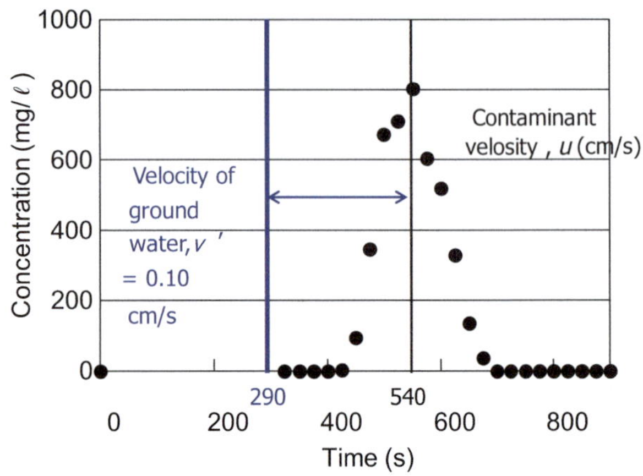

Fig. 11.10 Variation of contaminant concentration and water velocity

11.4.6 Advection and Dispersion Equation and Retardation Factor (R)

Advection and dispersion of contaminants is expressed by Eq. 11.9

$$\frac{\partial}{\partial t} C_T = D \frac{\partial^2}{\partial x^2} C_T - u \frac{\partial}{\partial x} C_T \tag{11.9}$$

Where

C_T: Total concentration of concentration in groundwater and concentration in soil (mg/L)
D: Dispersion coefficient of the contaminant in water (cm²/sec) (no sorption of soil is considered, or no soil is existed.)
u: Velocity of the contaminant in water (cm/sec)

$$C_T = n\, C_W + \rho\, C_S \tag{11.10}$$

Where

ρ: Density of soil (1.8 kg/L)
C_W: Concentration in groundwater (mg/L)
C_S: Concentration in soil (mg/kg)

Equation 11.10 is substituted to Eq. 11.9, then Eq. 11.11 can be gained.

$$\frac{\partial}{\partial t}(nC_W + \rho C_S) = D \frac{\partial^2}{\partial x^2}(nC_W + \rho C_S) - u \frac{\partial}{\partial x}(nC_W + \rho C_S) \tag{11.11}$$

Where contaminant in soil is absorbed, and advection and diffusion of contaminant in soil can be neglected. Therefore, Eq. 11.11 is expressed as Eq. 11.12.

$$\frac{\partial}{\partial t}(nC_W + \rho C_S) = D\frac{\partial^2}{\partial x^2}(nC_W) - u\frac{\partial}{\partial x}(nC_W) \quad (11.12)$$

Concentration of the contaminant in soil can be expressed as Eq. 11.13 using sorption coefficient, *Kd* (L/kg).

$$C_S[\text{mg/kg}] = Kd[\text{L/kg}] \times C_W[\text{mg/L}] \quad (11.13)$$

When Eq. 11.13 is substituted to Eq. 11.12,

Equation 11.12 can be expressed as Eq. 11.14.

$$\frac{\partial}{\partial t}\left(1 + \frac{\rho \cdot Kd}{n}\right)C_W = D\frac{\partial^2}{\partial x^2}C_W - u\frac{\partial}{\partial x}C_W \quad (11.14)$$

Here, $R = \left(1 + \frac{\rho \cdot Kd}{n}\right)$ is substituted to Eq. 11.14. Then, Eq. 11.15 is got.

$$R\frac{\partial}{\partial t}C_W = D\frac{\partial^2}{\partial x^2}C_W - u\frac{\partial}{\partial x}C_W \quad (11.15a)$$

or

$$\frac{\partial}{\partial t}C_W = \frac{D}{R}\frac{\partial^2}{\partial x^2}C_W - \frac{u}{R}\frac{\partial}{\partial x}C_W \quad (11.15b)$$

Dispersion coefficient (*D*) and velocity (*u*) of contaminant is expressed as $\frac{D}{R}$ and $\frac{u}{R}$. In Eq. 11.15b. However, if the velocity (*u*) of the contaminant is calculated using a peak of concentration in Fig. 11.10, the *u* will contain 1/*R*.

11.4.7 Mathematical Model

A mathematical model of partial differential equation using first-dimensional advection and three-dimension dispersion can be expressed in Eq. 11.16.

$$\frac{\partial C}{\partial t} = Dx\frac{\partial^2 C}{\partial x^2} + Dy\frac{\partial^2 C}{\partial y^2} + Dz\frac{\partial^2 C}{\partial z^2} - u'\frac{\partial C}{\partial x} \quad (11.16)$$

Analytical solution for point source and instantaneous (Baetslé 1969) is expressed as Eq. 11.17.

$$C(x,t) = \frac{C_0 \cdot V_0}{8(\pi \cdot t)^{\frac{3}{2}} \cdot (Dx \cdot Dy \cdot Dz)^{\frac{1}{2}}} e^{-\frac{\left(\left(x-u'\right)\cdot t\right)^2}{4 \cdot Dx \cdot t}} \tag{11.17}$$

Where

t: Time after injection of contaminant (second)
C_0: Concentration of Contaminant (mg/ℓ)
V_0: Inject volume of C_0 (cm^3)
x: Soil layer (ℓ) in the soil column (cm) or distance from the source of the contaminant.
u': Velocity of contaminant (cm/s) ($u/\underline{u}' = R$)
R: Retardation factor (−)
Dx, Dy, Dz : Hydrodynamic dispersion along flow path (x), transverse to flow path (y and z) (m^2/s) (Dy or $Dz = 0.1 \times Dx$ (USEPA 2002))

In Eq. 11.17, only Dx is unknown value. Dx can be estimated by the least square method using data of soil column experiment. As a result, calculated and measured concentration of the contaminant is shown in Fig. 11.11.

Next step is the application of the mathematical model to a contaminated site. Then, the model is used for prediction of concentration at a real contaminated site. Surveillance at the site or monitoring data can help to predict the concentration.

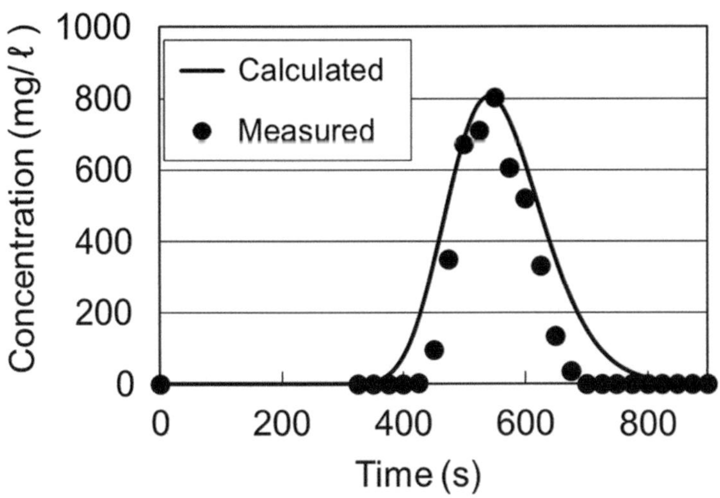

Fig. 11.11 Variation of the calculated and measured concentrations

11.5 Risk Management at a Contaminated Site

In this section, the benefit of risk management using RBCs is evaluated. *In situ* treatment is an affordable method to manage the contaminated site. In this section, the calculated RBCs were applied to the site for risk management, and the performance of management method using the RBCs was evaluated.

11.5.1 Case Study at a Model Contaminated Site

In a model contaminated site, soil and groundwater are contaminated by Cr(VI). The area is 40 m × 50 m and depth 5 m. Figure 11.12 shows the image of the model contaminated site. If contamination is found at a site, the contaminated soil/groundwater has to be treated. Otherwise, the contaminated site should be controlled under risk management.

In this section, procedure after detection of a contaminated site is shown. At first, if the site is a residential area for lease or sell, the site should be cleaned up in order to provide safe drinking groundwater. Otherwise, the site owner should explain the contamination level and the rental rate or land price will be decreased. The owner may negotiate with potential customers who take on or buy the site. Here, two cases are compared as a case study; in Case 1, the site is completely treated by excavation and disposal of contaminated soil. The soil is transported to a disposal site or containment site. In Case 2, the site is treated *in situ* (containment and pavement). Therefore, the contaminated soil and groundwater of the site is left, and monitoring is necessary for at least 2 years in the Japanese soil law. In addition, *in situ* pump and treatment can be executed in this case although the treatment period is not clear. The predicted period by the simulation may be changed after execution of the treatment.

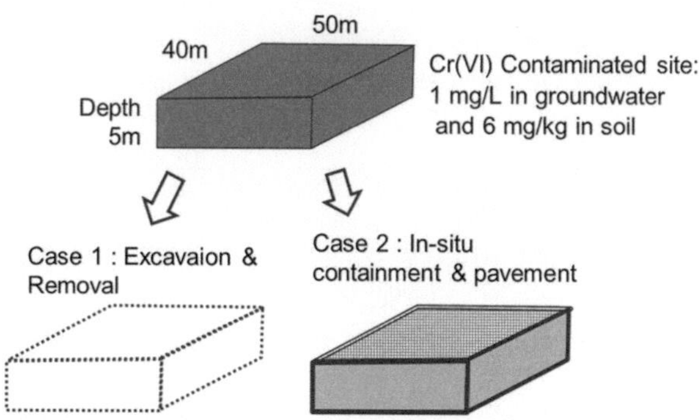

Fig. 11.12 Image of the model contaminated site

11.5.2 Choice of Countermeasures for Land Use

This section shows how to choose countermeasures (Fig. 11.12). RBCs in Section 11.3 are applied to the site for each condition of land use. If the concentrations of the contaminant in groundwater at the model site is 1 mg/L (and the concentration of the contaminant in soil is 6 mg/kg, which is calculated by Eq. 11.5, *Kd* of Cr (VI) is 6 kg/L), Table 11.5 shows the target RBCs for the model site. If groundwater is used for drinking purpose, RBCs are 0.0011 and 0.0024 mg/L for residential and occupational sites, respectively. Therefore, complete remediation using excavation and removal should be chosen (Case 1). On the other hand, if groundwater is not used or there is no well in the site, only *in situ* containment and pavement is needed (Case 2) (Res. 3, 5 and 11). For occupational area, RBCs of soil without using groundwater (Occ. 4 and 10) is 2.6 mg/L in consider of soil exposure from the ground. 1 mg/L in groundwater at the site is under 2.6 mg/L of the RBC; therefore, the site does not need any treatment.

11.5.3 Cost-Benefit Analysis

Cost-benefit analysis is an objective method to choose the kind of countermeasure for a contaminated site. Sasamoto et al. (2004) compare cost and benefit of combinations of countermeasures (excavation, *in situ* remediation, no treatment) and land uses (residential, occupational and parking). Especially, Japanese governments often use cost-benefit analysis for public projects. In this case, benefit has two means. One is decreasing risk, and the other is the future benefits from the treated site. Even if cost-benefit analysis is not applied to the site, the cost of the countermeasure is a big factor to choose it.

Table 11.6 shows treatment expenditure ($) for Case 1 and Case 2. Unit price for treatment in the table was based on data in Japan (Yamaki and Morishima 2013).

Table 11.6 Treatment expenditure ($) for case 1 and case 2

Case 1: Excavaiton and removal		Case 2: *In situ* containment and pavement	
Excavation (steel sheeting)	10,000 m^3 @$50/m^3 = $500,000	Impermeable wall Surround the area(40 m + 50 m) × 2 × 5 m = 1800 m^2	1800 m^2@ $300/m^2 = $540,000
Transportation & disposal	10,000 m^3 × 1.8 ton/m^3 @ $200/ton = $3,600,000	Pavement	2000 m^2@$50/m^2 = $100,000
Landfill & ground leveling	10,000 m^3 @$20/m^3 = $200,000	Monitoring wells:4	4 @$6000 = $24,000
Sum	$4,300,000		$664,000
Cost per 1 ton	$240/ton		$37/ton

It shows that if removal or purification (Case 1) is not necessary and *in situ* containment and pavement (Case 2) is acceptable, Case 2 is less expensive than case 1 (1/6 of the expenditure). Therefore, Case 2 is easy to choose as a point of initial cost. Owners have also to estimate annual benefits from the remediated site.

11.5.4 Discussion

In this chapter, a methodology to set RBCs has been introduced. The RBC was then applied to the model contaminated site. The results demonstrated that *in situ* treatment method, such *in-situ* containment and pavement, in which contaminated soil remains under the pavement, can be used for risk management by applying the RBC. Therefore, *in situ* treatment can be applied more readily, and the problem of brownfields can be solved gradually. The risk of contaminants to human health must be effectively decreased; however, excessive requirements for remediation do not promote the recovery of contaminated sites.

In Japan, the RBC of each contaminated site is not used for site management, and it may be difficult for Japanese residents to accept RBCs in place of the current environmental standards. However, if residents accept risk management using RBCs, it will be beneficial in promoting the recovery of contaminated sites. Therefore, this method provides useful information for risk management, which can promote the use of the contaminated sites and prevent health risks economically.

1. *Risk management using RBCs*

Table 11.7 shows merit and demerit of standards, risk assessment and RBCs. Environmental standard for contaminated site is the only one value for one contaminant. Therefore, it is easy to judge contamination by the value. However, the environmental standards in Japan were set based on drinking water containing the contaminant. Therefore, the values are too severe and extra-countermeasures may be required, if groundwater is not used for drinking purpose. On the other hand, risk assessment (RA) evaluates the risk of the site reasonably. However, RA needs to be

Table 11.7 Merit and demerit of standards, risk assessment and RBCs

	Environmental standard	Risk assessment on each site (RA)	Risk-based concentrations (RBCs)
Merit	The one value is easy to judge contamination	RA evaluates the risk of the site reasonably	RBC is reasonable value for risk management
Demerit	Extra-countermeasures may be required	RA is required for every site, and additional survey and analysis are often required	None (However, proposed RBC needs agreement with residents.)

done at every site, and the owner of the site needs to pay the cost of RA with severance.

RBCs, which are proposed in this chapter, have merits of environmental standard and RA. RBCs also compensate the demerit of the standard and RA. Once the proposed RBCs were agreed with residents, RBCs' table gives RBCs for each scenario at the site. In order to make the agreement with residents, communication between residents and governments, which is called risk communication, is very important. As a result, RBCs are reasonable for risk management, and required remediation can be performed at a reasonable cost. Therefore, the number of the abandon sites that are named 'brawn field' can be fewer.

2. *Problem of the Prediction by the Simulation*

Analytical solution of an advection and diffusion model has been used to simulate the decreasing concentrations of the contaminants. However, it is not easy to predict the variation of the contaminants' concentration correctly. The simulation of Sect. 11.4 should be considered carefully to apply it to a real contaminated site. In order to manage the risk of a contaminated site using RBCs, predicting concentrations are very useful. However, the calculations do not give perfect predictions. Therefore, monitoring is strongly recommended to compensate for the time lag between the calculated concentrations and the measured concentrations.

If there are not enough monitoring data at the site, the column experiment can compensate the data for the contaminated site. Therefore, variation of concentration can be predicted reasonably for risk management. The most effective method is the use of monitoring data at the real contaminated site and calculating the contaminants' concentration again. This cycle should be repeated. Monitoring data for a year would help the precision. In addition, the purpose of using the simulation is important. The prediction of the simulation has a range of error, which depends on the site condition. For the risk management, higher concentration, which estimates the risk higher, should be chosen. If the predicted concentration is higher than a real concentration, it is not a problem for risk management. However, if the concentration is lower than the real value, risk management is not appropriate.

If a risk of metal contamination is managed using RBCs, natural attenuation and *in situ* containment, which are inexpensive but cannot achieve reductions meeting environmental standards, can be used as countermeasure.

3. *Risk Management and Risk Communication*

RBCs are calculated and set scientifically. However, actual risk management is case by case. Therefore, risk communication is important. The residents should agree about the countermeasure. If the residents do not agree, treatment activity would not succeed.

On the other hand, even little contaminant remains in residential area, if the resident understand owners' will and plan for the countermeasure, there is no trouble about the contaminated area. The residents can accept the countermeasure. Therefore, landowners and governments have to understand that risk communication is very important for risk management.

References

ASTM (2000) Standard guide for risk-based corrective action, Table X3.2 Site and scenario-specific parameters, E 2081–00

Baetslé LH (1969) Migration of radionuclides in porous media. Progress in nuclear energy series XII, Health Physics. Pergamon Press, London, the United Kingdom, pp 707–730

Berg R (1994) Human exposure to soil contamination: a quantitative and quantitative analysis towards proposals for human toxicological intervention values (partly revised edition), National Institute of Public Health and Environmental Protection, No. 725201011, Annex 1.7, The Netherlands

Fujinaga A et al (2012) Risk-based concentration setting methodology for management of contaminated soil/groundwater, risk analysis. Int J 32(1):122–137. doi:10.1111/j.1539-6924.2011.01677.x

GEPC (2003) Explanation of technical method of surveillance and countermeasure based on soil contamination countermeasures Act, Geo-Environmental Protection Center, Appendix 2_7–2_13

Hayamizu T (1993) Revised environmental standard and standard of water-purity for tap water. J Jpn Soc Water Environ 16(4):224–242

JMHLW (1997) Table 2–7 average height and weight. Statistics database: Japanese nutrition condition. Survey of Japanese nutrition on fiscal 1997. (in Japanese) http://www.mhlw.go.jp/bunya/kenkou/kenkou_eiyou_chousa.html. Accessed 1 Dec 2003

JMHLW (1999) Japanese average duration of life on 1997: 3 Average duration of life, international comparison. Minister's Secretariat, Statistics and Information Department, Japanese Ministry of Health, Labor and Welfare. (in Japanese) http://wwwl.mhlw.go.jp/toukei/h9-ablif/life-3.html. Accessed 1 Dec 2003

JMIAC (2001) Document: stochastic population. (in Japanese) Statistical Department. http://www.stat.go.jp/data/jinsui/2001np/index.htm. Accessed 1 Dec 2003

JMOE (2001) Commission of risk assessment of contaminants in soil, report on risk assessment for intake of soil. (in Japanese). http://www.env.go.jp/water/report/h13-01/index.html. Accessed Jan 2010

JMOE (2003) Urgent countermeasure for organic arsine compounds at Kamisu, Ibaraki prefecture. (in Japanese). https://www.env.go.jp/chemi/gas_inform/pdfs/cd_h150606.pdf. Accessed 26 Aug 2014

JMOE (2009) Results of the survey on the enforcement status of the soil contamination countermeasures law and soil contamination investigations and countermeasures in FY 2009 (in Japanese). http://www.env.go.jp/water/dojo/chosa.html. Accessed 1 July 2014

Kammen DM, Hassenzahl DM (1999) Chapter 5: Toxicology. Should we risk it?: exploring environmental, health, and technological problem solving. Princeton University Press, New Jersey, USA, pp 154–158

Kobayashi E (1987) Countermeasures of contaminated groundwater – case study on Hyogo prefecture – report form a real site. J Environ Pollut Control 23(10):51–47

MDEQ (2012) Proposed soil criteria: ambient air inhalation component, remediation division, Michigan DEQ, USA. http://www.michigan.gov/documents/deq/deq-rrd-CSI-SoilToAmbient_airPathway_379820_7.pdf? 20131230015457. Accessed 26 Aug 2014

Nakanishi J, Ono K (2008) Risk assessment documents 21, chromium (VI), Maruzen. (in Japanese). https://unit.aist.go.jp/riss/crm/mainmenu/1-25.html. Accessed 1 July 2014

ODEQ (1998) Guidance for conduct of deterministic human health risk assessments, Table B-1 reasonable maximum estimates: Ed_A, Ed_C, Af_A, Af_C, A.52 (Cp), Waste Management & Cleanup Division, Oregon DEQ, USA

ODEQ (2003) Risk-based decision making for the remediation of petroleum-contaminated sites, Appendix C. Waste Management & Cleanup Division, Oregon DEQ, USA

RIEMAM (2009) Guide book of water quality standard for drinking water. Research Institute of Environmental Management, Administration and Maintenance of JAPAN, Maruzen

Sasamoto Y et al (2004) Proposal of land use of soil-contaminated site by application of risk assessment on health. GEPC Tech News 9:30–35 (in Japanese)

USEPA (1989) Risk assessment guidance for superfund vol 1: human health evaluation manual (Part A). Interim final, EPA/540/1-89/002. Office of Emergency and Remedial Response, Washington, DC

USEPA (1997) Exposure factors handbook, Table 3–30, Table 5–23, Table 6–2, Table 6–6, Table 6–8, Office of Research and Development National Center for Environmental Assessment, EPA/600/P-95/002Fa

USEPA (2002) BIOCHLOR 2.2. http://www.epa.gov/ada/csmos/biochlor.html. Accessed 1 Dec 2003

USEPA (2005) Partition coefficients for metals in surface water, soil, and waste, EPA/600/R-05/074

USEPA (2014) Regional screening levels tables, chemical specific parameters, Region 9 of the U.S. EPA. http://www.epa.gov/region9/superfund/prg/. Accessed 1 Aug 2014

WHO (1993) 10. Acceptable risk. guidelines for drinking-water quality, 2nd ed, vol 1, Recommendation

WHO (2001) 3. Chemical aspects. Water quality. guidelines, standards and health: Assessment of risk and risk management for water-related infectious disease

WHO (2013) Inorganic chromium(VI) compounds, concise international chemical assessment document 78

Yamaki J, Morishima Y (2013) Revised soil contamination countermeasures Act, Toyo Keizai Inc

MIX
Papier aus verantwortungsvollen Quellen
Paper from responsible sources
FSC® C105338

If you have any concerns about our products,
you can contact us on
ProductSafety@springernature.com

In case Publisher is established outside the EU,
the EU authorized representative is:
Springer Nature Customer Service Center GmbH
Europaplatz 3, 69115 Heidelberg, Germany

Printed by Libri Plureos GmbH
in Hamburg, Germany